Springer Series in
Surface Sciences

Editor: Robert Gomer

31

Springer Series in **Surface Sciences**

Editors: G. Ertl, R. Gomer and D. L. Mills   Managing Editor: H. K. V. Lotsch

A. R. Burns   E. B. Stechel
D. R. Jennison (Eds.)

Desorption Induced by Electronic Transitions DIET V

Proceedings of the Fifth International Workshop,
Taos, NM, USA, April 1–4, 1992

With 173 Figures

Springer-Verlag
Berlin Heidelberg New York
London Paris Tokyo
Hong Kong Barcelona
Budapest

Dr. Alan R. Burns
Dr. Ellen B. Stechel
Dr. Dwight R. Jennison

Sandia National Laboratories, Albuquerque, P.O. Box 5800,
Albuquerque, NM 87185, USA

Series Editors

Professor Dr. Gerhard Ertl

Fritz-Haber-Institut der Max-Planck-Gesellschaft, Faradayweg 4–6,
1000 Berlin 33, Germany

Professor Robert Gomer, Ph.D.

The James Franck Institute, The University of Chicago, 5640 Ellis Avenue,
Chicago, IL 60637, USA

Professor Douglas L. Mills, Ph.D.

Department of Physics, University of California,
Irvine, CA 92717, USA

Managing Editor: Dr. Helmut K. V. Lotsch

Springer-Verlag, Tiergartenstrasse 17,
W-6900 Heidelberg, Germany

ISBN 3-540-56473-X Springer-Verlag Berlin Heidelberg New York
ISBN 0-387-56473-X Springer-Verlag New York Berlin Heidelberg

Typesetting: Camera ready copy from the author/editor

54/3140-5 4 3 2 1 0 – Printed on acid-free paper

Preface

This volume in the Springer Series on Surface Sciences presents a recent account of advances in the ever-broadening field of electron- and photon-stimulated surface processes. As in previous volumes, these advances are presented as the proceedings of the International Workshop on Desorption Induced by Electronic Transitions; the fifth workshop (DIET V) was held in Taos, New Mexico, April 1–4, 1992. It will be abundantly clear to the reader that "DIET" is not restricted to desorption, but has for several years included photochemistry, non-thermal surface modification, exciton self-trapping, and many other phenomena that are induced by electron or photon bombardment. However, most stimulated surface processes do share a common physics: initial electronic excitation, localization of the excitation, and conversion of electronic energy into nuclear kinetic energy. It is the rich variation of this theme which makes the field so interesting and fruitful.

We have divided the book into eleven parts in order to emphasize the wide range of materials that are examined and to highlight recent experimental and theoretical advances. Naturally, there is considerable overlap between sections, and many papers would be appropriate in more than one part. Part I focuses on perhaps the most active area in the field today: *electron attachment*. Here the detection and characterization of negative ions formed by attachment of electrons supplied externally from the vacuum are discussed. In addition, the first observations of negative ions formed by substrate photoelectrons are presented. In Part II, the attachment of excited substrate electrons is shown to be involved in *photodesorption and photochemistry*. The use of femtosecond laser pulses not only provides a direct time window on the dynamics, but also has identified multiple excitations in the course of one desorption event. In Part III, *laser-induced ion desorption* is studied, which in many cases is relevant for the control and understanding of plasma ablation.

High-energy processes such as core-excitation-induced dissociation of adsorbates and gas-phase molecules are discussed in Part IV. This work has provided new insight into fast fragmentation and specific excitation channels. In addition, the relative importance of electronic excitation and resonant neutralization versus momentum transfer in the scattering of high-energy neutrals and ions at surfaces is discussed. In Part V, current efforts in DIET *theory* focus on excited-state lifetime calculations, Antoniewicz-type dynamics in view of electron attachment processes, and models for the desorption of physisorbed species. *Electron-stimulated desorption and chemistry* of adsorbates on metals is discussed in Part VI. The reader will find much overlap here with Part II, particularly in the importance of adsorbate internal dynamics in desorption yields.

An exciting new area that promises to expand rapidly is that of *spatially-resolved* DIET in Part VII. There it is shown that selective bond-breaking and dissociation of individual molecules can be performed with the tip of a scanning tunneling microscope. *Semiconductor* surfaces is the topic of Part VIII. Evidence for defect-mediated desorption is presented and the important role of mid-gap surface states is elucidated. Stimulated desorption also continues to be an effective probe of processes relevant to etching and growth technologies. DIET on *oxides*, Part IX, is also of great interest in materials applications, particularly in view of the large desorption and dissociation cross-sections. In the area of *alkali halides*, the subject of Part X, the resolution of some long-standing mysteries is emerging, such as the origin of excited alkali atoms above the surface. Finally, in Part XI, low-probability excited states are observed in *cryogenic overlayers*, where the excitation can be mobile and self-trap. Desorption from condensed layers on cooled beam tubes is also a potential problem in future accelerators such as the proposed Superconducting Super Collider.

For the generous support which allowed both the preparation of these proceedings and the organization of the DIET V workshop, we would like to thank Dr. Larry Cooper of the Office of Naval Research. We thank Sandia National Laboratories (operated for the Department of Energy) for administrative support. We are indebted to Dr. Fred Vook (Sandia) for making it possible to pursue these efforts. Finally, we thank Sharon Voccio (Sandia) for her considerable secretarial help before and during the workshop.

Albuquerque, NM   *A.R. Burns*
September 1992   *E.B. Stechel*
                                            D.R. Jennison

Contents

Part III   Laser-Induced Ion Desorption

Part IV   High Energy Processes

Part V   **Theory**

Part VI   **Electron Stimulated Desorption
              and Chemistry of Adsorbates on Metals**

Part IX Oxides

Part X Alkali Halides

Part XI   **Cryogenic Overlayers**

Part I

Electron Attachment

Surface Reactions and Desorption by Electron Attachment

L. Sanche

MRC Group in the Radiation Sciences and Canadian Centre
of Excellence in Molecular and Interfacial Dynamics, Faculty of Medicine,
Université de Sherbrooke, Sherbrooke, QC, Canada J1H 5N4

1.   ELECTRON ATTACHMENT

Electrons can attach to atoms and molecules in different ways depending on the charge transfer and charge stabilization mechanism involved [1]. The additional electron may be captured as it collides with the target or it may be supplied by an anion or metastable atom or molecule interacting with the target. The initial anion formed in this electron transfer process is usually in a transient short-lived (10^{-10}-10^{-16} sec) state which underdoes subsequent modifications. The formation of transient anions can be divided into two broad categories [2]. If the additional electron occupies a previously unfilled orbital of a target atom or molecule in its ground state, the transitory state is referred to as a single-particle resonance. The term "shape" resonance applies more specifically when temporary trapping of the electron is due to the shape of the electron-target potential. When the transitory anion is formed by two electrons occupying previously unfilled orbitals, the resonance is called "core-excited" or may be referred to as a two-particle, one-hole state. In the case of the latter, the electron is captured by the positive electron affinity of an electronically excited state of the target with or without the assistance of an angular momentum barrier. In a single-particle resonance the centrifugal barrier set by the colliding electron is usually responsible for the capture, but the additional electron may also be temporarily trapped by exciting nuclear motion in a molecule. In all cases, a potential well is formed which temporarily captures the additional electron. The dimensions of this well are such that it accommodates electrons having wavefunctions corresponding to low energies.

In the present context, electron attachment is considered to occur for a molecule condensed on a surface which can either be that of a metal, a semi-conductor or a dielectric. The process is shown schematically in Fig. 1 where the additional electron is supplied either from the solid or from vacuum. A unidimensional extra-electron potential is drawn to describe the electron-molecule interaction at the surface of a metal responsible for the formation of a single-particle anion state at an energy E_o or that of a 2-electron, 1-hole state at energies E_1-$|E_o|$ or E_2-$|E_o|$. This hypothetical representation of the 3-D problem is not exact but is given here to provide a simple picture. The height of the barrier may differ according to the nature of the substrate and rise above the vacuum level for dielectrics; on the vacuum side, another barrier may be added to simulate "shape" resonances. In the case of core-excited resonances (2-electron, 1-hole), a free electron of energy E_K coming from vacuum or a photoelectron with kinetic energy $h\nu$ may excite electronic energy levels (e.g., E_1 and E_2, respectively) of the surface molecule and find itself trapped in the potential well of the electronic state if this latter possesses at least one energy level. As shown in the diagram, a hot electron may be directly produced in the metal at the energy of a single particle resonance and jump into the electron-molecule potential well. Hot electrons, photoelectrons, and secondary electrons produced by higher energy particles, may all lose energy by scattering within the solid until they arrive at the resonance energy at which they have a high probability of being transferred to the adsorbed molecule. So, whatever the initial mechanism for producing low-energy electrons, the process is basically

Springer Series in Surface Sciences, Vol. 31
Desorption Induced by Electron Transitions DIET V
Editors: A.R. Burns · E.B. Stechel · D.R. Jennison   © Springer-Verlag Berlin Heidelberg 1993

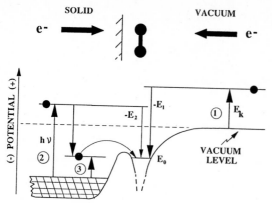

Fig. 1   Unidimensional potential representing the interaction of a low-energy electron with a molecule adsorbed on a metal surface. The electron either arrives from the vacuum side with kinetic energy E_k (1) or it is created in the metal as a secondary electron by a high-energy particle or by a photon as a photo- or hot electron (2 and 3, respectively). The molecular potential, which has one extra electron level at energy E_o, may represent a single-particle anion state or a 2-particle, 1-hole state if an energy equivalent to the electronic state involved (e.g., E_1 or E_2) is subtracted from the initial kinetic energy of the electron.

the same: electrons must be "tuned" to the resonance energy in order to transfer to the adsorbate. Many examples now exist in the literature which illustrate that resonance electron attachment and subsequent reactions occur with hot electrons [3,4] photoelectrons [3,4,5,6], secondary electrons [6] and monochromatic electrons injected from vacuum [2].

2.   DECAY OF TRANSIENT ANIONS

A transient molecular anion may decay in different ways depending on its lifetime and m-olecular orbital characteristics [2]. The additional electron may depart and leave the molecule in a vibrationally or electronically excited state. Because of the force existing between the induced and/or permanent dipoles at the surface during the lifetime of the anion, this latter is attracted toward the surface. If enough kinetic energy is imparted to the molecule by this motion, surface-molecule vibrational modes can also be excited [7]. The entire molecule may even desorb (i.e., by an Antoniewicz type mechanism) when this energy lies above the dissociation limit of the molecule-surface binding potential [8]. For a molecule within a solid, phonon modes of the host lattice are excited by this mechanism [9] as the electron leaves a given molecular site. Desorption of only neutral fragments is possible when the transient anion decays to an electronic excited state which is dissociative in the Franck-Condon (F-C) region [2]. For a diatomic molecule AB, the possible reactions may be represented as $e + AB \rightarrow (AB^-)^* \rightarrow AB^* + e \rightarrow A + B + e$ or $A + B^* + e$ leaving the fragment B in the ground or an excited electronic state. When the lifetime of the resonance is of the order of a vibrational period or longer, the $(AB^-)^*$ state is dissociative in the F-C region and one of the possible fragments has a positive electron affinity, then the anion may also dissociate into a stable anion and a neutral fragment in the ground or an excited state, giving rise to the dissociative electron attachment (DEA) reac-

4

tion $e + AB \rightarrow (AB^-)^* \rightarrow A^- + B$ or $A^- + B^*$. Finally, if the adsorbed molecule has a positive electron affinity and the lifetime of the anion is sufficiently long to allow transfer of its intramolecular vibrational energy to the surface-molecule bound and/or the phonon bath of the substrate or allow spontaneous photon emission, then the additional electron may be stabilized on the molecular target forming a permanent anion [10]. This process is unlikely for adsorbates on metals because the anion lifetime is expected to be short (10^{-15}-10^{-16} sec) due to tunnelling to the metal for potentials of the type shown in Fig. 1; but for dielectrics, tunnelling is greatly reduced and stabilization becomes much more probable.

In the following sections experimental evidence is provided for reactions giving rise to metastable molecule and anion desorption induced by monochromatic electrons incident from vacuum. It is also shown experimentally that transient anions can stabilize on surfaces, exchange their charge and energy with surface adsorbates and induce anion-molecule surface reactions.

3. EXPERIMENTS

The reported data were obtained with an electron beam (0.03-0.3 eV full width at half maximum) incident on a clean Pt(lll) or polycrystalline Pt substrate held at ~20 K which was covered with submonolayer to multilayer amount of atoms and/or molecules depending on the phenomenon being investigated. The desorbing metastable species were detected by an array of microchannel plates shielded against the arrival of positive and negative charges by suitable potentials as described in the experimental arrangement of Leclerc et al. [11]. The anion yields were measured by a mass spectrometer located near the substrate at an angle of 20° from the normal [12]. Charge trapping by molecules on the surface of rare gas multilayer films was monitored by measuring the energy shift in the energy onset of current measured at the Pt substrate using the technique of Marsolais et al. [13]. In the anion electron-stimulated desorption (ESD) experiment electrons were incident at 20° and 40° from the target plane. The other experiments were performed at normal incidence. Since the mean free path of an energetic (i.e., of the order of a few eV) metastable species or an anion is very low within a solid, the signal measured in the reported investigations is expected to emanate from the surface of the solid.

4. DESORPTION OF ANIONS AND METASTABLE SPECIES

The DEA reactions can be identified by monitoring the anion ESD signal from a molecular adsorbate as a function of electron energy (i.e., the anion yield function). Since a given molecular configuration of a transient negative ion occurs at a well-defined energy, each peak or maximum in a yield function identifies the energy of a particular transient anion state. The anion yield functions for condensed hydrocarbons [12], N_2O [14] and O_2 [15] are given as examples in Figs 2, 3(a) and 4(a) and (b), respectively.

The H^- yield functions from all saturated hydrocarbon films studied to date exhibit a single DEA peak whereas in those for the other anion fragments from multilayer films of saturated and unsaturated hydrocarbons, one or two well-defined transient anions contribute to the signal as shown in Fig. 2. The broad peaks due to dissociating anions are not always so well resolved as can be seen from the O^- ESD yield function from a four monolayer (ML) N_2O films of Fig. 3(a). Here, the shoulder at 7.0 eV and the maxima at 9.0 and 15.5 eV identify the positions of three dissociative N_2O^- states. The curve in Fig. 3(b) represents the metastable desorption signal produced by 5-25 eV electrons impinging on a four-layer N_2O target deposited on Pt(lll) [16]. The metastable signal is believed to be produced essentially by N_2^* metastable molecules arriving at the detector. This signal

5

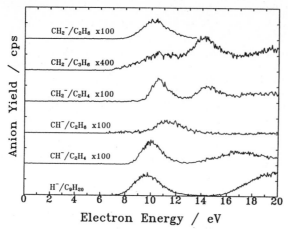

Fig. 2   Energy dependence of the H⁻, CH⁻ and CH_2^- electron stimulated desorption (ESD) yields produced from multilayer films of C_2H_6, C_3H_6, C_2H_4 and C_9H_{20} condensed on a Pt substrate. Each division of the vertical scale corresponds to 5,000 cps. The gain factor is indicated for each CH_2^- or CH⁻ yield function.

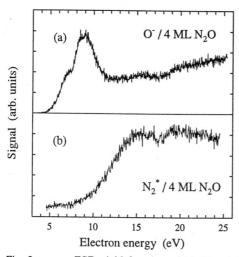

Fig. 3   ESD yield function of (a) O⁻ and (b) metastable species from a 4-monolayer (ML) film of N_2O condensed on Pt.

may, a priori, arise from direct excitation of a dissociative electronic state (i.e., $e + N_2O \rightarrow N_2O^* + e \rightarrow N_2^* + O + e$) or excitation via resonance scattering (i.e., $e + N_2O \rightarrow (N_2O^-)^* \rightarrow N_2O^* + e \rightarrow N_2^* + O + e$), but DEA is also a possible source of N_2^* (i.e., $e + N_2O \rightarrow (N_2O^-)^* \rightarrow N_2^* + O^-$). Comparison of the set of data in Fig. 3 indicates that the maximum at 15.1 eV in the metastable signal is probably due to N_2^* arising from the decay of an $(N_2O^-)^*$ state into the dissociation limit $N_2^* + O^-(^2P)$. Accordingly, the higher-energy maximum in Fig. 3(a) would correspond to the latter reaction and the other features to the

6

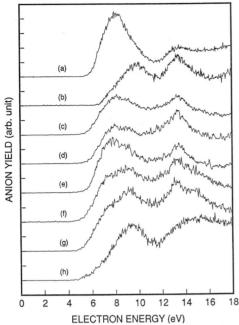

ANION YIELD (arb. unit)

ELECTRON ENERGY (eV)

Fig. 4   (a) and (b): energy dependence of ESD O$^-$ yields for a 4-ML O$_2$ coverage of Pt. Energy dependence of ESD OH$^-$ yields from 1 layer of (c) C$_8$H$_{18}$ and (d) C$_5$H$_{12}$ adsorbed on 4 ML of O$_2$ and 5 ML of a mixture of O$_2$ at 25% vol. in (e) C$_5$H$_{12}$, (f) C$_4$H$_8$, (g) C$_3$H$_6$ and (h) C$_2$H$_4$.

decay of two N$_2$O$^-$ states into the ground state dissociation limit N$_2$($^1\Sigma_g^+$) + O$^-$(^{2}P). Interestingly, the energy difference between the 9.0 and 15.5 eV maxima in Fig. 3(a) is close to that required to produce N$_2^*$($^3\Pi_g$) metastables.

The effect of discriminating on the energy of the desorbing anions is shown in Fig. 4(a) and (b) where the signal arises from O$^-$ produced by 4-16 eV electron impact on a 3 ML O$_2$ film. The curve in (b) was recorded with a potential of 1.5 volts retarding O$^-$ ions. In the gas-phase [1] and in the dilute matrices [17], the only anion state which has the right characteristics to produce DEA is the $^2\Pi_u$ state of O$_2^-$ at 7 eV. Two other states, $^2\Sigma_g^+$ at 9 eV and $^2\Sigma_u^+$ at 13 eV, which could also produce O$^-$ by DEA, cannot be formed in the single-target-electron frame of reference because of the cylindrical symmetry selection rule which excludes $\Sigma^- \leftrightarrow \Sigma^+$ transitions [18]. However in pure O$_2$ films, the isolated O$_2$ symmetry is broken and the DEA signal from the Σ states produces the two well-resolved peaks around 9 and 13 eV in the yield function of Fig. 4(b). The $^2\Pi_u$ O$_2^-$ state produces O$^-$ ions of lower kinetic energy and therefore produces a stronger signal in the unretarded data of Fig. 4(a) giving rise to a broad peak which is a convolution of the signals from the $^2\Pi_u$ and $^2\Sigma_g^+$ states.

5.   REACTIONS INDUCED BY DISSOCIATIVE ATTACHMENT

The radicals and anions produced with several eV of energy by DEA are expected to be highly reactive and lead to the formation of new products as they react with neighboring

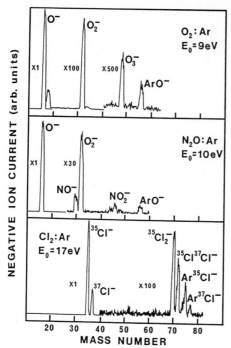

Fig. 5   Anion mass spectra produced by bombarding with 9, 10, and 17 eV electrons Argon multilayer films containing 10% vol. O_2, 30% vol. N2O and 20% vol. Cl_2, respectively. The numbers on the left of major peaks indicate the gain relative to the most abundant mass in each spectrum.

atoms or molecules located within or at the surface of a solid. Negative ions produced in reactions within rare gas matrices are shown in Fig. 5, which exhibits the mass spectra of anions desorbing from the surface of 80 Å thick matrix films containing 10%, 30% and 20% volume of O_2, N_2O and Cl_2, respectively [19]. The incident electron energy was 9, 10 and 17 eV, respectively. A detailed study [17] of the yield functions of all species produced in the upper portion of Fig. 5 indicated that the initial process responsible for the formation of O_2^- and O_3^- anions is the formation of O_2^- in the dimeric configuration $O_2 \cdot O_2^-$ which could decay to $O + O_2 \cdot O^-$ or convert electronically to O_4^- dissociating into O_2^- $+ O_2$ and $O_3^- + O$. The $Ar \cdot O^-$ van-der-Waals-complex anion could be formed from the initial configuration $Ar \cdot O_2 \cdot O_2^-$ dissociating into $Ar \cdot O^- + O_2 \cdot O$.

Reactions induced in O_2-hydrocarbon mixtures deposited on Pt [20] are shown in Fig. 4: (c) and (d) represent the OH^- yields arising from electron impact on a single ML of n-C_8H_{18} and n-C_5H_{12}, respectively, adsorbed on 4 ML of O_2 and (e), (f), (g) and (h) represent the OH^- yields for a 5 ML film composed of a mixture of O_2 at 25% vol. in n-C_5H_{12}, $1-C_4H_8$, C_3H_6, and C_2H_4, respectively. Since no compound at the film surface contains both hydrogen and oxygen, the OH^- ions cannot be produced by direct electron impact. However, they can result from surface reactions between ground-state molecules (i.e., O_2, C_nH_{2n+2} and C_nH_{2n}) and anions produced directly by the electron beam (i.e., O^-, H^- and CH_n^-). OH^- can be formed via the reactions

$$O^- + C_nH_{2n+2} \rightarrow OH^- + C_nH_{2n+1} \quad (1) \quad \text{or} \quad O^- + C_nH_{2n} \rightarrow OH^- + C_nH_{2n-1} \quad (2)$$

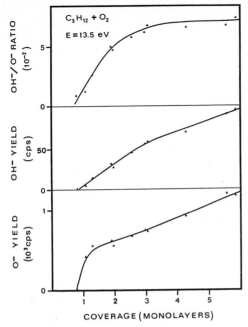

Fig. 6   Thickness dependence of the O^- and OH^- yields and OH^-/O^- intensity ratio for a film containing O_2 at 25% vol. in $n\text{-}C_5H_{12}$ deposited on Pt. The incident electron energy is 13.5 eV.

in the case of saturated or unsaturated hydrocarbon adsorbates, respectively. <u>A priori</u>, the reactions

$$H^- + O_2 \rightarrow OH^- + O \quad (3) \quad \text{and} \quad CH_n^- + O_2 \rightarrow OH^- + CH_{n-1}O \quad (4)$$

can also lead to ESD of OH^- since, as shown in Fig. 2, transient anions formed from condensed hydrocarbons can dissociate into H^- or some CH_n^- fragments. However, the OH^- yield functions can be correlated to the line shape of the O^- signal in Fig. 4(b) and to the O_2^- states producing that signal. Any OH^- formed by reactions (3) and (4) would bear the "signature" of the yield function for H^- and CH_n^- production (e.g., see Fig. 2), but this is not the case. *The OH^- signal bears the "signature" of DEA from O_2.* From such comparisons and studies as a function of O_2 concentration [20], the OH^- signal can be ascribed to reactions (1) and (2). The $^2\Sigma_u^+$ state of O_2^- is clearly visible as a strong peak at 13 eV in curves (c) to (g) of Fig. 4 and the $^2\Sigma_g^+$ and $^2\Pi_u$ states at 7 and 9 eV, respectively, are even resolved in curves (f) and (g) of Fig. 4.

Details on substrate coverage dependence of the O^- and OH^- yields measured from the mixture film of 25% vol O_2 in $n\text{-}C_5H_{12}$ are given as an example in Fig. 6 for an incident energy of 13.5 eV. The ratio of the OH^- signal to that of O^- (OH^-/O^-), defined as the observable efficiency of the reaction leading to OH^- formation, has a maximum value of 0.08 ± 0.01 above 4 ML; at 7.8 eV, the efficiency has the value of 0.016 ± 0.004 for the same coverage. The efficiencies of these electron-beam-induced surface reactions are of the order of 1% to 8% for a multilayer film (i.e., for an insulating surface) at low energies [20]. At submonolayer coverage of the reactants (i.e., on a metal surface), this efficiency

9

is reduced by at least an order of magnitude and O⁻ formation via DEA by more than two orders of magnitude. Thus, it appears that insulators and possibly semiconductors are well-suited surfaces for this type of electron-beam chemistry. Metal surfaces covered by a single layer of the reactants also provide significant reaction yields. These results indicate the possibility of inducing specific chemical modifications with low-energy electrons.

6. CHARGE AND ENERGY TRANSFER LEADING TO DISSOCIATIVE ATTACHMENT

When a transitory anion is formed in or at the surface of a well-ordered molecular solid composed of only one type of atom or molecule, it can, in principle, move in the solid or at its surface with a finite well-defined wave vector in a manner similar to that of an exciton. In the case of a core-excited resonance, we therefore expect the charge and electronic excitation energy to move from one site to another in the form of an electron-exciton complex. Such a simultaneous energy and charge transfer is difficult to demonstrate for interaction between similar species, but recently it has been shown [21] that such a complex formed in a rare gas multilayer film can migrate to the film's surface and under certain conditions transfer its energy and charge to a molecular adsorbate. To illustrate this phenomenon, the D⁻ yield functions are shown in Fig. 7 for (a) a pure C_2D_6 8-ML film, (b) 0.2 ML C_2D_6 deposited on a 80 ML Kr spacer film and, (c) 0.05 ML C_2D_6 deposited on a 50 ML Xe spacer film [21]. The D⁻ yield distribution measured from pure multilayer C_2D_6 shown in (a) has a single broad peak at 9.9 eV which arises from the DEA process $e + C_2D_6 \rightarrow C_2D_6^- \rightarrow C_2D_5 + D^-$. When deposited on Kr and Xe substrates,

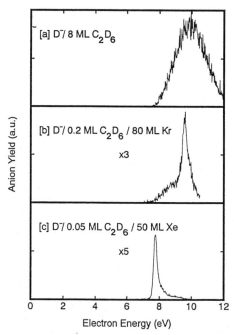

Fig. 7   D⁻ desorption yield obtained for (a) a 8 ML film of C_2D_6, (b) 0.2 ML C_2D_6 deposited on a 80 ML Kr spacer, and (c) 0.05 ML C_2D_6 deposited on 50 ML of Xe.

strong and narrow enhancements are observed at energies slightly below that needed to create an exciton in the rare gas (Fig. 7(b) and (c)). The normal ESD yield from the C_2D_6 is significantly reduced at incident electron energies well above and below that of the enhancement. For fixed quantities of C_2D_6, the absolute enhancement intensity increases as the rare-gas substrate is made thicker; the intensity of the sharp peak relative to the broad D^- distribution increases as the adsorbate coverage is decreased for a fixed substrate thickness. These trends suggest that the initial process occurs in the bulk rather than at the surface, and that it has significant mobility. Enhanced anion yields have also been obtained with an Ar substrate and with other molecular adsorbates [21]. In each of these cases, the enhanced ESD yields are related to electronic excitations of the rare-gas, however the energies of the observed enhancements are *below* the energy of the lowest optically accessible exciton states of the rare gas solids. The sharp enhancements whose energies were found to be independent of the incidence angle of the impinging electrons, could be explained by involving the formation of a core-excited resonance below the energy of the lowest exciton in the rare gas substrates (i.e., a transient anion formed by trapping an electron in the positive electron affinity well of the lowest excited state of the rare gas solid). Such states are well known for isolated rare-gas atoms (e.g., $Kr^-[4p^55s^2]$ $^2P_{3/2}$ at 9.52 eV and Xe^- $[5p^56s^2]$ $^2P_{3/2}$ at 7.90 eV [22]), where they are observed ~0.5 eV below the lowest excited neutral state of the atoms. Briefly stated, the core excited anion state or "electron-exciton complex" moves to the surface of the rare gas where it exchanges its charge and energy by forming a core-excited anion of the molecular adsorbate (e.g., $C_2D_6^-$) which afterwards dissociates thus producing the sharp enhancement in the ESD yield function (e.g. in the D^- signal of Fig. 7b and c) precisely at the energy of the complex formation.

7. SURFACE CHARGE TRAPPING

When an anion is produced by DEA, it does not necessarily escape in vacuum, but can remain trapped in the solid or at its surface where it can participate in ion-molecule reactions. Surface charging by DEA for O_2 and H_2O molecules deposited on a solid rare gas surface has been measured [10,23] with the method mentioned in Section 3. The surface charging coefficient A_s for 0.1 ML O_2 on a 20 ML Kr film is represented by the middle curve (c) in Figure 8, as a function of the energy of the electron beam causing the charge. This coefficient is directly proportional to the trapping cross section. Pure Kr films exhibit negligible charge accumulation in this energy range. The result in Fig. 8(c) is comparable to the energy dependence of the anion yields derived from the gas-phase attachment rate coefficient for stable O_2^- production (Fig. 8a) [24] and the gas-phase DEA cross section (Fig. 8b) [1]. Curve (d) represents the ESD signal from a 20 ML Kr film covered with 0.1 ML of O_2.

From the correspondence between the curves of Fig. 8, charge trapping at the surface of the Kr film above 4 eV has been ascribed to dissociation of the $^2\Pi_u$ state of O_2^- into the limit $O(^3P) + O^-(^2P)$ [10]. Desorbing $O^-(^2P)$ ions could arise from either the $^2\Pi_u$ or the $^2\Sigma_g^+$ state of O_2^-, but only a small fraction of the signal probably arises from the Σ state since this latter must be formed near another O_2 molecule in order to violate the $\Sigma^- \leftrightarrow \Sigma^+$ selection rule. The maximum in the charging coefficient A_s is found about 0.6 eV below the maximum in the gas-phase O^- yield due to lowering of the $^2\Pi_u$ state by the polarization potential of the Kr surface. The DEA maximum in the ESD yield arises from the same O_2^- surface state as that in the charging signal. However, the DEA maximum lies at higher energy, since in order to overcome the polarization force of the Kr film, desorbing O^- ions must arise, on the average, from a higher energy portion in the F-C region of the O_2^- $^2\Pi_u$ state.

11

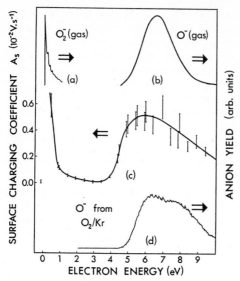

Fig. 8   Anion yields produced by 0 to 10 eV electron impact on (a and b) gaseous
O_2 and (d) on a 20 ML Kr film covered with 0.1 ML O_2. The electron
energy dependence of the surface charging coefficient A_s for 0.1 ML O_2 on
a 20 ML Kr film is shown in (c).

        In the 0 to 1.2 eV range, temporary electron attachment to gaseous O_2 leads to for-
mation of the $^2\Pi_g$ resonance state of O_2^- which occurs in vibrational levels v ≥4, since the
v <4 levels lie below ground state O_2 [24]. However, when during the lifetime of the $^2\Pi_g$
anion, vibrational energy is transferred to another particle, the v<4 levels can be reached
and the electron becomes permanently attached to the molecule. A similar stabilization
process is responsible for surface charge accumulation in the range 0 to 2 eV. The electron can either lose energy to vibrationally excite O_2 via the $^2\Pi_g$ O_2^- state and afterwards
stabilize at a trapping site, or the $^2\Pi_g$ state may itself stabilize by energy transfer to phonons. The estimated lifetime of the $^2\Pi_g$ O_2^- state within clusters is of the order of 10^{-12} s
[25] which is comparable to the phonon vibrational period of the Kr lattice (5×10^{-13}s) [26].
Hence, the additional electron resides a sufficiently long time at an O_2 site to polarize the
phonon modes of the Kr lattice.

8.   DIFFERENCES BETWEEN ISOLATED AND SURFACE TRANSIENT ANIONS

Although considerable progress has recently been made in the field of resonance electron
scattering from molecular adsorbates [2] and molecular solids [27], knowledge on transient
anion formation is still much more abundant in the gas phase [1]. Very often this is the
only information available to explain surface reactions which may involve electron capture
by an adsorbate. So, guidelines are needed in attempting to apply gas-phase information to
condensed phase experiments.
        When compared to the gas-phase data, transient anions formed at surfaces by electron
capture by a physisorbed molecule are found to have the following characteristics: (1) their
energies are usually lowered by about 0.5 to 2.0 eV depending on the polarizability of the

surface and/or the lowering of the anion's symmetry; (2) they usually have a lower number of intra-molecular decay channels due to their lower energies but new channels appear for electron emission into the surface; (3) lifetimes are either shorter or longer depending on the change in the number of decay channels and symmetry; (4) probabilities for initial electron capture, decay into inter- and intramolecular excitations and decay into DEA may vary by orders of magnitude depending on energy, symmetry and orientation. It may not be easy to predict the condensed phase behavior from the gas-phase information because of the interrelationship between these various parameters. In the case of DEA, the lifetime is often a crucial parameter which may affect exponentially the magnitude of the anion yields, but breakdown of molecular symmetry (e.g., Σ^- or Σ^+) may also have drastic effects [28]. Chemisorption changes the molecular characteristic of the target itself so that any comparison with gas-phase data is even more hazardous in this case.

The author is indebted to P. Rowntree, L. Parenteau, M. Michaud and A. Mann for constructive comments and assistance. This work is sponsored by the Medical Research Council of Canada and the Canadian Centres of Excellence in Molecular and Interfacial Dynamics.

REFERENCES

1. For a review on anion formation see H.S.W. Massey, *Negative Ions*, University Press, London (1976); L.G. Christophorou, *Electron-molecule Interactions and their Applications*, Vol. 1 and 2, Academic Press, Orlando, 1984.
2. For review see L. Sanche, J. Phys. B *23*, 1597 (1990).
3. For a review see X.-L. Zhou, X.-Y. Zhu and J.M. White, Surf. Sci. Rep. *13*, 73 (1991).
4. W. Ho, SPIE Vol. 1056, *Photochemistry in Thin Films*, 157 (1989).
5. Z. Lu, M.T. Schmidt, D.V. Podlesnik, C.F. Yu, and R.M. Osgood, Jr., J. Chem. Phys. *93*, 7951 (1990) and citations therein.
6. J.M. Seo, S.E. Harvey, Y. Chen, and J.H. Weaver, Phys. Rev. B *43*, 11893 (1991) and citations therein.
7. K. Jacobi, M. Bertolo, P. Geng, W. Hansen and C. Astaldi, Chem. Phys. Lett. *173*, 97 (1990).
8. J.W. Gadzuk, Phys. Rev. B *44*, 13466 (1991).
9. L. Sanche and M. Michaud, Phys. Rev. Lett. *59*, 645 (1987).
10. L. Sanche and M. Deschênes, Phys. Rev. Lett. *61*, 2096 (1988).
11. G. Leclerc, A.D. Bass, M. Michaud, and L. Sanche, J. Electr. Spectr. Rel. Phenom. *52*, 725 (1990).
12. P. Rowntree, L. Parenteau, and L. Sanche, J. Phys. Chem. *95*, 4902 (1991).
13. R.M. Marsolais, M. Deschênes, and L. Sanche. Rev. Sci. Instrum. *60*, 2724 (1989).
14. L. Sanche and L. Parenteau, to be published.
15. R. Azria, L. Parenteau, and L. Sanche, Phys. Rev. Lett. *59*, 638 (1987).
16. A. Mann, G. Leclerc, and L. Sanche, to be published.
17. L. Sanche, L. Parenteau, and P. Cloutier, J. Chem. Phys. *91*, 2664 (1989).
18. H. Sambe and D.E. Ramaker, Chem. Phys. Lett. *139*, 386 (1987).
19. L. Sanche and L. Parenteau, J. Chem. Phys. *90*, 3402 (1989).
20. L. Sanche and L. Parenteau, J. Chem. Phys. *93*, 7476 (1990).
21. P. Rowntree, H. Sambe, L. Parenteau, and L. Sanche, to be published.
22. L. Sanche and G.J. Schulz, Phys. Rev. A *5*, 1672 (1972).
23. A.D. Bass and L. Sanche, J. Chem. Phys. *95*, 2910 (1991).

24. D. Spence and G.J. Schulz, Phys. Rev. A *5*, 724 (1972).
25. Y. Hatano, in *Electronic and Atomic Collisions*, Elsevier, New York (1986).
26. K. Kern, P. Zeppenfield, R. David, and G. Comsa, Phys. Rev. B *35*, 886 (1987).
27. For a review see L. Sanche, in *Excess Electrons in Dielectric Media*, C. Ferradini and J.-P. Jay-Gerin, eds., CRC Press, Boston (1991).
28. L. Sanche, Comments At. Mol. Phys. *26*, 321 (1991).

Desorption of H$^-$, O$^-$ and OH$^-$ Ions During Impact of Very Low Energy Electrons on Surfaces

*M. Bernheim and T.-D. Wu**

Laboratoire de Physique des Solides, Bât. 510,
Université Paris-Sud, F-91405 Orsay, France
*Visitor from Xiamen University, People's Rep. of China

Abstract. Large variations in H$^-$, O$^-$ and OH$^-$ desorption yields, from the same aluminum crystal saturated with water molecule, suggest that additional charge exchanges followed the initial bond breaking. Ejected with similar kinetic energies, the heavy ions encounter increased neutralizations in spite of their high electron affinities. In addition the reported experiments show that preliminary electron bombardment often remain essential to raise the efficiency of the low energy desorption process. A variation of the surface work function suggests that important reorganizations occur during the electron impact on the investigated surfaces.

1. Introduction

Only a few experiments dealt with the desorption of negative ions during low energy electron collisions on chemisorbed molecules. However, below *15 eV*, negative ion desorptions might occur for the high electron affinity elements of adsorbed molecules [1,2,3]. Here, the resonant desorption is attributed to dissociation of anti-bonding states as is the case for the dissociative electron attachments on free molecules [4].

Modifications of surface compositions during low energy electron bombardments may influence many experimental investigations, electron loss spectroscopies or scanning tunneling microscopies [5], for instance.

The main purposes of the reported experiments were
- to characterize the relevant desorption process,
- to study both sample and chemisorbed molecule influences,
- to assess a possible surface analytical microscopy based on controlled selective desorption from heterogeneous samples [2].

Further on, a better knowledge of processes at play during very low energy electron interactions with surfaces might contribute to a better understanding of heterogeneous catalytic reactions.

2. Main features of the experimental set-up

The main part of the experimental set-up consisted of an electrostatic immersion lens whose cathode was the investigated surface. Both incident electron and desorbed ion beams crossed this lens along the main optical axis but in opposite directions.
- A *4 keV* electron beam was decelerated in the very last part of its path to a final energy *ev*, directely controlled by the difference *v* of voltages applied to the sample and to the electron source filament. In such a configuration, the constant size

Springer Series in Surface Sciences, Vol. 31
Desorption Induced by Electron Transitions DIET V
Editors: A.R. Burns · E.B. Stechel · D.R. Jennison   © Springer-Verlag Berlin Heidelberg 1993

of the electron beam on the investigated surface kept very constant the incident electron density for any incident electron energy [6, 7].

- Ion collection occured on an area (125 μm diameter) optically defined by a field stop. In addition, an aperture stop set both energy and angular discriminations for the desorbed ion collection [8, 9].

For any ion species, a computer controlled mass spectrometer enabled us to record the desorption yield variations with incident electron energy as well as the ion energy distributions for any electron energy.

Slight differences in work function between sample surface and filament caused variations of the final electron energies also. The assessment of the onset of electron beam intensity i collected on the sample yielded an *in situ* determination of the surface work function variations. Thus, a precise calibration of incident electron energy was reached for any state of the adsorbed layer.

It is worth saying that, for increasing incident energies, intensity i variations gave additional information concerning the surface density of states between Fermi and vacuum levels (according to the *Total Current Spectrometry* method [10]) as well as variations of the secondary electron emission yield.

A mass filtered Ar^+ ion beam was used to sputter all impurities off the investigated surface ; the surface cleanliness was monitored by the usual secondary ion emission. A controlled molecular beam was directed on the surface afterwards for a time long enough to build thin saturated chemisorbed layers. Obviously, the real experimentation was performed after such a delay to get rid of most of unfixed molecules. During each of these steps the sample remained at room temperature, isolated in ultra high vacuum.

A detailed description of the whole experimental set-up, of its practical adjustments and of most experimental results, was reported elsewhere [11].

3. Some experimental results and discussion

The experiments established mainly the occurrence of transformations in the adsorbed layers during electron bombardment even when the incident energy was kept very low.

For instance figure 1 reports the variations of H^- desorption yield and collected intensity i (curve **B**) for an (111) aluminum surface saturated with chemisorbed water molecules. Both curves were recorded simultaneously for increasing incident energies on a freshly prepared surface never yet exposed to electrons. According to the reference curve **A**, recorded on the bare surface, the water vapour adsorption was followed by a *1,1 eV* lowering of the surface work function and by a *20%* reduction of collected intensity, due to increased electron reflection from the surface.

About $6 \ 10^{11}$ electrons per second reached the investigated area. Thus for *6 eV* energy electrons, the H^- desorption probability could be estimated about 10^{-5} ion per electrons in conditions of a full ion collection. During these recordings, the total electron dose received by the surface was $3,6 \ 10^{14}$ electrons corresponding to an integrated flux at about $0.25 \ C/cm^2$, and to an arrival of about $7 \ 10^3$ electrons per adsorbed molecule. The mean ion desorption yield remained much too low to induce significant variation in superficial composition. Nevertheless the next recording gave

16

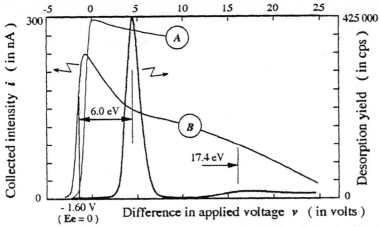

Fig. 1 Recordings relative to the H⁻ ion desorption from an (111) Al crystal surface saturated with chemisorbed water molecules. Electron intensity i covers an area larger by a factor 3 than the area where the usefull ion collection occurs. The origin of energy scale ($v = -1.6\ V$) corresponds to the collection of electrons ejected from the filament with the most probable kinetic energy.

evidence of a *0.9 eV* increase in the work function attributed to a reorganization of the adsorbed molecules. Then as the electron bombardment lasted the work function increased further and a round hump grew progressively for the i intensity curve. The H⁻ ion intensity decreased gradually. The resonant desorption energy was shifted just slightly towards higher energies (*6.2 eV*).

The yields for O⁻ and OH⁻ ions desorbed from the same aluminum surface saturated with water molecules were much lower than the yield for H⁻ ions (relative factors *1 / 200* and *1 / 2000* respectively) in spite of high electron affinities of O and OH (*1.46* and *1.83 eV* respectively, compared to *0.75 eV* for H). Since the energy distributions of H⁻, O⁻ and OH⁻ ions were very similar, these yield variations could hardly be attributed to a mass dependend energy transfer in the electron collision process. Let us add that the image force influence is the same for any ion species.

Surprisingly, for a freshly prepared surface, the O⁻ and OH⁻ ion desorptions never occured under thresholds of about *20 eV* (Fig.2). However the electron bombardment renders the desorption process efficient for low incident energies (Fig.3). It is known that a dissociation in H and OH radicals occurs as soon as water molecules were adsorbed on aluminum [12]. The reported experiments show that the electron bombardment triggers further transformations of the sample surface.

Besides similar preliminary electron bombardment was invariably necessary to cause efficient O⁻ ion desorption out of chemisorbed oxygen or carbon monoxide. In addition the O⁻ desorption features scarcely depend on the nature of the adsorbed species (CO, O₂, or H₂O) and varies much more with the nature of the investigated metal. Thus we assume that the electron impacts might build similar surface structures that controll most of the desorption process.

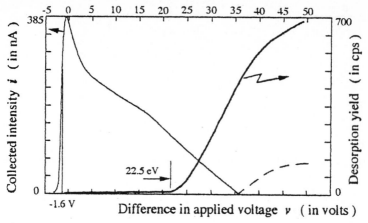

Fig. 2 Variations of the OH⁻ desorption yield measured on an (111) Al crystal in conditions similar to figure 1. For the investigated energy range, the growth of secondary electron emission renders negative the *i* intensity for energy above *38 eV*.

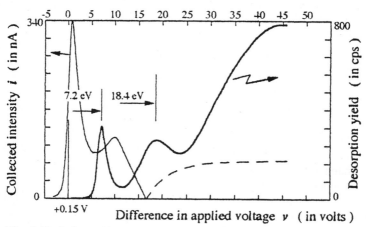

Fig. 3 Variations of OH⁻ desorption yield and of intensity *i* once an equilibrium was build after several sucessive recordings. A clear resonant desorption of OH⁻ is now centered at *7.2 eV*. At about *18.4 eV* a wide additional desorption peak might be related to electrons having experienced a *10 eV* energy loss during excitation of surface plasmon.

Intense negative ion desorptions might also occur for many metallic or semiconductor surfaces exposed to air as well as for other compound samples. Tunneling microscope investigations established that a 2x1 surface reconstruction was triggered for a Si (111):H-1x1 single crystal as soon as the negative voltage applied to the tip passed *3.5 V*. The H⁻ ion desorption recorded for such a sample immediately confirmed the assumed hydrogen desorption for the same said energy range (fig. 4).

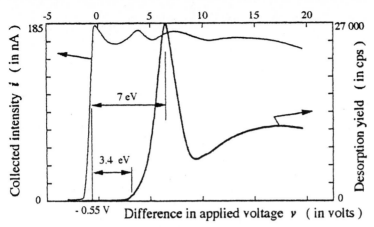

Fig. 4 H⁻ ion desorption from a Si (111) : H - 1x1 sample. This sample was treated in an ammonia fluoride solution to build a stable surface, free of usual reconstructions. Indeed in-site chemisorbed hydrogen atoms were used to saturate the surface dangling bonds [13]. The intensity of H⁻ ions decreased as the surface exposure to the electron increased. The structures reported on the intensity i curve depend on the electron surface exposure. These structures vary with the crystal orientation and surface preparation as well.

For several metal surfaces saturated with water molecules, Nickel for instance, the H⁻, O⁻ and OH⁻ desorption efficiencies remain particularly low. However modifications in the work function and in collected intensity hint that the water vapour adsorption was followed by significant electron induced transformations.

Here we presume that the high work function of Nickel surfaces increased the neutralization of the desorbed particles notabily. A similar assumption may explain the weak desorption yields for heavy ions (O⁻ and OH⁻) out of the aluminum surfaces saturated with water vapour molecules

Hence in many cases a quenching process, due to additional electron transitions to the surface, is supposed to follow the initial bond breaking :

The ion survival dominates the efficiency of the ion desorption process. As for the usual secondary ion emission [14], a velocity and work function dependant neutralization reduces the efficiency for the desorption of heavy negative ions especially.

Acknowledgement The authors would like to thank **P. Dumas** for providing the relevant silicon samples.

References

[1] Liu Zhen Xian and D. Lichtman, *Surface Science* **114** , 287 (1982).
[2] M. Bernheim, M. Chaintreau, R. Dennebouy and G. Slodzian, *Surface Science* **126** , 610 (1983).

[3]L. Sanche and L. Parenteau, *J. Chem. Phys.* **93**, 7476 (1990).

 P. Rowntree, L. Parenteau and L. Sanche, *J. Chem. Phys.* **94**, 8570 (1991)

[4] L. G. Christophorou, *Atomic and Molecular Radiation Physics*, Wiley (1971).

[5] R. S. Becker, G. S. Higashi, Y. J. Chabal and A. J. Becker *Phys. Rev. Letters* **65**, 1917 (1990).

[6] M. Bernheim and T.-D. Wu, *Desorption Induced by Electronic Transitions, DIET IV*, Vienna.(1989), *Springer series in Surface Sciences* **19**, 208(1990).

[7] T.-D. Wu and M. Bernheim, *Inelastic Ion Surface Collision, IISC 8*, Vienna (1990), *Nucl. Instr. Meth.* ser. B **58**, 496(1991).

[8] R. Castaing and G. Slodzian, *J. Phys. E* **14**, 1119 (1981).

[9] G. Slodzian, *Optik* **77**(4), 148 (1987).

[10] S. A. Komolov and L. T. Chadderton, *Surface Science* **90**, 379 (1979).

[11] Ting Di Wu, *Thesis* (Orsay, 1992).

[12] P. A. Theil and Th. E. Madey, *Surface Science Reports* **7**, 211 (1897)

[13] P. Dumas ,Y.J. Chabal and G.S. Higashi, *Phys. Rev. Letters* **65** 1124 (1990)

[14] N.D. Lang and J.K. Nørskov, *Physica Scripta* **T6**, 15 (1983).

Electron and Photon Stimulated Desorption of Negative Ions from Oriented Physisorbed Molecules

R.J. Guest[1], R.A. Bennett[1], L.A. Silva[1], R.G. Sharpe[1], J.C. Barnard[1], R.E. Palmer[1], and M.A. MacDonald[2]

[1]Cavendish Laboratory, University of Cambridge,
 Madingley Road, Cambridge CB3 0HE, UK
[2]SERC Daresbury Laboratory, Warrington, Cheshire WA4 4AD, UK

We have studied electron and photon stimulated desorption of O^- from oriented O_2 physisorbed on graphite. Two different dissociation processes have been observed in electron stimulated desorption, dissociative attachment (DA) and dipolar dissociation (DD). Dissociative attachment is shown to be sensitive to the molecular orientation on the surface, whereas in the case of dipolar dissociation, the yield is less orientation dependent, a result we attribute to the rotation of the molecule on the surface during the dissociation process. Photon stimulated desorption of O^- ions from O_2/graphite may also proceed via substrate excitation and photoelectron impact on the adsorbate.

1. Introduction

It has now been established that two molecular dissociation processes, dissociative electron attachment (DA: $e^- + AB \rightarrow AB^- \rightarrow A + B^-$) and dipolar dissociation (DD: $e^- + AB \rightarrow A^+ + B^- + e^-$) observed in gas phase electron scattering are preserved when molecules are physisorbed onto a surface [1]. Indeed, more DA resonances may be seen in the condensed phase than in the gas phase. This has been attributed to the breaking of the molecular symmetry in the presence of the substrate and neighbouring molecules, which allows electronic transitions forbidden in the gas phase to take place. This phenomenon has been clearly seen in the case of O_2; only one resonance is seen in the gas phase (6.5eV) whereas four resonances (6, 8, 9 and 13eV) have been invoked to explain studies of O_2 condensed on polycrystalline Pt, or on rare gas films [2].

An alternative mechanism for electron stimulated desorption (ESD) of molecules and ions is via an indirect process involving charge transfer from the substrate [3-6]; such charge transfer can be induced by photoemission of electrons from the substrate. The recent observation of negative ion desorption by Dixon-Warren et al [7] gives perhaps the most direct evidence for such a process. In general, there will always be competition between direct photodissociation and the indirect charge transfer mechanism; hence determination of the energy dependence of the relevant cross-sections is a vital signature of the mechanism operating in a given system. For example, the cross-section for photon stimulated desorption (PSD) from a semiconductor surface may display a threshold equal to the optical band gap of the substrate, the photon energy above which hot carriers can be produced [3,8].

We present the results of electron and photon stimulated desorption studies of physisorbed oxygen on graphite. Oxygen forms a number of different phases on graphite, dependant on the coverage and temperature [9]; we exploit this by varying the O_2 coverage to produce molecules in different orientations on the same surface.

Springer Series in Surface Sciences, Vol. 31
Desorption Induced by Electron Transitions DIET V
Editors: A.R. Burns · E.B. Stechel · D.R. Jennison   © Springer-Verlag Berlin Heidelberg 1993

At low temperatures this system exhibits a submonolayer phase transition between the low coverage, lying down δ phase (20° mean tilt), and the higher coverage standing up ζ phase (35° mean tilt). In the multilayer regime, the molecule remains standing up on the surface, but the mean tilt is reduced. In all phases, the molecule is thought to be librating on the surface, and the mean tilt indicates an instantaneous average, obtained with NEXAFS [10]. Thus by selecting our oxygen coverage between the saturated δ and ζ phases, we are able to select the molecular orientation, and hence compare the yield and ion angular distributions for the different alignments.

A hemispherical monochromator was used to produce an electron beam to induce ESD. The beam resolution was 35meV and current 5nA; this current was sufficiently low to keep the ESD yield almost constant over a period of a few hours. The ions desorbing from the surface were detected with a quadrupole mass spectrometer (Hiden SIMS probe) mounted at 90° to the electron beam. Angular distribution measurements were performed by rotating the sample between the fixed monochromator and mass spectrometer. The PSD measurements were taken at beamline 3.1 at the Daresbury Synchrotron Radiation Source using tuneable vacuum-ultra-violet (VUV) radiation in the energy range 15-35eV. All spectra were calibrated to the photon flux, as measured by the photomultiplier current viewing a sodium salicylate coated window under irradiation. The photons were incident at an angle of 45° to the surface with the desorbing ions measured normal to the surface. An extraction voltage of 2V was applied to the mass spectrometer to improve signal intensity. During both experiments the highly oriented pyrolytic graphite sample was mounted in good thermal contact with, but electrically isolated from a liquid helium cryostat so that its temperature remained at less than 30K. The electrical isolation allowed the incident electron current to be measured during ESD and the photoemitted current during PSD. Correction for the electron background when measuring negative ions was made by subtracting the mass 15 or mass 17 signal.

2. Results & Discussion
2.1 Electron Stimulated Desorption

We have studied the O^- yield as a function of electron energy both in the negative ion resonance region (5-17eV) and also at higher energies where dipolar dissociation occurs. In a recent paper [11], Silva and Palmer reported observing three dissociative attachment resonances in the ζ phase. The angular distribution of O^- ions desorbed via the three resonances was found to be identical. Here, we compare resonant and non-resonant dissociation, and so focus on one resonance (at 13.5eV), and compare this with the dipolar process.

The O^- yield as a function of electron energy is shown in figure 1 for different O_2 doses corresponding to δ phase and ζ phase coverages. The ζ phase exists as both a monolayer and a bilayer structure on graphite; at the present time we are not able to determine where in the ζ phase regime our results were taken. In addition we show the O^+ yield for comparison. The O^+ yield shows a threshold below which no ions are detected and above which the yield rises with electron energy; this is characteristic of dipolar dissociation. In the O^- spectra however, ions are detected below this threshold due to dissociative attachment. The intensity in this region is very strongly dependent on coverage as at high O_2 coverages, resonances are clearly seen, which are absent at low coverages. We have measured the angular distribution of the

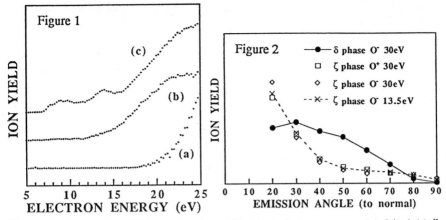

Figure 1. Oxygen ion yield as a function of incident electron energy. (a) O^+ yield, ζ phase, (b) O^- yield, δ phase, (c) O^- yield, ζ phase. The spectra were recorded at an ion emission angle of 20° from the surface normal.

Figure 2. Oxygen ion yield as a function of sample angle

emitted ions by rotating the sample between (close to) normal and grazing ion emission. Angular profiles for DA and DD in the lying down (δ) and standing up (ζ phase) are shown in figure 2. It can be seen that in the ζ phase the O^- yield increases towards normal emission from the surface. The angular distributions for positive and negative ions look identical at dipolar dissociation energies, as expected. Also, the 13.5eV resonant dissociation profile in the ζ phase closely resembles that for dipolar dissociation. However, the low coverage (lying down molecule) dipolar dissociation profile is far broader than that taken at higher O_2 dose, and peaks at 30° from the surface normal.

To explain these results we have to account for the following observations: (i) the 13.5eV resonance appears quenched at low O_2 coverages; (ii) the dissociative attachment and dipolar dissociation angular distributions appear the same at high coverages, and (iii) the angular distribution at low coverages (when the molecules are expected to be lying flat on the surface) is peaked well away from grazing emission. Looking first at the (standing up) ζ phase results: a clear DA resonance is seen at 13.5eV. This has been assigned to the $^2\Sigma_u^+$ resonance of molecular O_2^- [12]; the transition from the oxygen $^3\Sigma_g^-$ ground state to this resonance is forbidden in the gas phase due to the violation of the $\Sigma^+ \leftrightarrow \Sigma^-$ selection rule. However it has been proposed that that this resonance can be seen in ESD from the physisorbed state due to the surface and neighbouring molecules perturbing the symmetry of the molecular wavefunctions. At higher energies, dipolar dissociation occurs, as expected. The angular profiles observed in the ζ phase (figure 2) all increase towards normal emission. The cut-off at 20° from the normal occurs due to geometric constraints. It is expected that during the dissociation process, the molecules will dissociate along the bond axis. Since in this phase the molecules stand up on the surface, the ion yield will peak along the surface normal; this is consistent with what is observed. The broadness of the distribution can be accounted for by two factors, firstly the tilting of the molecules on the surface. NEXAFS studies [10] have shown that there is a considerable molecular tilting from the surface normal (~35°) in the ζ phase; this tilt will tend to broaden the ion angular distribution away from the surface normal. Secondly, the negative ions, once produced, will be attracted by their image charge induced in the surface [13]. Again, this will tend to increase the polar angle of desorption away from the surface normal.

As expected, the O^+ and O^- dipolar dissociation angular distribution yields look virtually identical. It is perhaps more surprising that the dissociative attachment yield of negative ions shows the same form. The observed angular distribution depends on two factors: the angular dependence of the excitation probability of the ground state O_2, and the emission profile of the negative ions. Although it is not expected that the excitation probability will have a simple angular dependence eitherin DA or in the DD regime, due to multiple scattering of the low energy electrons in the adsorbate overlayer [14], it will be different for the two mechanisms. Thus we can deduce from the similarity in the observed angular distributions that they are dominated by the emission profile of the ions; i.e. by the orientation of the molecules on the surface.

From a similar argument, one would expect an ion produced in the lying down δ phase to be emitted parallel to the surface. This would cause it to be strongly attracted to the surface by its image, and so no O^- signal would be detected. Of course due to its librational motion [15], the oxygen molecular axis is not exactly parallel to the surface; NEXAFS [10] measurements have indicated a mean instantaneous molecular tilt of ~20°. Thus some molecular fragments will be projected at low angles away from the surface and may be able to escape. There exists a critical angle of emission [13], depending on the ion kinetic energy and the strength of the interaction with its image, for an ion to be able to escape from the surface. We estimate a critical angle of 35° from the surface at 13.5eV and 25° at 30eV. This is consistent with the data in figure 1 since at 30eV a greater proportion of the O_2 molecules will be tilted above the critical angle than at 13.5eV.

These notions, however, cannot explain why the angular distribution from the δ phase at 30eV peaks so far away from the surface (i.e. at 30° from the normal). This can be accounted for by assuming that the molecule rotates during the dipolar dissociation process, (but not during dissociative attachment). During DD, the molecule exists for some time as a rigid dipole on the surface. Classical calculations show that the optimum orientation of a dipole on the surface is vertical; thus one would expect the molecule to rotate on the surface due to a torque arising from its image in the substrate. No such rotation is possible in the DA mechanism, so for example the 13.5eV feature is not observed in the δ phase.

To develop this idea further, we compare, in a qualitative way, the timescales for dissociation and rotation. The period of the O_2 librational motion is ~10^{-13}s [15]. As the nuclei of the O_2 molecule separate, the formation of the image dipole in the surface, the molecule-surface interaction will be stronger, and hence the librational frequency will be increased. Thus the time period for rotation will be reduced from 10^{-13}sec. The dissociation timescale can be estimated from the potential energy curves of O_2, as it corresponds to the time taken to traverse the appropriate potential energy curve. Dehmer and Chupka [16] show that the predissociation curves for dipolar dissociation are much broader than the potential curve of ground state O_2; thus the dissociation time is longer than the vibrational period of ground state O_2 (10^{-14}sec). Thus the reduced period for libration approaches the increased intramolecular vibrational period; the two values become comparable suggesting there is sufficient time for molecular rotation during dissociation. We have modelled the dynamics of the molecule on the surface [17], using a classical image potential for the molecule-surface interaction, and a Morse potential for the intramolecular interaction [16]. These simulations yield a time for rotation of the order of 10^{-13} sec, indicating the plausibility of molecular rotation on the timescale of dissociation.

2.2 Photon Stimulated Desorption

The photodesorption yield of O^- ions from 2 monolayers of O_2 on graphite is shown in Figure 3. In the energy range studied (15-35eV), two broad resonant structures of approximate width 3 eV are observed at 25 and 30 eV. Above 32 eV a steeply rising signal is seen. No significant O^- desorption was measured below 19 eV. The gas phase ion yield curve [18], also plotted in Figure 3, is shown for comparison. Since the perturbation of the O_2 electronic structure upon physisorption is relatively small, one might expect the gas and condensed phase cross sections, and hence the ion yield curves to be similar. From Figure 3, this is clearly not the case; most of the O^- signal from the O_2 layer is observed in the region where the gas phase signal is weak and structureless. Thus we ascribe the dominant photodesorption mechanism to an indirect excitation process, as described in more detail elsewhere [19]. Photoelectrons are generated in the graphite and then propagate through the oxygen overlayer inducing dipolar dissociation. The structure apparent in the yield spectra is due to strong inter band transitions in the graphite, modulating the photoelectron yield, and the requirement that the photoelectron has at least 17 eV kinetic energy to induce DD. Additional photoelectrons are generated in the oxygen overlayer, giving rise to the increase in ion yield above 32eV. Unfortunately in the present experiment we could not energy resolve the photoelectrons and hence cannot show the resonances in the yield of photoelectrons of greater than 17 eV kinetic energy.

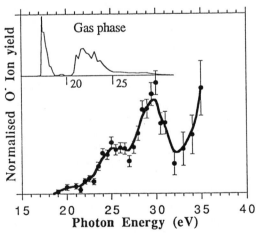

Figure 3. Negative ion yield from O_2/graphite as a function of incident photon energy at 4L dosage. The gas phase photodissociation cross-section from Dujardin et al [18] is plotted for comparison.

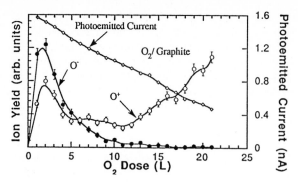

Figure 4. Dependence of O^+ and O^- ion yields and photoemmitted current upon dosage, and hence thickness, of oxygen.

There is no clear evidence of the direct gas phase process in Fig.3 which has a large cross section at energies below 25 eV. This may be due firstly to the relative strengths of the direct and indirect processes with the indirect substrate mediated mechanism dominating. A second possibility is that the predissociative Rydberg states of O_2 may be quenched, either by the substrate or neighbouring molecules in the overlayer.

Since we expect the desorbing ions to be generated in the outermost layers of the condensed film and the mean free paths for electrons propagating through the O_2 film is low in the energy range studied, there should be a strong coverage dependence in the ion yield. In order to test this hypothesis the ion yields and photoemitted electron current were investigated as a function of O_2 coverage; the results are shown in Figure 4. Oxygen was deposited sequentially in small doses and the ion yields and photoemitted current were measured at each point with a photon energy of 24eV. Both the O^- and O^+ ion yields rise to a maximum intensity at a relatively low dosage (2L, corresponding to about 1 monolayer), and then fall sharply with increasing overlayer thickness. The ion yield increases at low coverage since the number of O_2 molecules on the surface is increasing. The similarity in the form of the O^+ and O^- yields at low O_2 dose suggests that dipolar dissociation is the dominant mechanism for ion production , furthermore preliminary studies of the energy dependence at low coverages indicate that both O^+ and O^- exhibit very similar crossections of the same magnitude which deviate significantly from the gas phase dissociation. The decreasing signal intensity at larger coverages arises from the attenuation of electrons propagating from the substrate to the outer surface of the O_2 layer, from where photodissociated fragments may be desorbed.. This phenomenon is reflected in the measured photoemitted current, also shown in figure 4, which decays monotonically with increasing film thickness.

The increase in O^+ yield at high O_2 coverages (figure 4) can be accounted for by a second mechanism, the dissociation of O_2^+ ions generated by direct photoionisation. This is the primary mechanism by which O^+ is generated in the gas phase. These ions typically have a low kinetic energy [20], and in order to leave the surface, they have to escape from the attractive image potential induced in the graphite surface, which falls with distance from the surface. This desorption channel therefore becomes 'switched on' when the film thickness exceeds a critical value. From classical electrostatic calculations we estimate that the threshold occurs at a dosage of 8L in the case of O^+ ions ejected normal to the surface. O^+ ions emitted away from the normal as a consequence of molecular libration will not be able to escape until a higher film thickness is reached and then may account for the increase in the yield as the film thickness is further increased. Complicating factors in this model, the finite angular acceptance of the mass spectrometer and the lower detection efficiency for low energy ions in the experiment at present hinder any quantitative evaluation of the data.

3. Conclusion

We have observed O^- desorption from O_2 condensed on the surface of a graphite sample irradiated with either low energy electrons or synchrotron radiation in the vacuum ultra violet. Both dipolar dissociation and dissociative attachment mechanisms have been observed in ESD whereas in PSD, the dominant mechanism for the production of desorbed O^- ions appears to be dipolar dissociation induced by photoelectrons from the graphite substrate. The PSD yield of O^+ has been shown to be due to two mechanisms, a photoelectron induced process (as for O^-) at low coverage, and a direct process from the dissociation of O_2^+ ions. The direct process is 'switched on' at higher coverages when the O^+ ion can escape from its image in the substrate.

The yield of O^- from dissociative attachment has been shown to be sensitive to the molecular orientation on the surface. The resonances observed when the molecule is standing up are not seen when the molecule is lying down. This is accounted for by the desorbing ions being attracted back to the surface by their image charge in the lying down case. Dipolar dissociation is seen for both the standing up molecules and for the lying down species, with the angular distribution peaked close to normal emission. This is attributed to molecular rotation during desorption due to the image charge interaction of the dipole with the substrate, resulting in the molecular axis rotating towards the surface normal.

Acknowledgements

We thank Andrew Hopkirk and Richard Hall for helpful discussions. This work was funded by the United Kingdom S.E.R.C. and the Royal Society. R.J.G. and R.A.B. are grateful for CASE studentships from ICI and VSW Scientific instruments, respectively

References

[1] L.Sanche, J. Phys. B 23 1597 (1990)
[2] H.Sambe & D.E.Ramaker, Surface Science 269/270 444 (1992)
[3] L.J.Richter, S.A.Buntin, D.S.King, R.R.Cavanaugh, Phys. Rev. Lett. 65 1057 (1990).
[4] S.R.Hatch, X.-Y.Zhu, J.M.White, A Campion J. Chem. Phys. 92 2681 (1990).
[5] J.W.Gadzuk, L J.Richter, S.A.Buntin, D.S.King, R.R.Cavanaugh, Surface Science 235 317 (1990).

[6]   E.P.Marsh, T.L.Gilton, W.Meier, M.R.Schneider, J.P.Cowin, Phys. Rev. Lett. 61 2725 (1988).

[7]   St.-J.Dixon-Warren, E.T.Jensen, J.C.Polanyi, Phys. Rev. Lett. 67 2395 (1991).

[8]   Z.Ying, W.Ho, Phys. Rev. Lett. 60 57 (1988).

[9]   M.F.Toney and S.C.Fain, Jr., Phys. Rev. B 30 1248 (1984)

[10] R.J.Guest, A.Nilsson, O.Björneholm, B.Hernnäs, A.Sandell, R.E.Palmer and N.Mårtensson, Surface Science 269/270 432 (1992)

[11] L.A.Silva and R.E.Palmer, Surface Science 272 313 (1992)

[12] R.Azria, L.Parenteau and L.Sanche, Phys. Rev. Lett. 59 638 (1987)

[13] Z.Miscovic, J.Vukanic and T.E.Madey, Surface Science 141 285 (1984)

[14] R.E.Palmer and P.J.Rous, Rev. Mod. Physics in press

[15] R.D.Etters and K.Kobashi, J.Chem. Phys 81 6249 (1984)

[16] P.M.Dehmer and W.A.Chupka, J. Chem. Phys 62 4525 (1975)

[17] R.J.Guest, R.A.Bennett, L.A.Silva and R.E.Palmer, to be submitted

[18] G.Dujardin, L.Hellner, L.Phillipe, R.Azria, M.J.Besnard-Ramage, Phys. Rev. Lett. 67 1844 (1991)

[19] R.A.Bennett, R.G.Sharpe, R.J.Guest, J.C.Barnard, R.E.Palmer and M.A.MacDonald, submitted to Chem. Phys. Lett.

[20] M.Ukai, K.Kameta, K.Shinsaka, Y.Hatano, T.Hirayama, S.Nagaoka and K.Kimura, Chem. Phys. Lett 167 334 (1990)

Resonance Electron Scattering by O_2 Monolayers on Graphite: Reinterpreted

H. Sambe and D.E. Ramaker

Chemistry Department, George Washington University,
Washington, DC 20052, USA

1. Introduction

Both electron stimulated desorption (ESD) and inelastic electron scattering (IES) are enhanced strongly by anion resonant states. Occasionally, peak features observed in ESD and IES arise from the same anion resonant states. IED, however, unlike ESD, provides information on the symmetry of the anion resonant states through the angular distributions of the inelastically scattered electrons. Such electron angular distributions from O_2 on graphite have been studied extensively by Palmer and co-workers. Their results have been published in more than eight papers over the past four years [1-8] and summarized in two recent review articles [7,8].

On graphite, O_2 can be prepared in either the δ or ζ phase, where the molecules lie either parallel (δ phase) or perpendicular (ζ phase) to the graphite surface [9]. The excitation profile of the $X(^3\Sigma^-_g) \rightarrow a(^1\Delta_g)$ electronic transition exhibits two peaks at 6 and 14 eV for the δ phase and just one peak at 8 eV for the ζ phase [7]. Palmer and co-workers assigned these 6 and 8 eV peaks to the same $^2\Pi_g$ resonant state and attributed the 2-eV shift of the peak energies in the δ and ζ phases to multiple electron scattering in the solid [3]. No assignment was made for the 14 eV feature. In this paper, we propose an alternative interpretation for the origin of the 6 and 8 eV peaks and assign the 14 eV peak.

The 6, 8, and 14 eV peak features have been observed also in the ESD of O^- ions from O_2 films and O_2 on rare gas films, and attributed to the $^2\Pi_u$(6eV), $^2\Sigma^+_g$(8eV), and $^2\Sigma^+_u$(14eV) resonant states, respectively [10]. Excitation to the $^2\Sigma^+_g$ and $^2\Sigma^+_u$ states from the O_2 ground state is symmetry forbidden in the gas phase but is allowed in the condensed phase when the cylindrical symmetry of the O_2 is relaxed by interaction with the surrounding [11]. Recent ESD studies for O_2 clusters clearly demonstrated that the 8 and 14 eV peaks arise from the "symmetry-forbidden" states [12]. Furthermore, in the gas phase, the symmetries of $^2\Pi_u$ and $^2\Sigma^+_g$ for the 6 and 8 eV peaks have been established experimentally based on the angular distribution of O^- ions resulting from electron impact on O_2 in the ground state and the $^1\Delta_g$ excited state respectively [13,14]. These assignments are completely in line with theoretical calculations [15] and analysis [11].

2. Origin of the 6, 8, and 14 eV peaks in the X→a excitation

The question here is: Do the 6, 8, and 14 eV peaks observed in the X→a excitation profile arise from the same $^2\Pi_u$(6eV), $^2\Sigma^+_g$(8eV), and $^2\Sigma^+_u$(14eV) resonant states as those observed in

Springer Series in Surface Sciences, Vol. 31
Desorption Induced by Electron Transitions DIET V
Editors: A.R. Burns · E.B. Stechel · D.R. Jennison   © Springer-Verlag Berlin Heidelberg 1993

the above experiments? We shall answer this question by examining the angular distributions of inelastically scattered electrons and the electron attachment selectivity observed by Palmer and co-workers [1-8].

$^2\Pi_u$(6eV): The $^2\Pi_u$ state was observed at 6 eV in the v=0→1 vibrational excitation profile [7]. The angular distribution of inelastically scattered electrons at incident electron energy E_i=6.5eV for the X→a excitation is almost identical to that for the v=0→1 excitation [7]. This agreement is observed for both the δ and ζ phases, suggesting strongly that the 6 eV peaks in the X→a and v=0→1 excitation profiles arise from the same $^2\Pi_u$ state. This may be the most important finding of the present paper.

$^2\Sigma^+_g$(8eV): The emitted electrons from the $^2\Sigma^+_g$ state decaying into the a($^1\Delta_g$) state have δ_g symmetry. Therefore, the angular distribution of emitted electrons from this state for the ζ phase must be $\sin^4\theta$ for small θ, where θ is the angle from the direction normal to the graphite surface. This $\sin^4\theta$ dependence agrees with experimental results [5]. Further supporting evidence: The 8 eV feature appears in the X→a excitation profile for both the δ and ζ phases. (Although there is no discernible peak at 8 eV in the δ phase, there is a broad shoulder which cannot be accounted for by the 6-eV resonance alone. The 6-eV resonance, which was observed in the v=0→1 excitation, does not have such an extended tail [7].) The 8 eV intensity for the ζ phase is stronger than that for the δ phase [7]. This non-selectivity and relative intensity can be explained, if the 8 eV feature arises from the $^2\Sigma^+_g$ resonant state.

$^2\Sigma^+_u$(14eV): The angular distribution of the inelastically scattered electrons at 14 eV is consistent with the $^2\Sigma^+_u$ symmetry [16]. Further supporting evidence: The 14 eV feature observed in the X→a excitation for the δ phase disappears for the ζ phase. Likewise, the $^4\Sigma^-_u(3\sigma_u)$ feature observed in the v=0→1 excitation for the δ phase disappears for the ζ phase [7]. This parallel behavior is expected, if the 14 eV feature arises from the $^2\Sigma^+_u$ resonant state.

Based on the above, we conclude that the 6, 8, and 14 eV peaks observed in the X→a excitation profile indeed arise from the same $^2\Pi_u$(6eV), $^2\Sigma^+_g$(8eV), and $^2\Sigma^+_u$(14eV) states as those observed in ESD and other experiments. This conclusion disagrees with the assignment made by Palmer and co-workers,who attributed the 6 and 8 eV peaks to the $^2\Pi_g$ state.

3. The $^2\Pi_g$ intermediate state in the solid

What happened to the $^2\Pi_g$ state in the solid? In the gas phase, the X→a electronic excitation proceeds via the $O_2^-(^2\Pi_g)$ intermediate state at least for incident electron energies less than 5 eV [17]. This intermediate state should be differentiated from the $O_2^-(1\pi_g)\,^2\Pi_g$ shape resonance peaked around 0.2 eV [18]. The latter has the $1\pi_g$ component predominant within the molecular region and a negligible amplitude outside this region, while the former has comparable amplitudes in both regions. Theoretical calculations [19] predict that this non-resonant $^2\Pi_g$ intermediate state produces a very broad peak (FWHM≈10eV), if there is one, in the X→a excitation function.

The non-resonant $^2\Pi_g$ scattering, unlike the resonant $^2\Pi_g$ scattering, is expected to be perturbed strongly when O_2 is

condensed. In fact, there is ample evidence for quenching of this non-resonant $^2\Pi_g$ intermediate state in condensed phase:
 (a) The observed threshold energy (3.0 eV) is delayed by 2.1 eV from the expected threshold energy of 0.9 eV [7,20].
 (b) The angular distribution of inelastically scattered electrons at E_i=6 eV indicates $^2\Pi_u$ symmetry rather than $^2\Pi_g$ symmetry, as was argued in Sec. 2.
 (c) The observed features are too sharp (FWHM$\approx$4eV) to arise from non-resonant $^2\Pi_g$ scattering. The latter should have either no peak or a very broad peak with FWHM$\approx$10 eV.
To sum up, the non-resonant $^2\Pi_g$ intermediate state is quenched upon condensation and hence cannot be assigned to the 6 and 8 eV features.

4. Resonance energy shift due to multiple electron scattering

Palmer and co-workers offered two reasons for the 2-eV shift of the "$^2\Pi_g$" resonance peaks [3,7]:
 (1) theoretical calculations assuming the $^2\Pi_g$ symmetry and
 (2) the similarity of the angular distributions for E_i=6.5 and 8.5 eV.
Item (1): The assumption of the $^2\Pi_g$ symmetry is not valid as argued in Sections 2 and 3. This explains partly why their theoretical calculations did not agree well with the experimental data [3,7].
Item (2): Although the angular distributions for E_i=6.5 and 8.5 eV are similar to each other for the δ phase, those for the ζ phase are quite different. The only common feature between them is that both decrease toward the normal direction.
In short, there is no evidence to support the 2-eV shift of the "$^2\Pi_g$" resonance.

5. A concluding remark

Palmer and co-workers assigned the 6 and 8 eV peaks, observed in the X$\rightarrow$a excitation, to the $^2\Pi_g$ resonance and attributed the 2-eV shift of the peak energies in the δ and ζ phases to multiple electron scattering in the solid. Based on this single investigation, they claimed that multiple electron scattering can substantially shift resonance energies and warned electron spectroscopists who study adsorbed molecules to be aware of this possibility [3,7]. Moreover, these authors repeatedly suggested this possibility to explain experimental results obtained by others [3,8]. Because of the Palmer and co-workers' claim, other authors considered (although did not adopt) this possibility for an explanation of their results [21]. This paper has demonstrated clearly that the Palmer and co-workers' claim is unfounded at least for the cases studied so far.

Acknowledgment

This work was supported by the Office of Naval Research.

References

[1] R.E. Palmer, P.J. Rous, J.L. Wilkes and R.F. Willis, Phys. Rev. Lett. **60** (1988) 329.
[2] P.J. Rous, R.E. Palmer and R.F. Willis, Phys. Rev. **B 39** (1989) 7552.
[3] P.J. Rous, E.T. Jensen and R.E. Palmer, Phys. Rev. Lett. **63** (1989) 2496.
[4] E.T. Jensen, R.E. Palmer and P.J. Rous, Phys. Rev. Lett. **64** (1990) 1301.
[5] E.T. Jensen, R.E. Palmer and P.J. Rous, Chem, Phys. Lett. **69** (1990) 204.
[6] P.J. Rous, R.E. Palmer and E.T. Jensen, Phys. Rev. **B 41** (1990) 4793.
[7] E.T. Jensen, R.E. Palmer and P.J. Rous, Surf. Sci. **237** (1990) 153.
[8] R.E. Palmer and P.J. Rous, Rev. Modern Phys. April **64** (1992) 383.
[9] M.F. Toney and S.C. Fain, Phys. Rev. **B 30** (1984) 1115; H. You and S.C. Fain, Phys. Rev. **B 33** (1986) 5886; M.F. Toney and S.C. Fain, Phys. Rev. **B 36** (1987) 1248.
[10] H. Sambe and D.E. Ramaker, Surf. Sci. (in press).
[11] H. Sambe and D.E. Ramaker, Phys. Rev. **A 40** (1989) 3651.
[12] R. Hashemi and E. Illenberger, Chem. Phys. Lett. **187** (1991) 623.
[13] R.J. Van Brunt and L.J. Kieffer, Phys. Rev. **A 2** (1970) 1899.
[14] D.S. Belic and R.I. Hall, J. Phys. **B 14** (1981) 365.
[15] H.H. Michels, Adv. Chem. Phys. **45** (1981) part 2, 225.
[16] R.E. Palmer, private communication.
[17] E.S. Chang, J. Phys. **B 10** (1977) L677.
[18] G.J. Schulz, Rev. Modern Phys. **45** (1973) 423; G.J. Schulz, in Principle of Laser Plasmas edited by G. Bekefi (John Wiley & Sons, New York, 1976) p.33.
[19] C.J. Noble and P.G. Burke, J. Phys. **B 19** (1986) L35; D. Teillet-Billy, L. Malegat and J.P. Ganyacq, J. Phys. **B 20** (1987) 3201.
[20] L. Sanche and M. Michaud, Phys. Rev. Lett. **47** (1981) 1008.
[21] W. Hansen, M. Bertolo and K. Jacobi, Surf. Sci. **253** (1991) 1; T.S. Jones, M.R. Ashton, M.Q. Ding, and N.V. Richardson, Chem. Phys. Lett. **161** (1989) 467.

Non-Adiabatic Curve Crossings
in Dissociative Electron Attachment of Condensed O_2

*H. Sambe and D.E. Ramaker**

Chemistry Department, George Washington University,
Washington, DC 20052, USA

Abstract. Previously published data on electron stimulated desorption (ESD) from condensed O_2 or O_2 in solid matrices have been analyzed. Dissociative electron attachment (DEA) via the $^2\Sigma_g{}^+$ resonant state can result in two different dissociation products (i.e. by proceeding to the dissociation limits $O(^3P) + O^-(^2P)$ and $O^*(^1D) + O^-(^2P)$). The $O^*(^1D)$ product requires dissociation via a non-adiabatic curve crossing. $O^*(^1D)$ is the dominant DEA product in gaseous O_2. Upon condensation, $O(^3P)$ becomes the dominant product; however, the branching ratio depends strongly on the kinetic energy of the dissociating atoms.

1. Introduction

Our previous theoretical findings concerning DEA of O_2 are summarized in Table 1. Of the many O_2^- states below 10 eV, only two states, i.e. the $^2\Pi_u$ and $^2\Sigma_g{}^+$ states, live long enough to dissociate yielding O^- ions. The

Table 1   Dissociative electron attachment mechanisms below 10 eV.

CHARACTERISTIC	MECHANISM		
	I	II	III
Symmetry	$^2\Pi_u$	$^2\Sigma_g{}^+$	$^2\Sigma_g{}^+$
Elect. config.	$1\pi_u{}^{-1}1\pi_g{}^2$	$3\sigma_g{}^{-1}1\pi_g{}^2$	$3\sigma_g{}^{-1}1\pi_g{}^2$
Vert. E (eV)	6.0	7.7	9.0
DEA from g.s.	Yes	No	No.
Rel. Lifetime	smaller	larger	larger
Dissoc. limit[a]	1st	1st	2nd
Init. trajectory[b]	In	Out	In

[a]The 1st limit is $O(^3P) + O^-(^2P)$ and the 2nd is $O(^1D) + O^-(^2P)$.
[b]"In" and "Out" refer to the initial movement of the O^- ion toward or away from the surface.

Springer Series in Surface Sciences, Vol. 31
Desorption Induced by Electron Transitions DIET V
Editors: A.R. Burns · E.B. Stechel · D.R. Jennison   © Springer-Verlag Berlin Heidelberg 1993

33

other O_2^- states autoionize too fast [1]. The autoionization lifetime of the $^2\Sigma_g^+$ state is predicted to be longer than that of the $^2\Pi_u$ state [1].

The σ^- selection rule [2] forbids electron attachment to the $^2\Sigma_g^+$ state directly from the ground state ($^3\Sigma_g^-$). To remove this restriction, the cylindrical symmetry of the O_2 molecule must be perturbed by its neighbors. The $^2\Sigma_g^+$ state can however be reached directly from the first excited state ($^1\Delta_g$) even for gaseous O_2. Such a mechanism,

$$ e^- + O_2^*(^1\Delta_g) \rightarrow O_2^{-*}(^2\Sigma_g^+) \rightarrow O^*(^1D) + O^-(^2P), \qquad (1) $$

has been observed for gaseous O_2 and identified as such by Belic and Hall [3]. The $^2\Sigma_g^+$ state is observed at 8.5 eV vertically above the O_2 ground state consistent with its predicted vertical energy of 8.1 ± 0.9 eV [1].

Belic and Hall [3] not only identified the $^2\Sigma_g^+$ symmetry, but also found that the $^2\Sigma_g^+$ state dissociates into the second lowest limit almost exclusively as indicated in Eq. (1) and Fig. 1. This implies a very high efficiency for a non-adiabatic curve crossing, since the lowest $^2\Sigma_g^+$ state adiabatically dissociates to the lowest limit. The O_2^- adiabatic potential curves (see Fig. 1) calculated by Michels utilizing the MCSCF method [4], show an avoided curve crossing for the lowest $^2\Sigma_g^+$ state and suggests a high efficiency for the non-adiabatic curve crossing because of the relative slopes of the two interacting curves.

Now, the question arises, what happens when the O_2 molecules are condensed or placed in a matrix? Does the lowest $^2\Sigma_g^+$ state still dissociate predominantly to the second lowest limit, as it does in the gas phase?

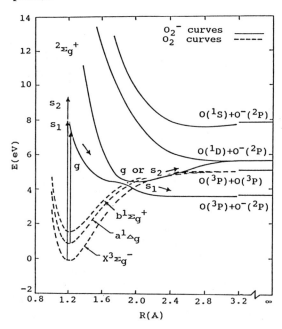

Fig 1. Calculated potential energy curves for $^2\Sigma_g^+$ states of O_2^-[4]. The arrows indicate the dominate dissociation path in the gas phase (g) and in the solid (s) or condensed phase at 8 eV (s_1) and 9 eV (s_2).

2. The role of the surface

In the gas phase, the $O_a + O_b^-$ and $O_a^- + O_b$ dissociation limits of the homonuclear O_2^- molecule are degenerate, where subscripts (a and b) are used to differentiate the identical oxygen atoms. This degeneracy will be removed when an O_2^- is located on a polarizable surface and its symmetry axis is not parallel to the surface. For the non-parallel case, the O_2^- state can dissociate into either $O + O^-/S$ or $O^- + O/S$ where 'S' indicates the surface. The $O + O^-/S$ limit is lower than the $O^- + O/S$ limit by a surface polarization energy. Regardless of the degeneracy, each O_2^- state must be connected uniquely to one of these limits. This connection between the O_2^- state and the dissociation limit can be determined by applying the Wigner-Witmer and non-curve-crossing rules [5]. The results obtained are shown in Table 1. Dissociation into the $O + O^-/S$ ($O^- + O/S$) limit implies that the O^- ion moves in (out) from the surface, and that some ions follow a "bounce" trajectory off the surface [6].

Within the Landau-Zener theory [7], the branching ratio at a curve crossing is determined by the relative slopes of the two curves at the crossing and the velocity of the dissociating ions at the crossing. Since at the surface, the two dissociation limits have different energy, the relative slopes of the two curves will be changed, and thus the branching ratio will be changed. In the sub-surface region, the polarization is expected to be more homogenous, in which case the energy spacing between the dissociation will not change significantly. However, in this case the dissociating fragments must break out of a "cage" which will slow down the dissociating fragments. Thus we expect the branching ratio to be altered in the condensed phase by both the relative slopes of the potential curves and the velocity of the dissociating fragments.

As seen in Table 1, three different mechanisms are identified yielding O^- with different electronic excitations, kinetic energies, and initial directions of motion. The singlet and triplet oxygen atoms [that is, $O^*(^1D)$ and $O(^3P)$] generally react with neighboring molecules quite differently because of the spin-conservation rule. This allows us to determine which product was formed upon dissociation, i.e. to answer the question whether the dissociation was adiabatic or nonadiabatic.

3. Results and Conclusions

We have examined a wide range of previously published and unpublished data involving O_2 condensed on rare gases and on Pt, and mixed gases (e.g. O_2 and CO, O_2 and N_2, $^{18}O_2$ and $C^{16}O$ and O_2 with C_nH_n) of differing relative concentrations and thicknesses [8-11] with this question in mind. Application of retardation voltages gives some indication of the relative kinetic energies of the ions. These studies, much too lengthy to be described here, have produced the following conclusions:

35

1) The DEA of gaseous O_2 proceeds predominantly via the allowed mechanism I. Upon condensation of O_2, this mechanism appears to be suppressed with increasing thicknesses of O_2, or increasing amounts of CO in the mixed gases. Since O^- anions via this mechanism initially move towards the surface of the O_2 film, requiring a "bounce" off the surface to be observed, the suppression is attributed to the increased probability of O^-/O_2 or O^-/CO interactions during the bounce.

2) Furthermore, the increased O_2-O_2 or O_2-CO molecular interaction breaks down the σ_2^- selection rule increasingly allowing the forbidden excitation to the $^2\Sigma_g^+$ state. The forbidden mechanisms II and III become the principal contributors of O^- from solid O_2 or mixed gases.

3) The dissociation process of the $O_2^-(^2\Sigma_g^+)$ resonant state is significantly modified upon condensation. On the surface, the O_2 dissociates predominantly adiabatically into the lowest limit. This reversal from the gas phase is attributed to a 1 eV reduction in the energy difference of the two dissociation limits and resultant change in the relative slopes of the two curves.

4) In the sub-surface region, the $O_2^-(^2\Sigma_g^+)$ apparently also dissociates predominantly adiabatically to the lowest limit, but this is now attributed to the cage effect, since the polarization is more homogeneous. However, the O^- ions in this case are not desorbed. At higher excitation energy, as expected by the Landau-Zener theory, the branching ratio shifts toward the second dissociation limit, where the production of $O^*(^1D)$ plays a key role. Evidence exist that the $O^*(^1D)$ participates in the formation of energetic complexes with the conversion of electronic energy to kinetic energy and the desorption of O^- ions at higher kinetic energy.

We suspect that this is the first instance of where the effects of an adiabatic curve crossing have been so clearly elucidated in the ESD yields from a surface. The strong dependency of the branching ratio on the matrix polarization and cage effects are qualitatively consistent with that expected from the Landau-Zener theory of curve crossings [7]. Further work is required here to determine if this theory can semi-quantitatively explain the data; however, this is a formidable tasks in light of the complex matrix effects involved.

References

1. H. Sambe and D.E. Ramaker, Phys. Rev. **40**, 3651 (1989).
2. H. Sambe and D.E. Ramaker, Chem. Phys. Letters **139**, 386 (1987).
3. D.S. Belic and R.I. Hall, J. Phys. **B14**, 365 (1981).
4. H.H. Michels, Adv. in Chem. Phys. **45**, 225 (1981).
5. E. Wigner and E.E. Witmer, Z. Phys. **51**, 859 (1928).

6. H. Sambe and D.E. Ramaker, in Desorption Induced by Electronic Transitions DIET IV, Eds. G. Betz and P. Varga (Springer-Verlag, Berlin, 1990), p. 251.

7. K.J. Laidler and K.E. Shuler, Chem. Rev. **48**, 153 (1951).

8. R. Azria, L. Parenteau, and L. Sanche, Phys. Rev. Letters 59, 638 (1987);Chem Phys. Letters, **171**, 229 (1987); **156**, 606 (1989).

9. L. Sanche and L. Parenteau, Phys. Rev. Letters **59**, 136 (1987); J. Chem. Phys. **93**, 7476 (1990).

10. L. Sanche and M. Michaud, Phys. Rev. **B30**, 6078 (1984).

11. H. Sambe and D.E. Ramaker, to be published in Surf. Sci.

*Supported in Part by the Office of Naval Research.

Charge-Transfer Photodissociation and Photoreaction in Adsorbates at a Metal Surface

St.J. Dixon-Warren, E.T. Jensen, and J.C. Polanyi

Department of Chemistry, University of Toronto,
Toronto, Ontario, Canada M5S 1A1

The yield of desorbing Cl^- ions produced by charge-transfer (CT) photo-dissociation of CCl_4 adsorbed on a Ag(111) substrate, $CCl_4/Ag + hv \rightarrow CCl_3/Ag(111) + Cl^-$, is reported as a function of coverage at 193 and 248nm and near the threshold for photoelectron emission at 280nm. The ion emission cross-section is found to be enhanced for the first molecular layer in direct contact with the substrate. The cross-section for ion emission is very low; however the total CT photodissociation cross-section is shown to be large by measuring the yield of CH_3Cl produced in the reaction of the Cl^- with an overlayer of CH_3Br. The CH_3Br overlayer lowers the work function and so increases the Cl^- emission, confirming the CT mechanism.

1. Introduction

Photochemistry at surfaces is a rapidly developing field, that is appearing for the second time in this series of DIET meetings [1]. It is becoming well established that the photochemistry of molecules adsorbed at metal surfaces can be mediated by the presence of photoexcited electrons in the metal, as well as by direct absorption of the light in the adsorbate. The photoelectrons induce dissociation in the adsorbate by a CT mechanism that is similar to gas-phase dissociative electron attachment (DA), $AB + e^- \rightarrow A + B^-$. The dissociation products formed at the surface can desorb or they may react with coadsorbed species or with the substrate.

The results presented here are part of an ongoing effort in this laboratory to understand the mechanism of photoinduced CT processes at metal surfaces [2]. Recent developments in the understanding of CT photochemistry have been reviewed by several authors [3].

2. Experimental

The experiments were performed in an ultra-high vacuum chamber that has been described previously [4]. The chamber was equipped with a quadrupole mass spectrometer, an Auger electron spectrometer and a negative ion time-of-flight spectrometer, all of which faced a rotatable sample manipulator. An Ag(111) sample that could be cooled to 100 K and heated to 800 K was mounted on the sample manipulator. The sample was irradiated at 77° glancing incidence with a UV laser. Both an excimer laser (for 193 and 248nm light) and the frequency-doubled output of a Nd:YAG pumped dye laser (for 280nm light) were used.

Springer Series in Surface Sciences, Vol. 31
Desorption Induced by Electron Transitions DIET V
Editors: A.R. Burns · E.B. Stechel · D.R. Jennison   © Springer-Verlag Berlin Heidelberg 1993

3. Results and Discussion

The results presented here form part of an investigation into the photoinduced CT processes in chloromethanes, RCl, adsorbed on Ag(111). In an earlier paper [4], we gave direct evidence for CT photodissociation of CCl_4 adsorbed on Ag(111). The evidence was the detection of desorbing Cl^- ions produced by CT photodissociation,

$$CCl_4/Ag(111) + hv \rightarrow [CCl_4^-]^{\ddagger}/Ag(111) \rightarrow CCl_3/Ag(111) + Cl^-(g) \qquad (1)$$

The temporary negative ion, $[CCl_4^-]^{\ddagger}$, is formed by the capture of a photoexcited electron from the substrate. Photoinduced CT dissociation yielding Cl^- emission has also been shown to occur in several other chloromethanes adsorbed on Ag(111), including $CHCl_3$, CH_2Cl_2, CH_3Cl and CCl_3Br [5].

3.1 Charge-Transfer Photodissociation of CCl_4 on Ag(111)

The yield of Cl^- ions as a function of CCl_4 coverage on Ag(111) at 193 and 248nm are presented in figs. 1(a) and (b) (filled circles). The laser power was very low so as to minimize space charge effects and reduce the depletion of the adsorbate. The shape of the Cl^- emission versus coverage curves is similar at both wavelengths. The coverage, in monolayers (ML), was determined independently using thermal desorption [5]. The TPD and AES spectra showed that the CCl_4 reacted with the Ag(111) during the first ~1 ML dose, consequently, the rate of increase in the Cl^- emission between 0 and 1 ML was lower since the amount of molecular CCl_4 was

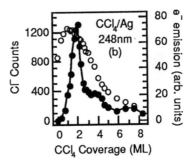

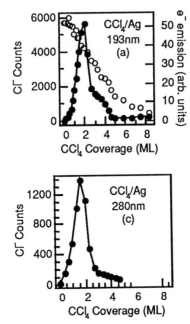

Figure 1. Yield of Cl^- ions (filled circles) and photoelectrons (open circles) from CCl_4/Ag(111) as a function of the CCl_4 coverage at (a) 193nm (0.05 $\mu J/cm^2$); (b) 248nm (1.0 $\mu J/cm^2$); (c) 280nm (180 $\mu J/cm^2$).

less. The Cl⁻ emission, figs. 1(a) and (b), increases steeply from ~1 ML dose to a sharp maximum at ~2 ML dose, corresponding to the completion of the first *molecular* layer of CCl_4. A similar sharp maximum in the Cl⁻ emission was observed with $CHCl_3$ and CH_2Cl_2 on Ag(111) at 193nm; however, in these two cases the thermal reaction did not occur and the sharp maximum was observed at ~1 ML [5]. The yield of Cl⁻ decreased rapidly as the second molecular layer of chloromethane was added, indicating that the second layer trapped ions formed in the first layer and had a lower cross-section for Cl⁻ emission than the first layer.

The photoelectron emission current was measured concurrently with a charge-sensitive preamplifier connected to the sample, and is shown in open circles of figs 1(a) and (b). At 193nm the photoelectron emission was observed to decrease monotonically with increasing dose. The decrease was ascribed to scattering of the photoelectrons back to the metal by the adsorbed layers. The photoemission of electrons at 248nm showed an increase during the first ~1 ML dose, suggesting a small decrease in the work function.

The cross-section for Cl⁻ emission depended on the photon energy [4]. The Cl⁻ emission commenced at about 4.1eV (288nm) and increased monotonically with increasing photon energy. The cross-section, which increased by about 10x between 5.0eV (248nm) and 6.4eV (193nm) photon energies, was proportional to the amount of photoelectron emission estimated with Fowler's equation [6].

The yield of Cl⁻ ions as a function of the CCl_4 coverage was measured in the ion emission threshold region, at 280nm, as shown in fig. 1 (c). The laser power at 280nm was less stable than with the excimer laser, thus each point has been power normalized. Photoelectron emission was not detectable with the charge-sensitive preamplifier, suggesting that the Cl⁻ ions resulted from the capture of sub-vacuum level photoelectrons. The curve in fig. 1 (c) shows the sharp maximum at ~2 ML, corresponding to the completion of the first molecular layer of CCl_4.

3.2 Enhanced Ion Emission from the First Molecular Layer

The most striking feature in figs. 1 (a)-(c) is the steep decline in the Cl⁻ emission cross-section as the CCl_4 coverage was increased above 2ML. The decline in the Cl⁻ emission was much greater than that of the electron emission, and hence is not due to a decrease in the supply of photoelectrons. The steep decline in the ion emission yield from the second and higher molecular layers was observed for all the chloromethanes studied [5], suggesting that the ion emission cross-section was enhanced for molecules in *direct* contact with the substrate as compared with subsequent layers. Evidence that contact with the surface was required to give the enhanced cross-section was obtained by adsorbing the CCl_4 on top of a spacer layer of $CHCl_3$. Chloroform was chosen as a spacer layer, because its surface chemistry was well characterized [5] and because it has an ion emission cross-section ~100x smaller than that for CCl_4. Figure 2 shows the yield of Cl⁻ ions as a function of the coverage of CCl_4 adsorbed on top of ~2 ML of $CHCl_3$. There is no evidence for the steep decline in the Cl⁻ emission yield following the completion of the first molecular layer of CCl_4 on the $CHCl_3$ spacer layer. The Cl⁻ emission does decrease slowly, probably as a result of attenuation of the photoelectron current reaching the top layer of CCl_4 by the layers beneath. Such attenuation effects with spacer layers have been observed before [7].

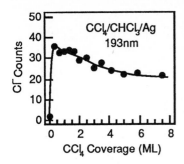

Figure 2. Yield of Cl^- ions from CCl_4 dosed on top of $CHCl_3$ precovered Ag(111) at 193nm.

The enhanced cross-section for Cl^- emission from the first molecular layer of RCl is unexpected. Quenching the temporary negative ion, $[RCl^-]^\ddagger$, would be expected to be higher for molecules in direct contact with the surface, thus the enhancement must be due to a compensating effect. The enhancement cannot be due to changes in the the work function, since the photoelectron emission decreases steadily with increasing coverage. Furthermore, if changes in the work function were involved then a larger enhancement would be expected at 280 nm, near threshold for photoelectron emission, than at higher photon energies [5]. It is unlikely that phase changes in the RCl layers are responsible, since the enhanced yield was observed for all the RCl studied. Finally, the sharp maximum was observed even at the photoelectron emission threshold, where the photoelectron energy distribution is dramatically different from that at shorter wavelengths, thus it is unlikely that changes in the photoelectron energy distribution, induced by the adsorbed layers, are responsible.

The enhancement could, however, be due to chemisorption interactions of RCl with the metal substrate. There are two possibilities for enhanced Cl^- emission by chemisorption, one involving reaction dynamics and the other involving thermodynamics: (i) A reaction of the radical R with the Ag(111) as RCl dissociates could impart more energy to the Cl^- ions as the new bond is formed. In gaseous exothermic reactions it is common for 50% of the energy released in bond formation to be channelled into product translation. (ii) Static weakening of the R-Cl bonds due to the chemisorption of RCl will allow more energy to be available in Cl^- translation after bond scission.

The maximum cross-section for Cl^- emission is estimated to be $\sim 10^{-25} cm^2$ from ~ 2 ML CCl_4 on Ag(111) at 193nm [5]. This cross-section is extremely small; it is $\sim 10^6 x$ less than the gas-phase photolysis cross-section of CCl_4 [8]. Comparison with the gas-phase DA cross-section [9], taking account of the quantum efficiency for photoelectron emission, shows that only 1 in every $\sim 10^8$ photoelectrons was able to induce DA to yield a desorbing Cl^- ion [5]. The low emission cross-section is likely to be due to energetic and geometric constraints that confine the majority of the Cl^- ions to the surface. The measured peak translational energy of the Cl^- ions in the gas-phase DA of CCl_4 is $\sim 0.1eV$ [10], which is less than the $\sim 1eV$ [11] image attraction of an ion to the surface. The observed Cl^- emission represents the small fraction of Cl^- ions emitted in a favourable direction and in the high energy tail of the initial Cl^- ion energy distribution. In the next section the total CT photodissociation cross-section will be shown to be large, even though the yield of $Cl^-(g)$ is small.

41

3.3 Effect of an Overlayer of CH_3Br

A measure of the total CT photodissociation cross-section was obtained by using an overlayer of CH_3Br as an *in situ* detector for the Cl^- ions formed at the surface by CT photodissociation of CCl_4. The Cl^- ions reacted with the CH_3Br in a process resembling the well-known S_N2 substitution reaction [12], $Cl^- + CH_3Br \rightarrow CH_3Cl + Br^-$. The low kinetic energy release of this reaction [12] resulted in the CH_3Cl photoproduct being retained at the surface. The CH_3Cl was detected by TPD following the UV laser irradiation. The cross-section for formation of CH_3Cl at 248nm was $\sim 2 \times 10^{-19} cm^2$ [5]. This cross-section is more that $\sim 10^6$ times larger than that for $Cl^-(g)$ emission; it is also larger than the gas-phase photolysis cross-section of both CCl_4 [8] and CH_3Br [13].

A second effect of a CH_3Br overlayer was to lower the surface work function [5,14]. This resulted in the initial increase in the amount photoelectron emission and hence of $Cl^-(g)$ emission, as shown in fig. 3. The inverse correlation of the $Cl^-(g)$ ion yield with the work function confirms the CT photodissociation mechanism. The $Cl^-(g)$ emission decreased at higher CH_3Br dose because the overlayer trapped the Cl^- ions at the surface.

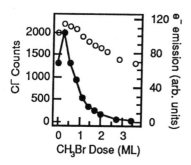

Figure 3. Yield of Cl^- ions (filled circles) and electrons (open circles) from a fixed dose of $CCl_4/Ag(111)$ as CH_3Br is dosed on top (193nm radiation).

4. Future Directions

The detection of Cl^- ions desorbing from a chloromethane covered Ag(111) surface constitutes the first direct probe of the CT photodissociation mechanism. The effect of varying the nature of the substrate on ion emission has not been studied. The use of a substrate with a lower work function would be of particular interest, since it should be possible to red-shift the dissociation to below the RCl bond energy, as has been previously observed by less direct means [3]. The use of a dielectric spacer layer, such as Xe, would give information on the effect of varying the image stabilization [11] of the anion, $[RCl^-]^{\ddagger}$, and also the effect of variation in the photoelectron energy distribution as a result of scattering in the spacer layer.

Acknowledgments

This work was supported by the National Sciences and Engineering Research Council of Canada (NSERC), the Ontario Laser and Lightwave Research Centre, and the National Networks of Centres of Excellences program. ETJ thanks NSERC for the award of a Postdoctoral fellowship.

References

[1] *Desorption Induced by Electronic Transitions*, DIET IV, Eds. G. Betz and P. Varga, (Berlin: Springer-Verlag, 1990)

[2] St.J. Dixon-Warren, E.T. Jensen, J.C. Polanyi, G.-Q. Xu, S.H. Yang and H.C. Zeng, Faraday Discuss. Chem. Soc. **91**, 451 (1991) and references therein.

[3] (a) X.-L. Zhou, X.-Y. Zhu and J.M. White, Surface. Sci. Rep. **13**, 73 (1991); (b) J.C. Polanyi and H. Rieley, in *Dyanamics of Gas-Surface Interactions*, Eds. C.T. Rettner and M.N.R. Ashfold, (London: Royal Society of Chemistry, 1991), p. 329.

[4] St.J. Dixon-Warren, E.T. Jensen and J.C. Polanyi, Phys. Rev. Lett. **67**, 2395 (1991).

[5] St.J. Dixon-Warren, E.T. Jensen, J.C. Polanyi, J. Chem. Phys. *in preparation.*

[6] R. H. Fowler, Phys. Rev. **47**, 45 (1931).

[7] T.L.Gilton, C.P Dehnbostel and J.P Cowin, J. Chem. Phys. **91**, 1937 (1989).

[8] C. Hubrich and F. Stuhl J. Photochem. **12**, 93 (1980).

[9] S.C. Chu and P.D. Burrow, Chem. Phys. Lett. **172**, 17 (1990).

[10] C.W. Walter, K.A. Smith and F.B. Dunning, J. Chem. Phys. **90**, 1652 (1989).

[11] M. Michaud and L. Sanche, J. Elect. Spectr. and Rel. Phen., **51**, 237 (1990).

[12] S.T. Graul and M.T. Bowers, J. Am. Chem. Soc. **113**, 9696 (1991).

[13] D.E. Robbins, Geophys. Res. Lett. **3**, 213 (1976).

[14] X.L. Zhou, F. Solymosi, P.M. Blass, K.C. Cannon and J.M. White, Surf. Sci. **219**, 294 (1989).

Part II

Photodesorption
and Photochemistry

Desorption Induced by Femtosecond Laser Pulses: Surface Dynamics at High Electronic Temperatures

T.F. Heinz, J.A. Misewich, M.M.T. Loy, D.M. Newns, and H. Zacharias*

IBM Research Division, Thomas J. Watson Research Center,
Yorktown Heights, NY 10598-0218, USA
*Permanent address: Universität-GHS Essen, W-4300 Essen 1, Germany

Abstract. The desorption of NO/Pd(111) induced by femtosecond laser pulses has been investigated. The unique experimental signature of these studies indicates that the high degree of electronic excitation present in a metal under femtosecond illumination is responsible for desorption. A model for this process involving desorption induced by multiple electronic excitations (DIMET) is presented.

1. Introduction

Desorption induced by electronic transitions (DIET) continues to be a subject of great fundamental and practical interest, as this series of publications attests [1]. One of the critical aspects of DIET events is the energy flow between adsorbate and substrate. For chemisorbed species on metal surfaces the rate of energy flow is rapid [2, 3], typically occurring on a time scale of femtoseconds for electronic degrees of freedom and picoseconds for vibrational degrees of freedom. With the advent of ultrafast laser technology an examination of these processes in the time domain has become feasible. These capabilities have recently been exploited in elegant measurements of the rate of vibrational energy relaxation [4]. The picosecond time scale of vibrational energy relaxation revealed in these experiments leads one to speculate about the behavior that might be observed for excitation on a subpicosecond time scale. In this case, strong departures from equilibrium behavior may be anticipated, leading to the activation of otherwise inaccessible reaction pathways and new insights into energy flow in particle/surface interactions.

Indeed, our initial studies of molecular desorption from a metal substrate induced by femtosecond laser pulses revealed some unique features in the final-state distributions [5]. The process was characterized by a high desorption yield with a nonlinear fluence dependence and distinctive internal energy distributions. These investigations were followed by direct time-domain measurements in two laboratories [6,7] that demonstrated a subpicosecond time scale for this type of desorption process, indicating that an electronic mechanism is operative. In this paper, we review some key experimental findings of our study [5,6] of the desorption of nitric oxide (NO) molecules from Pd(111) surfaces induced by irradiation with femtosecond visible light pulses; after giving a brief account of the nature of the resulting substrate excitation, we then present a model [8] for desorption occurring through multiple electronic transitions that may arise in the presence of the high densities of electronic excitation produced by femtosecond laser pulses.

2. Experimental

The experiments were conducted using an ultrahigh-vacuum chamber with a base pressure below 1×10^{-10} Torr. The Pd(111) sample was prepared with cycles of argon-ion sputtering, oxidation, and annealing. The cleanliness of the sample was checked with Auger and thermal desorption spectroscopy, while the order of the single-crystal surface was verified by low-energy electron diffraction (LEED). The sample was generally held at a temperature of 400 K. At this temperature the

NO molecules, which adsorb nondissociatively, occupy bridge sites with a binding energy of ~ 1 eV [9]. Saturation coverage of NO was maintained throughout the experiment by dosing with a pulsed molecular beam.

The desorbed NO molecules were detected in the gas-phase by a (1+1) multiphoton ionization process via the $A^2\Sigma^+$-state. By scanning the frequency of the ionizing laser (in the vicinity of 225 nm) we were able to determine rotational, vibrational and spin-orbit distributions of the desorbed molecules. Velocity distributions were obtained from time-of-flight spectra by varying the delay time between the desorbing and ionizing laser pulses. The femtosecond laser pulses were formed by amplifying the output of a colliding pulse mode-locked (CPM) laser in a four-stage dye amplifier at a 10-Hz repetition rate. The output of the system has a pulse width of 400 fs and a spectrum centered at 620 nm.

3. Results

Using the approach described above, we have characterized yield and energy distributions of the NO molecules desorbed by femtosecond excitation pulses [5]. Although the mean translational energy of the desorbed species ($<E_t> = 0.1$ eV) is similar to that observed in thermal desorption experiments, the rotational-state distribution had a pronounced high energy tail and the vibrational temperature ($T_v > 2000$ K) was much greater than the peak lattice temperature. One of the striking features of the initial investigations was the magnitude of the desorption yield, which was much larger than that expected based on a conventional thermal or photochemical desorption mechanism. Finally, the fluence dependence of the yield, in contrast to conventional photo-induced DIET processes, was highly nonlinear.

In order to obtain time-resolved data we have applied a two-pulse correlation scheme [6]. In this approach, the femtosecond laser pulse is split into two orthogonally polarized pulses with a well-defined separation in time. Then the desorption yield is monitored as a function of the time delay between the two excitation pulses. Although the detection by ionization is slow, femtosecond resolution is obtained by controlling the separation of the femtosecond pulses. In the correlation studies, the ionization laser was tuned to the $Q_{11} + P_{21}$ bandhead and the signal is taken as representative of the total desorption yield for the NO molecules. Data from a correlation experiment using equal excitation pulses is shown in Fig. 1. The correlation feature observed has a width of ~600 fs (FWHM).

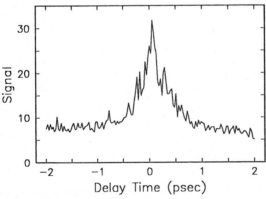

Fig. 1. Total desorption yield as a function of delay time between a pair of femto-second laser pulses.

4. Discussion and Model

A description of the nature of the excitation of the Pd substrate is the first step in interpreting the process of desorption induced by femtosecond laser pulses. The response of metals to femtosecond light pulses has been examined extensively in the literature [10]. When the light is absorbed, it produces electronic excitation in the form of electron-hole pairs. Because the rate of coupling to phonons is finite, the electrons may be heated out of equilibrium with the lattice. This situation has been modeled by introducing separate electronic and lattice temperatures, which are taken to be governed approximately by a set of coupled diffusion equations. As a result of the low electronic heat capacity compared with that of the lattice, a significant increase in the electronic temperature can be attained with only modest lattice heating. With the fluences used in our experiments, for example, we obtain peak electronic temperatures of thousands of degrees, but lattice temperature rises of only a few hundred degrees. In regard to the temporal profile, the electronic temperature increases with the absorbed energy from the laser pulse and then decays in ~ 1 ps as energy flows to the phonons. The lattice temperature climbs on the time scale of 1 ps as the two systems come into equilibrium; the lattice temperature then remains nearly constant for tens of picoseconds until cooling occurs by the diffusion of heat away from the surface. As a result of the differences in temporal profiles, a process driven by substrate phonons is expected to exhibit a correlation feature with a width of tens of picoseconds, whereas an electronic process should show a correlation width on the order of 1 ps.

The narrow feature observed in the femtosecond correlation studies indicates that desorption must be caused by the electronic excitation, rather than the lattice excitation of the substrate. This conclusion is supported by the vibrational temperature of the desorbed NO molecules, which far exceeds the peak lattice temperature of the substrate, and the high desorption yield mentioned above. (It should be noted that direct excitation of the adsorbed species can be ruled out on several grounds [5,6].) While the substrate degree of freedom responsible for desorption can be identified, the microscopic mechanism by which the high density of electronic excitation couples to the adsorbate center-of-mass motion remains to be established. This issue is, of course, the long-standing question of nonadiabatic couplings in particle/surface interactions [11].

We have recently suggested a mechanism involving desorption induced by multiple electronic transitions (DIMET) [8]. This mechanism can be considered as a generalization of the classical Menzel-Gomer-Redhead (MGR) model for DIET [12]. In this scheme, motion on an excited electronic state converts potential energy into adsorbate kinetic energy; desorption then follows upon deactivation whenever the adsorbate remains on the excited potential surface for a time $t > \tau_c$, the critical time to escape from the well. For the case of metal surfaces, however, the short lifetimes of excited electronic states imply that most of the excited molecules will deactivate before $t = \tau_c$. Such an excitation-deexcitation cycle can result in energy in the molecule-surface vibration of the ground electronic state. In the usual DIET process, this vibrational excitation is inconsequential. If, however, the adsorbate is re-excited before the vibrational excitation has relaxed, an enhanced desorption probability may result. This process of sequential pumping of vibrational excitation through excitation-deexcitation cycles is the essence of the DIMET mechanism.

The DIMET process is shown schematically for two excitation cycles in Fig. 2(a). As in the conventional DIET mechanism, the initial step is a transition to an excited potential. In this example, the adsorbate stays on the excited potential only for a time $t < \tau_c$. Thus it acquires some energy in the molecule-surface vibration, but cannot desorb. The second step in Fig. 2(a) shows how an appropriately phased excitation can lead to desorption even when the adsorbate remains on the excited potential for a time $t < \tau_c$. This behavior results from the lowering of the critical lifetime by the initial vibrational energy in the system. Of course, if the second excitation occurs during the wrong phase of the vibrational motion, then the effective critical time for the second excitation may be increased.

49

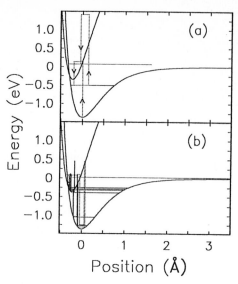

Fig. 2. (a) DIMET $n_{ex} = 2$ excitations. (b) DIMET with $n_{ex} = 20$ from a stochastic trajectory simulation.

To examine the DIMET mechanism under conditions appropriate for the experiments, we have performed semiclassical stochastic trajectory simulations [13]. The potential surfaces used are those shown in Fig. 2, with the upper surface corresponding to a simple model of a negative ion resonance [14]. The deactivation rate is assumed to fall exponentially with distance, the activation rate is then determined from the electronic temperature and potential energy separation by using a detailed balance relation [8]. For an absorbed fluence of 6.5 mJ/cm² and an excited-state lifetime at the well minimum of 2 fs, we find the DIMET desorption probability to be $P_{DIMET} = 9 \times 10^{-4}$, whereas the conventional DIET probability under the same conditions is only $P_{DIET} = 2.8 \times 10^{-7}$. Examination of the stochastic trajectories reveals that the average adsorbate experiences 12 excitation-deexcitation cycles; however, among those that desorb, the average number of excitations is 21. A typical trajectory for a desorbing molecule is displayed in Fig. 2(b). In this example, the adsorbate experiences 20 cycles of excitation-deexcitation.

While the magnitude of the DIMET yield is much larger than that for conventional DIET, we find that the yield, although very parameter sensitive, is generally lower than experiment. (Note that the fluences in the calculation are significantly larger than those of the experiment.) A comparison can be made with other experimental observations. For example, the correlation feature predicted by the DIMET calculation has a width of 670 femtoseconds, in good agreement with experiment. Our trajectory simulations also provide results on the translational energy distribution of the desorbed molecules. We find the DIMET result of $< E_t > = 80$ meV to be close to the value reported above. A striking experimental observation is the highly nonlinear fluence dependence of the desorption yield. The DIMET desorption model also predicts a nonlinear fluence dependence, as illustrated in Fig. 3. An upper bound for desorption by a conventional DIET process is also plotted in the Figure. Although the simplified model used for the DIMET simulations is not quantitatively accurate, several of the distinctive characteristics can be reproduced at a qualitative level. More sophisticated treatments would include substrate coupling to the adsorbate-surface vibration on a single potential surface [15], which has already been considered extensively in the context of vibrational damping [16], and would also take into account the internal degrees of freedom of the adsorbate [17].

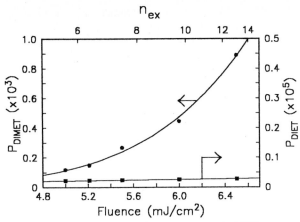

Fig. 3. DIET and DIMET desorption probabilities as a function of laser fluence. The average number of excitations of the molecules is shown on the top scale.

5. Conclusions

In summary, experiments on desorption induced by femtosecond laser pulses show several interesting features, most notably a large yield compared to that expected for conventional thermal or photochemical processes, high degree of vibrational excitation in the desorbed molecules, and a narrow correlation feature associated with a subpicosecond response. These observations imply a desorption mechanism driven by the high density of electronic excitation present under femtosecond illumination. The DIMET picture of multiple excitations is suggested as a mechanism capable of coupling this electronic excitation effectively to the adsorbate center-of-mass motion.

References

1. *Desorption Induced by Electronic Transitions - DIET IV,* edited by G. Betz and P. Varga (Springer, Berlin, 1990) and references therein.
2. Ph. Avouris and R. E. Walkup, Annu. Rev. Phys. Chem. **40**, 143 (1989).
3. E. J. Heilweil, M. P. Casassa, R. R. Cavanagh, and J. C. Stephenson, Annu. Rev. Phys. Chem. **40**, 143 (1989).
4. A. L. Harris, L. Rothberg, L. H. Dubois, N. J. Levinos, L. Dhar, Phys. Rev. Lett. **64**, 2086 (1990); J. D. Beckerle, M. P. Casassa, R. R. Cavanagh, E. J. Heilweil, and J. C. Stephenson, Phys. Rev. Lett. **64**, 2090 (1990).
5. J. A. Prybyla, T. F. Heinz, J. A. Misewich, M. M. T. Loy, and J. H. Glownia, Phys. Rev. Lett. **64**, 1537 (1990).
6. F. Budde, T. F. Heinz, M. M. T. Loy, J. A. Misewich. F. de Rougemont, and H. Zacharias, Phys. Rev. Lett. **66**, 3024 (1991).
7. J. A. Prybyla, H. W. K. Tom, and G. D. Aumiller, Phys. Rev. Lett. **68**, 503 (1992).
8. J. A. Misewich, T. F. Heinz, and D. M. Newns, Phys. Rev. Lett. **68**, 3737 (1992).
9. H. Conrad, G. Ertl, J. Küppers, and E. E. Latta, Surface Sci. **65**, 235 (1977); H.-D. Schmick and H.-W. Wassmuth, *ibid.* **123**, 471 (1982); M. Bertolo and K. Jacobi, *ibid.* **226**, 207 (1990); **236**, 143 (1990).
10. P. B. Corkum, F. Brunel, N. K. Sherman, and T. Srinivasan-Rao, Phys. Rev. Lett. **61**, 2886 (1988) and references therein.
11. *Many-Body Phenomena at Surfaces*, edited by D. Langreth and H. Suhl (Academic Press, New York, 1984) and references therein.

12. D. Menzel and R. Gomer, J. Chem. Phys. **41**, 3311 (1964); P. A. Redhead, Can. J. Phys. **42**, 886 (1964).
13. More sophisticated treatments of molecular dynamics with electronic transitions have recently been developed. See J. C. Tully, J. Chem. Phys. **93**, 1061 (1990) and references therein.
14. J. W. Gadzuk, L. J. Richter, S. A. Buntin, D. S. King, and R. R. Cavanagh, Surface Sci. **235**, 317 (1990).
15. D. M. Newns, T. F. Heinz, and J. A. Misewich, Prog. Theor. Phys. Suppl. **106**, 411 (1991) and to be published.
16. M. Head-Gordon and J. C. Tully, J. Chem. Phys. **96**, 3939 (1992) and references therein.
17. A. R. Burns, E. B. Stechel, D. R. Jennison, Phys. Rev. Lett. **58**, 250 (1987).

A New Way to Characterize the Excitation Mechanism in Surface Photochemistry

M. Wolf*, X.-Y. Zhu, and J.M. White

Department of Chemistry and Biochemistry, University of Texas,
Austin, TX 78712, USA
*Present address: Fritz-Haber-Institut, Faradayweg 4–6,
 W-1000 Berlin 33, Germany

Abstract. Direct and substrate-mediated excitation mechanisms in surface photochemistry can be discriminated based on the pronounced angular dependence of the evanescent electric fields at the vacuum interface of a thin metal film evaporated onto a transparent substrate. We can unambiguously establish the change from a direct to a substrate-mediated excitation mechanism in the photodesorption of SO_2 (hv = 6.4 eV) on a quartz prism, with and without a 300 Å Ag film.

1. Introduction

UV irradiation of an adsorbate-substrate system may result in excitation of the valence electrons of both the substrate and the adsorbate. As a consequence, surface photochemistry can be either initiated by direct adsorbate-localized excitation induced by the electric fields at the surface or by electronic excitations in the substrate which may be converted to adsorbate excitations, *e.g.*, by attachment of hot substrate carriers [1].

Provided there are pronounced features in the gas phase UV-absorption spectrum of the adsorbate or in the substrate absorbance, the wavelength dependence of the photochemical cross section helps to discriminate between these two mechanisms. For example, the photodissociation of $Mo(CO)_6$ on various substrates closely resembles the gas phase absorption, supporting the dominance of a direct excitation mechanism [2]. On the other hand, photodesorption of SO_2 from Ag(111) is believed to be substrate-mediated, as indicated by the correlation of the wavelength dependence with the bulk plasmon resonance of Ag [3].

Another approach to discriminating between substrate-mediated and direct excitation processes is based on the polarization and angle-of-incidence dependence of the photochemical cross section [4]. However, Richter *et. al.* [5] recently pointed out that this method fails if the transition dipole moment under consideration lies predominantly in the surface plane. To our knowledge there exists no method that can rigorously determine the excitation mechanism in all cases.

We present a new approach to this problem. A transparent substrate covered with a thin metal film is irradiated from the backside inducing photodesorption of molecules adsorbed at the vacuum-film interface. The pronounced angular dependence of the evanescent electric fields around the critical angle of total internal reflection allows discrimination between substrate-mediated and adsorbate-localized excitations.

2. Experimental

The experiments were conducted in a UHV chamber ($2 \cdot 10^{-10}$ torr) with a 10 x 10 mm rectangular quartz prism, which could be cooled with liquid nitrogen to 120 K and heated via IR irradiation to ~550 K. The Ag films were freshly evaporated in UHV at room temperature; the thickness was determined by a quartz crystal microbalance mounted on the shutter of the doser. Above 50-100 Å thickness continuous films with the dielectric properties of bulk silver will be formed [6]. SO_2 was dosed through a microcapillary array facing the front face (film side) of the prism at a distance of ~3 mm. A quadrupole mass spectrometer (QMS) was used for postirradiation thermal desorption (TDS), time-of-flight (TOF) spectroscopy and to monitor the yield of desorption products during irradiation. The prism was irradiated from the back side (see inset in Fig.1) by *p-polarized* light of an excimer laser ($h\upsilon$ = 6.4 eV, pulse energy ~1 mJ/cm^2). The initial coverage in all the experiments was 2 monolayers.

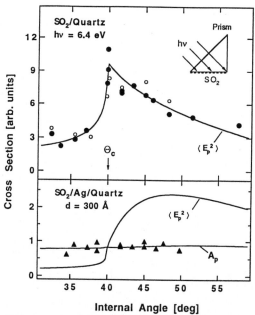

Figure 1 Relative cross sections for photodesorption of SO_2 ($h\upsilon$ = 6.4 eV) as a function of the internal angle in quartz (120 K, coverage ~2 ML). Upper panel: SO_2 desorption from the bare prism (obtained from desorption yield (●) and postirradiation TDS (O)). Lower panel: SO_2 depletion cross section on a 300Å Ag film (postirradiation TDS(▲)). The solid lines show the total electric field strength $\langle E_p^2 \rangle$ and the absorbance A_p in the Ag-film, respectively.

3. Results and Discussion

The upper panel of figure 1 shows the relative photodesorption cross section of SO_2 adsorbed on the bare quartz prism (120 K) for numerous internal angles. Molecular SO_2 is the predominant desorption product. The filled circles show cross sections obtained from the exponential decay of the SO_2 desorption yield as a function of photon fluence; cross sections derived from postirradiation TDS are shown as the open circles. At 45° an absolute cross section of $7 \cdot 10^{-19}$ cm^2 is found. The whole data set is nicely fit by the calculated total electric field strength $\langle E_p^2 \rangle$ at the quartz-vacuum interface (solid line), as expected for processes initiated by *direct excitation* of randomly orientated molecules. TOF measurements reveal that parent SO_2 desorbs with a mean translational energy, $\langle E_{trans} \rangle$, of 125 meV. The observed photodesorption of SO_2 from quartz can be explained by a desorption mechanism as suggested by Nishi *et. al.* for NH_3 and H_2O ice [7], and by Polanyi and co-workers for H_2S and OCS adsorbed on LiF [8]. It is believed that direct adsorbate excitation is followed by an electronic-to-translational energy transfer process, which might either involve direct exchange repulsion or inter-adsorbate quenching between neighboring molecules.

With a 300 Å Ag film evaporated on the quartz prism, the total cross section drops by nearly one order of magnitude ($8.5 \cdot 10^{-20}$ cm^2 at 45°) due to the efficient quenching in the presence of a metal surface. The cross sections (triangles in the lower panel of figure 1) show no dependence on the internal angle. The data can be fit by the calculated absorbance in the Ag film, A_p, which remains nearly constant, while the total p-polarized electric field strength, $\langle E_p^2 \rangle$, changes significantly within ~10° around the critical angle (solid lines) [9]. This provides direct support for the *dominance of substrate-mediated excitation.* [3]. In the range of 0.5 - 2.5 mJ/cm^2 the desorption cross section is found to be independent of the absorbed laser fluence.

We speculate that SO_2 desorption from Ag is induced by attachment of hot electrons (attachment to form gas phase SO_2^- is nondissociative below 2 eV [10]) followed by an Antoniewicz scenario for desorption [11]. Furthermore, the direct excitation process that dominates on the clean prism is almost completely quenched in the presence of the 300 Å silver film (the partial cross section of the direct process decreases at least several orders of magnitude). This change of the dominating excitation mechanism is also reflected in the desorption dynamics, where we find a considerably lower translational energy in the substrate-mediated case ($\langle E_{trans} \rangle = 70$ meV).

SO_2 on Ag is, therefore, a good example of why photochemical processes on metals are observed with significant cross section despite strong excited state quenching by the substrate [2]. The UV-absorption coefficient of a free gas phase molecule is typically $<10^{-3}$, while the UV-absorption of most metals is above 50%. The higher initial number of substrate excitations may therefore result in the dominance of a substrate-mediated (charge-transfer) process if similar quenching rates for the excited states are assumed[12].

55

Acknowledgment: This work was supported by the National Science Foundation, Grant CHE 90-15600 and by the Robert A. Welch Foundation. MW acknowledges a fellowship of the Alexander-von-Humboldt Society.

References

1   X.-L. Zhou, X.-Y. Zhu and J. M. White, *Surf. Sci. Rep.* **13**, 73 (1991).
2   W. Ho in "Desorption Induced by Electronic Transitions, DIET IV", eds G. Betz and P. Varga (Springer, Berlin, 1990).
3   M. E. Castro and J. M. White, *J. Chem. Phys.* **95**, 6057 (1991).
4   X.-Y. Zhu, J. M. White, M. Wolf, E. Hasselbrink and G. Ertl, *Chem. Phys. Lett.* **176**, 459 (1991).
5   L. J. Richter, S. A. Buntin, D. A. King and R. R. Cavanagh, *Chem. Phys. Lett.* **186**, 423 (1991).
6   M. Fukui and K. Oda, *Appl. Surf. Sci.* **33**, 882 (1988).
7   N. Nishi, H. Shinohara and T. Okuyama, *J. Chem. Phys.* **80**, 3898 (1984).
8   I. Harrison, J. C. Polanyi and P. A. Young, *J. Chem. Phys.* **89**, 1498 (1988). J. C. Polanyi and P. A. Young, *J. Chem. Phys.* **93**, 3673 (1990).
9   We assume a amooth Ag film with $n = 1.27$, and $k = 1.2$ (bulk values at 6.4 eV) an optical thin adsorbate layer (see [6]).
10   J. Rademacher, L. G. Christophorou, and R. P. Blaustein, *J. Chem. Soc. Faraday Trans.* II **71**, 1212 (1975).
11   P. R. Antoniewicz, *Phys. Rev.* **B 21**, 3811 (1980).
12   E. Hasselbrink, *Comments At. Mol. Phys* **27**, 265 (1992).

Photodesorption Dynamics and Structural Information for O$_2$ Adsorbed on Pd(111)

A. de Meijere, H. Hirayama*, F. Weik, and E. Hasselbrink

Fritz-Haber-Institut der Max-Planck-Gesellschaft,
Faradayweg 4–6, W-1000 Berlin 33, Germany
*Permanent address: NEC Corporation, Microelectronics Research
 Laboratories, 34 Miyukigaoka, Tsukuba 305, Japan

Abstract. UV irradiation (hv = 3.9 to 6.4 eV) of O$_2$ adsorbed on Pd(111) induces photodesorption and photodissociation. Both processes are of non-thermal character. Moreover, thermal conversion processes between adsorbate states are observed and are found to be coupled to a thermal desorption channel. As the incident photon energy is raised from 3.9 eV to 6.4 eV, the cross sections for all processes rise exponentially by a factor of 38. Angle resolved measurements with respect to the polar and azimuthal desorption angle show that desorption flux and translational energy carry information about the adsorbate binding geometry prior to desorption.

1. Introduction

Since the last conference in this series the UV photochemistry of adsorbed molecules has been subject of a surge of scientific efforts towards a fundamental understanding. The photochemistry of various O$_2$-substrate systems has attracted considerable interest: Pt(111) [1-3], Pd(111) [4-7] and Ag(110) [3, 8].

2. Experimental

A standard UHV-apparatus with a molecular beam for O$_2$ dosing was used. Sample order and orientation in the laboratory system were checked by LEED. The clean sample showed a sharp hexagonal pattern and after saturation with atomic oxygen a p(2x2) pattern. The azimuthal orientation of the sample could be established by referring to calculated LEED I-V-curves for the (01) and (11) spots.

For irradiation an excimer laser was employed running at 5.0 and 6.4 eV photon energy. Additional photon energies, in particular 3.9, 4.5, and 5.9 eV, were available by Raman shifting the excimer lines. Low laser fluences, typically ≤ 3 mJ/cm^2/pulse, were applied in order to prevent noticeable heating of the substrate. A quadrupole mass spectrometer (QMS) mounted on a rotatable platform served for recording thermal desorption spectra (TDS) and time-of-flight (TOF) spectra.

Using a set of crystal holders the azimuthal orientation of the sample could be varied in 30° steps with respect to the plane defined by the surface normal and the direction toward the detector. The intersection of this detection plane with the sample surface defines the azimuthal direction of desorption. Alignment of a crystallographic direction with this plane was checked by LEED with an accuracy of about 5°.

Springer Series in Surface Sciences, Vol. 31
Desorption Induced by Electron Transitions DIET V
Editors: A.R. Burns · E.B. Stechel · D.R. Jennison   © Springer-Verlag Berlin Heidelberg 1993

3. Results

On Pd(111) O_2 adsorbs molecularly at 100 K forming three distinct species: α_1-, α_2-, and α_3-O_2(a). These are observed in thermal desorption at 120, 155 and 200 K, respectively. α_1- and α_2-O_2(a) are bound in side-on geometry in bridge and atop sites, respectively. The α_3-O_2(a) adsorption geometry will be discussed later on. The photochemistry of O_2 is apparent in several processes: photodissociation to form atomic oxygen, photodesorption of O_2 with translational energies greatly exceeding equilibration with the surface temperature, conversion between different adsorbate states and a desorption process with translational energies indicating accommodation with the sample temperature [6, 7]. The first two processes are evidently of non-thermal character, in particular dissociation since it does not thermally occur at temperatures below 200 K:

$$O_2(a) \xrightarrow{h\nu} O_2(\uparrow)_{fast}$$
$$\xrightarrow{h\nu} 2\,O(a).$$

The latter processes have been interpreted as a consequence of the accumulation of atomic oxygen on the surface. Matsushima [9] had shown experimentally that the presence of atomic oxygen rapidly reduces the saturation coverage for molecular oxygen, most predominantly for the most tightly bound species, α_1-O_2(a). This can be understood in terms of competition for binding sites. In the same manner the photolytic formation of atomic oxygen will lead to competition between the nascent oxygen atoms and the residual molecules for binding sites, of course favouring the more tightly bound atomic oxygen. At high initial coverages this causes desorption via a displacement mechanism. This desorption is characterized by accommodation to the surface temperature since the excess energy is rapidly dissipated during the displacement process. The data also show that displacement can to some extend be compensated by conversion from α_1-O_2(a) to α_2- and α_3-O_2(a):

$$O(a) + O_2(a) \longrightarrow O(a) + O_{2,displaced} \longrightarrow O(a) + O_2(\uparrow)_{slow}$$
$$\longrightarrow O(a) + O_{2,converted}.$$

Detailed measurements of TOF-spectra, the desorption fluxes, and the state of the adsorbate as a function of irradiation allow us to model the kinetics of this system. This procedure yields cross sections for photodesorption and photodissociation of the different O_2-species.

Evaluation of data taken at 5 different photon energies, shows that all partial cross sections keep a constant ratio to each other independent of photon energy. They rise exponentially by a factor of 38 over the range from 3.9 to 6.4 eV indicating a single common excitation for all processes. It is attributed to attachment of hot electrons created by photon absorption within the substrate to an O_2 affinity level, most likely the $3\sigma_u$-level, which is antibonding with respect to the O-O-bond. Charge transfer into the $3\sigma_u$-level has also been identified as crucial for thermally activated dissociation [10]. The dependence on photon energy is then due to the rate of electron tunnelling through the surface barrier into this level. The data are consistent with a position of this level about 4 eV above the Fermi energy in agreement with common expectation. At the

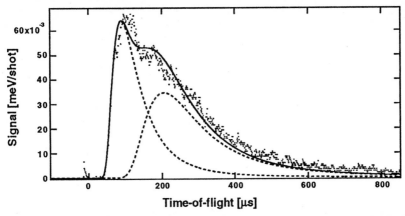

Fig.1: Time-of-flight spectra for photostimulated desorption of O_2 from Pd(111). The sample had been saturated with O_2 and irradiated at 95 K sample temperature. The bimodal spectrum indicates a "fast" and a "slow" desorption channel. The average translational energies $\langle E_{trans}\rangle/2k$ are 650 and 120 K, respectively.

same time the transient formation of a negative ion state causes attraction towards the surface by the image charge. This effect eventually leads to desorption via the Antoniewicz mechanism [11]. Upon reneutralization the molecule finds itself displaced towards the surface and stretched in the intramolecular bond. Branching into desorption and dissociation then depends on the position and momenta in these two coordinates upon reneutralization and the finer details of the potential energy surface.

We have also carried out angle resolved time-of-flight spectroscopy for conditions under which desorption is primarily due to a single one of the $O_2(a)$ binding states.:

1) The O_2 saturated surface is irradiated at 153 K sample temperature. This elevated temperature prevents adsorption of α_3- and α_2-$O_2(a)$. Hence, the signal should be entirely due to desorption from α_1-$O_2(a)$.

2) The surface was predosed at 400 K with oxygen resulting in about 0.2 ML coverage of atomic O. Then it was saturated with O_2 and irradiated at 130 K. This preparation results in predominant desorption from α_2-$O_2(a)$.

3) The surface was predosed at 400 K with oxygen as above. Then it was saturated at 95 K with O_2 and irradiated with $8.4\cdot10^{21}$ photons/cm^2 to remove most of the α_2-$O_2(a)$. Further irradiation causes predominant desorption from α_3-$O_2(a)$.

TOF-spectra have been collected as function of desorption angle. From the "fast" desorption peak observed we derive the average translational energy $\langle E_{trans}\rangle/2k$ of the photochemically desorbed molecules (Fig. 1). For the three preparations different variations of $\langle E_{trans}\rangle/2k$ with desorption angle are observed. For desorption in the plane containing the [$\bar{2}11$] azimuth the data for the α_1-state indicate that $\langle E_{trans}\rangle/2k$ decreases with increasing desorption angle from 630±40 K in normal direction to 400±60 K at 72° (Fig. 2). For the α_3-state the same value is found for normal desorption but then a strong increase leading to a maximum around 60° desorption angle with $\langle E_{trans}\rangle/2k = 1050±100$ K is observed. Finally $\langle E_{trans}\rangle/2k$ decreases again to 700±100 K at 72°. For

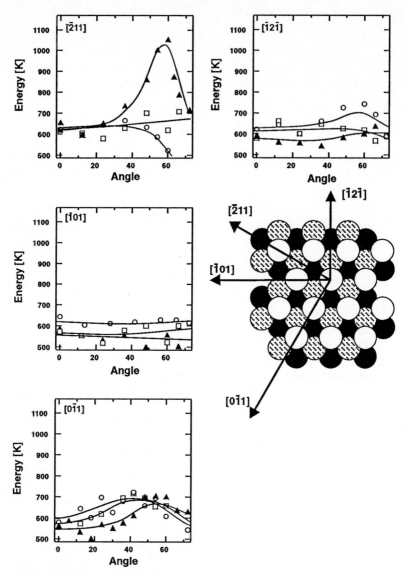

Fig.2: Azimuthaly resolved average translational energies $\langle E_{trans}\rangle/2k$ vs. desorption angle for desorption along the indicated azimuths (o: a_1-O_2; □: a_2-O_2; ▲: a_3-O_2). A marked maximum is observed for α_3-O_2 at a desorption angle of 60° in the [$\bar{2}$11] azimuth. This direction is from an on-top site towards a neighbouring *fcc*-site. It is pointig away from the most likely adsorption sites of atomic oxygen which is a next nearest neighbouring *fcc*-site. Lines are drawn to guide the eye.

α_2-O_2(a) the values lie in between. However, it should be kept in mind that none of the preparations achieves exclusive desorption from a single state. At best it results in preferential desorption from one state.

This kind of data was recorded for different azimuthal orientations of the sample. We find that the peak in the $\langle E_{trans} \rangle$ vs. angle data for α_3-O_2(a) has a *three*-fold and not a six-fold symmetry regarding rotation of the sample. In particular the peak in $\langle E_{trans} \rangle/2k$ is observed along the [$\bar{2}$11] and [1$\bar{2}$1] azimuth and not along the opposite [2$\bar{1}\bar{1}$] azimuth. The data for the other states are rather invariant with sample rotation.

4. Discussion

The top layer of a (111) surface has a six-fold symmetry upon rotation. It is well established in the literature that atomic oxygen forms a p(2x2) structure on these (111) surfaces. Oxygen atoms are expected to adsorb in the three-fold hollow sites. Hence, the 2x2 structure can align with the (111) surface by either populating exclusively *hcp* or *fcc* sites. On transition metals such as nickel and platinum the p(2x2) is located on the *fcc* sites [12, 13]. It is plausible that this also holds true for Pd(111). The population of the 2x2 grid of *fcc* sites by atomic oxygen reduces the six-fold symmetry to a three-fold symmetry in the same manner as taking into account the *abc*-stacking of a *fcc*-(111)-surface does. The orientation of the so imposed three-fold symmetry in real space does not change when going from one terrace to another. Whereas one would expect the second metal layer to have a negligible effect upon a weakly bound molecular species like O_2, the coadsorbed atomic oxygen will have a strong effect due to repulsive interaction between the dipoles resulting from the charge transfer involved in bonding.

α_3-O_2(a) is a superoxo species characterized by a vibrational mode of 1015 cm^{-1}. In analogy to organometallic complexes two binding geometries have been proposed for such a species: one with the molecular axis parallel to the surface, and another where the molecule is bound in an on-top site with its molecular axis inclined from the surface normal. With the atomic oxygen p(2x2) structure present a 1/4 ML of Pd surface atoms is left with no neighbouring oxygen atoms. These will be the preferred adsorption sites of the α_3-O_2(a) in order to minimize repulsion. Repulsion will further be effective in two ways. First, it might orient the tilted molecular axis of the O_2 molecule to point away from it, resulting in orientation towards a neighbouring empty *fcc* site. Second, in the excited state there might be an additional repulsive lateral force. The first argument is agreement with a report by Shinn and Madey [14] on the observation of a six-fold pattern of O^+-desorption beams around the surface normal in ESDIAD of O_2 from Cr(110) where O_2(a) is also characterized by its vibrational frequency as a superoxo-species. In the usual ESDIAD manner the six-fold pattern was interpreted to indicate an inclined molecular axis correlated with the surface symmetry. In the experiment reported here both mechanisms might contribute. However, in any case structural information is contained in the dynamics of photostimulated desorption of intact molecules.

The structures reported here are much more pronounced in the $\langle E_{trans} \rangle$ vs. angle data than in the flux vs. angle data where they only become obvious after deconvoluting the contributions from the different $O_2(a)$-states. The large differences in $\langle E_{trans} \rangle$ of the various O_2-species serve to discriminate between these contributions.

We thank Kurt Kolasinski for helpful suggestions regarding this manuscript. Herbert Over is thanked for carrying out the LEED-calculations which were essential to identify the crystallographic orientations and Patricia Thiel for helpful discussions.

References

1. X.-Y. Zhu, S.R. Hatch, A. Campion, J.M. White: J. Chem. Phys. **91,** 5011 (1989)
2. W.D. Mieher, W. Ho: J. Chem. Phys. **91,** 2755 (1989)
3. S.R. Hatch, X.-Y. Zhu, J.M. White, A. Campion: J. Phys. Chem. **95,** 1759 (1991)
4. X. Guo, L. Hanley, J.T. Yates, Jr.: J. Chem. Phys. **90,** 5200 (1989)
5. L. Hanley, X. Guo, J.T. Yates, Jr.: J. Chem. Phys. **91,** 7220 (1989)
6. M. Wolf, E. Hasselbrink, J.M. White, G. Ertl: J. Chem. Phys. **93,** 5327 (1990)
7. M. Wolf, E. Hasselbrink, G. Ertl, X.-Y. Zhu, J.M. White: Surf. Sci. **248,** L235 (1991)
8. S.R. Hatch, X.-Y. Zhu, J.M. White, A. Campion: J. Chem. Phys. **92,** 2681 (1990)
9. T. Matsushima: Surf. Sci. **157,** 297 (1985)
10. I. Panas, P. Siegbahn, U. Wahlgren: J. Chem. Phys. **90,** 6791 (1989)
11. P.R. Antoniewicz: Phys. Rev. **B 21,** 3811 (1980)
12. J. Haase, B. Hillert, L. Becker, and M. Pedio, Surf. Sci. **262,** 8 (1992)
13. K. Mortensen, C. Klink, F. Jensen, F. Besenbacher, and I. Stensgaard, Surf. Sci. **220,** L701 (1989)
14. N.D. Shinn, T.E. Madey: Surf. Sci. **176,** 635 (1986)

Dynamics of Ammonia Photodesorption from GaAs(100): A Vibration-Mediated Mechanism

X.-Y. Zhu, M. Wolf, T. Huett, and J.M. White

Department of Chemistry and Biochemistry, University of Texas, Austin, TX 78712, USA

Abstract. UV photodesorption of NH_3 from GaAs(100) is characterized by a strong isotope effect in the cross section, $\sigma_{NH3}/\sigma_{ND3} = 4.1 \pm 0.5$, a fact which cannot be accounted for by the mass difference of the leaving particles. We take this as the first evidence for UV-photodesorption from electronically quenched, but vibrationally hot ground state adsorbates.

1. Introduction

UV photon-driven desorption of adsorbates from metal and semiconductor surfaces is frequently described in terms of the Menzel-Gomer-Redhead (MGR) model [1] and its variations [2-3]. The MGR model predicts an isotope effect because a lighter particle is accelerated in a shorter time on the excited potential energy surface (PES) and, thus, has a higher probability in escaping strong quenching by the substrate. Quantitatively, for two isotopically labeled species with masses m_1 and m_2 for the leaving particles, the isotope effect is given by [1, 4, 5]:

$$\frac{\sigma_{m1}}{\sigma_{m2}} = \left(\frac{1}{P_{m1}}\right)^{(m_2/m_1)^{1/2} - 1} \tag{1},$$

where σ_{m1} and σ_{m2} are the desorption cross sections and P_{m1} is the probability of photodesorption. The latter is given by the ratio of σ_{m1} to σ_{ex}, the excitation cross section. It is important to note that equation (1) is independent of the details of the PES. Strong isotope effects, consistent with the MGR model, have been observed in ESD [5] and in surface photon-driven processes [4].

In this report, we demonstrate that the isotope effect in UV photodesorption of NH_3 from GaAs(100) is too large to be accounted for by equation (1). We propose that desorption from vibrationally excited ground state ammonia is important and, thus, that electronic-to-vibrational energy transfer is a central consideration in this system. This vibrational pumping mechanism is an alternative to the direct resonant intraadsorbate vibrational excitation in IR laser induced desorption [6]. Detailed accounts of this study are presented elsewhere [7].

2. Experimental

All experiments were conducted in a UHV chamber [7] and involved temperature programmed desorption (TPD) and time-of-flight (TOF) mass spectroscopy. The GaAs(100) substrate (15x10x1 mm, 10^{15}/cm^3 Si-doped) was cleaned by standard procedures to yield a (4x6) LEED pattern, which corresponds to a Ga-rich surface. Ammonia was dosed onto the front face of the sample at 102 K to give a saturation coverage of 0.5 ML [7]. The light source was a pulsed (11 ns) ArF excimer laser ($h\nu$ = 6.4 eV). A low pulse energy (1 mJ/cm^2) was used to minimize transient surface heating. The flight distance (sample to QMS ionizer) for TOF measurements was 6 cm.

3. Results and discussion

Photon irradiation of monolayer ammonia covered GaAs(100) leads to molecular photodesorption and a small amount of photodissociation ($\leq$10%) [7]. Time-of-flight distributions of photodesorbing ammonia, averaged over 1 x 10^{18} photons/cm^2, are shown in the inset of Fig. 1. The flux-weighted average translational energy is, within experimental error, independent of the isotopic composition: $<E_{trans}/2k>$ = 300 ± 30 K, higher than the substrate temperature (102 K). The nonthermal character of this process is provided by: (1) the linear relation between the TOF intensity and the laser pulse energy (0.5 to 11 mJ/cm^2) and (2) the independence of the translational energy on laser pulse energy.

While the *mean translational energy* is, within experimental error, independent of isotope substitution, the *flux* of photodesorbing NH$_3$ is four times that for ND$_3$, indicating a strong isotope effect. To confirm this, we have plotted, in Fig. 1, the molecular ammonia coverage as a function of

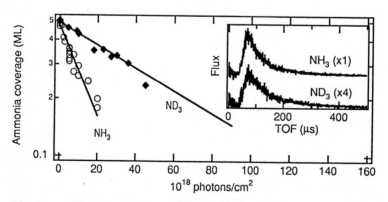

Fig. 1. Post-irradiation TPD areas for NH$_3$ (open circles) and ND$_3$ (solid diamonds) as a function of photon fluence at $h\nu$ = 6.4 eV. *Inset:* TOF spectra of NH$_3$ and ND$_3$ (x4) from saturated GaAs(100) surfaces at 102 K.

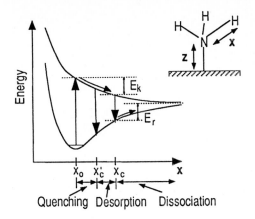

Fig. 2. Schematic diagram for UV photodesorption mediated by vibrational excitation. The x-axis corresponds to the N-H bond length and the z-corresponds to the surface-NH_3 distance.

photon fluence for both NH_3 and ND_3. Photodissociation channel makes a negligible contribution [7]; thus, the slopes measure the photodesorption cross sections. Least-square fits to the data yield, $\sigma_{NH_3} = 5.4 \times 10^{-20}$ cm^2, and $\sigma_{ND_3} = 1.3 \times 10^{-20}$ cm^2, i.e., $\sigma_{NH_3}/\sigma_{ND_3} = 4.1 \pm 0.5$, in agreement with the TOF results presented in the inset. The same result is also obtained for coadsorbed NH_3 and ND_3.

We now demonstrate the inadequacy of the MGR model for ammonia photodesorption. Using equation (1) with a mass ratio for the leaving group, $m_{NH_3}/m_{ND_3} = 1.176$, we have $\sigma_{ex} = \sigma_{NH_3}/P_{NH_3} = 10000$ Å^2, which is not physically sensible. Since isotope substitution has a much larger effect on the N-H coordinate than the surface-NH_3 coordinate, we now examine how N-H excitation, i.e., vibrational excitation, might lead to desorption.

The proposed variation of the MGR model, shown schematically in Fig. 2, involves UV photon excitation, either directly or through the substrate, to give an electronically excited molecule, assumed to be antibonding with respect to N-H (this naturally accounts for some photodissociation, as observed). Consistent with the MGR picture, most of the excited ammonia is quenched to give vibrationally excited NH_3 on the ground state PES. This vibrational energy in the N-H coordinate (x) is coupled to the N-surface coordinate (z) and, if sufficient energy transfers, desorption follows. The vibrational energy transfer process has already been demonstrated in IR induced desorption of NH_3 from solid surfaces, where resonant excitation of the ν_s mode of NH_3 leads to desorption [6b].

We introduce two critical distances in the stretch of N-H, x_c and x_c'. For de-excitation occurring at $x > x_c$, the kinetic energy, E_k, (in this case of the H atom) is greater than the barrier for dissociation, E_r, and the coordinate x is not bound. The reverse holds for $x < x_c$. We assume a *second* critical distance, x_c': de-excitation at $x > x_c'$ results in a ground state NH_3 adsorbate with sufficient *intramolecular* vibrational content to undergo the aforementioned vibrational energy transfer, and desorb. Neglecting

65

the low probability for dissociation, a modified form of equation (1) can be derived to give the isotope effect in ammonia photodesorption [7]

$$\frac{\sigma_{NH_3}}{\sigma_{ND_3}} = \left(\frac{1}{P_d(NH_3)}\right)(m_D/m_H)^{1/2} - 1 \tag{2}$$

It is important to note that the masses entering equation (2) are not those for the leaving particles, but the reduced masses in the N-H and N-D stretching coordinates which can be approximated by those of H and D atoms. From the measured isotope effect and the mass ratio of $m_D/m_H = 2$, we obtained a photodesorption probability of $P_d = 0.033$ for NH_3. Therefore, the excitation cross section is $\sigma_{ex} = \sigma_{NH_3}/P_d = 1.6 \times 10^{-18}$ cm^2.

Equation (2) can also be intuitively understood: since the energy is fed in through the N-H coordinate, it is the mass effect in this coordinate which is responsible for the isotope effect in NH_3 photodesorption.

Acknowledgement: This work was supported in part by the STC Program of the National Science Foundation, grant CHE 8920120.

References

[1] (a) D. Menzel and R. Gomer, *J. Chem. Phys.* **41**, 3311 (1964); (b) P. A. Redhead, *Can. J. Phys.* **42**, 886 (1964).
[2] Ph. Avouris and R. E. Walkup, *Ann. Rev. Phys. Chem.* **40**, 173 (1989).
[3] X.-L. Zhou, X.-Y. Zhu, and J. M. White, *Surf. Sci. Rep.* **13**, 73 (1991).
[4] M. Wolf, S. Nettesheim, J. M. White, E. Hasselbrink, and G. Ertl, *J. Chem. Phys.* **94**, 4609 (1991).
[5] T. E. Madey, J. T. Yates, Jr., D. A. King, and C. J. Uhlander, *J. Chem. Phys.* **52**, 5215 (1970).
[6] (a) T. J. Chuang, *Surf. Sci. Rep.* **3**, 1 (1983); (b) T. J. Chuang, H. Seki, and I. Hussla, *Surf. Sci.* **158**, 525 (1985).
[7] (a) X.-Y. Zhu, M. Wolf, T. Huett, and J. M. White, *J. Chem. Phys.* to be published; (b)X.-Y. Zhu and J. M. White, *Phys. Rev. Lett.* **68**, 3359 (1992).

Molecular Photochemistry on Surfaces: Role of Adsorbate–Substrate Structure

J.M. White

Department of Chemistry, Center for Materials Chemistry,
University of Texas, Austin, TX 78712, USA

Abstract. Recent surface photochemistry experiments - CF_3I and SO_2 on Ag(111), and AsH_3 on GaAs(100) - are briefly described, emphasizing the important role played by the local surface structure.

1. Introduction

Photochemistry at adsorbate-substrate interfaces is a relatively new field of surface chemical science [1]. As for the gas phase, there are fundamental questions of reaction mechanisms and dynamics, questions now being addressed in many laboratories. Compared to the gas phase, adding a substrate introduces new possibilities for excitation, relaxation and reaction. Research now underway in our laboratory deals with many of these; as outlined in the abstract, this paper overviews three systems.

2. SO$_2$ on Ag(111)

Sulfur dioxide has a rich photochemistry and photophysics [2] with fluorescence (lifetime $\sim 10^{-7}$ s) and phosphorescence (lifetime $\sim 10^{-4}$ s) dominating below 5 eV and direct photodissociation (predissociation $\sim 10^{-9}$ s) entering the picture at higher photon energies. Bimolecular reactions between excited and ground state SO_2 are well-known [2]. On a metal, where electronic state quenching is expected to compete effectively (10^{-13} - 10^{-15} s), we expect that all of these processes will be quenched unless accumulated vibrational excitation is sufficient to activate the molecule. Only processes occurring on a picosecond or femtosecond time scale will be competitive.

Indeed, between 193 and 437 nm (6.4 to 2.8 eV), the photon-driven chemistry of SO_2, weakly held through the S atom on Ag(111), is limited to parent desorption for coverages up to one monolayer [3]. For coverages between 10 and 60 ML, SO and SO_3 desorb during illumination at 193, 248 and 351 nm, and O desorbs at 193 nm. The presence of SO and SO_3 up to 351 nm, but not at 437 nm, is attributed to the bimolecular process described above. The presence of O atoms at 193 nm, is attributed to direct photodissociation.

Interestingly, within the first monolayer, the initial yields, measured by integrating SO_2 time-of-flight spectra following 15 ns excimer laser pulses at

Springer Series in Surface Sciences, Vol. 31
Desorption Induced by Electron Transitions DIET V
Editors: A.R. Burns · E.B. Stechel · D.R. Jennison © Springer-Verlag Berlin Heidelberg 1993

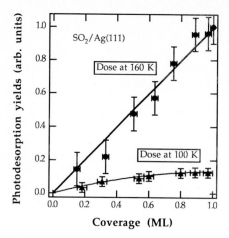

Figure 1. Pulsed excimer laser (248 nm) driven desorption of SO_2 from Ag(111) as a function of coverage between 0 and 1 ML. Yields were calculated from integration of TOF distributions (e.g., see Figure 2). Two preparation methods were used: (1) the Ag(111) was held at 160 K during dosing and then cooled to 100 K for photolysis (circles) and (2) the Ag(111) was held at 100 K for both dosing and photolysis (triangles).

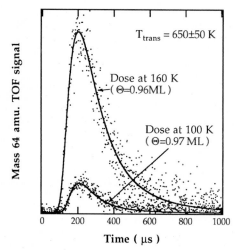

Figure 2. Pulsed excimer laser (248 nm) driven time-of-flight (TOF) distributions for 0.96±0.01 ML of SO_2 dosed on Ag(111) at 100 K (lower) and 160 K (upper). In both cases photolysis was done with the system at 100 K. Solid curves are least squares fits to a modified Boltzmann distribution [1b, 7]. Both curves are fit with a temperature of 650±50 K.

248 nm, depend strongly on the dosing method (**Figure 1**) [3]. Dosing at 160 K, which lies below the monolayer temperature programmed desorption (TPD) peak at 175 K, gives yields that vary linearly with coverage and are much larger than those prepared by dosing the same coverage at 100 K. Dosing at 100 K and heating to 160 K, also increases the yield. While the yields vary strongly , the translational temperatures, determined from fits to the ToF spectra (**Figure 2**), do not. This suggests that the transition states leading to desorption are the same, that the excitation is substrate mediated, and that the yield differences reflect variations of the excitation and/or quenching probabilities with local surface structure. While intuitively sensible, there is no direct spectroscopic evidence for different orientation distributions. Orbital overlap should be a key factor in excitation and quenching processes that involve substrate charge transfer. The notion of a common transition state for photon-driven desorption of submonolayer SO_2 is

confirmed by the constant translational temperature (650 ± 50 K), between 193 and 437 nm [3]. For even longer wavelengths, the translational temperature drops. These observations are consistent with resonant hot substrate electron attachment to SO_2, and, for sufficiently long wavelengths, a hot electron energy distribution that moves to energies below the resonance energy. The dominant role played by substrate-mediated desorption is confirmed by experiments which track the desorption yield as a function of surface electric field strength [4].

3. CF3I/Ag(111)

Unlike SO_2, alkyl halides, partially and fully halogenated, all show strong unstructured optical UV absorption spectra that reflect prompt (10 - 100 fs) dissociation on repulsive potential energy curves [2]. Under these conditions, quenching and dissociation occur on about the same time scales and we often observe, even for submonolayer coverages, dissociation with relatively high yields [1]. A fully halogenated example, CF3I on Pt(111) at 60 K has been photolyzed with a Hg arc (< 5.3 eV) [5]. For one monolayer, there is XPS evidence for the photodissociation of the C-I and the loss of both F and C, with parent desorption accounting for less than 20% of the loss. For CF3I/Ag(111), now under study in our laboratory [6], the same overall process occurs, and, as for SO_2, there are strong dynamical variations with structure (**Figure 3**). There is a persistent fast TOF peak near 100 µs, which is attributable to direct excitation, but the intensity and distribution of the signal towards longer times, attributed to substrate-

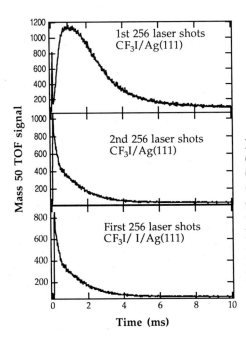

Figure 3. Pulsed excimer laser (248 nm) driven time-of-flight (TOF) distributions for CF3 (monitored as CF_2^+, which dominates the cracking pattern) desorption derived from the photolysis of 1 ML of CF3I on Ag(111) at 100 K. Upper panel: first 256 laser shots. Middle panel: second 256 laser shots. Bottom panel: first 256 laser shots with 0.3 ML of predosed I.

mediated excitation, varies significantly with the number of laser shots. That accumulation of I, and associated surface structural changes, is responsible for these dynamical changes is confirmed by the similarity of CF_3 TOF spectra in the presence of preadsorbed I (0.3 ML). Semi-quantitatively, the position and intensity of the fast peak remains constant, while the intensity of the slower portion decreases and the average velocity increases as the I coverage increases. Thus, as for numerous alkyl halide-substrate systems [1], direct excitation and substrate-mediated electron attachment and bond cleavage, that competes effectively with quenching, are important here.

4. AsH₃ on GaAs(100)

Another very interesting case involves arsine on Ga-rich GaAs(100) (4x6), which we have studied using LEED, TPD, HREELS and XPS [7]. Without light, saturation AsH_3 (1 ML = 1×10^{14} molecules/cm²) desorbs in two peaks; the first is reversible molecular AsH_3 desorption at 144 K (α-peak, 0.56 ML), while the second is recombinative desorption (AsH_2 + H) between 200 and 500 K (β-peak, 0.29 ML). There is some As-deposition (0.15 ML) and,

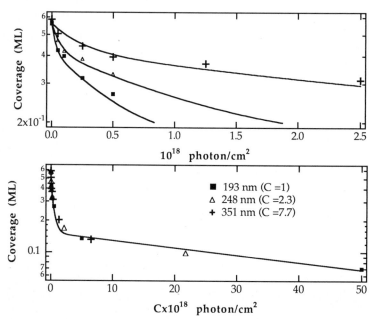

Figure 4. Semi-log plot of remaining α AsH₃ from Ga-rich GaAs(100) as a function of photon fluence. Upper panel: 0.0 to 2.5 x 10¹⁸ photons/cm². Lower panel: 0.0 to 50 x 10¹⁸ photons/cm². Initial α AsH₃ coverage is saturated (0.56 ML) and solid curves are fits to model with three kinds (cross sections) of α AsH₃ described in text. Lower panel shows "universal" fit using different factors, C, to scale the cross sections for three wavelengths.

between 400 and 600 K, H_2 desorption. The β-configuration is relatively photo-inactive, so we concentrate on the α-state. At 193 nm, all the evidence suggests that it decomposes in three distinct steps. In the first, one hydrogen is removed to the gas phase (only atomic hydrogen was observed during irradiation), and AsH_2 is formed. In the second step, hydrogen transfers from AsH_2 to surface Ga. In the third step, hydrogen is photochemically desorbed from both GaH and AsH. All the photodissociation products, AsH_x (x=1,2) and GaH, lead to H_2 desorption in post-irradiation TPD. At constant total photon fluence, the yield, based on TPD, was independent of pulse energy between 0.3 and 8 mJ/cm^2. We conclude that surface heating does not control the results. Since photodissociation occurs at wavelengths much redder (351 nm) than those capable of directly exciting arsine, and since the wavelength dependent data can all be put on a single "universal curve" (**Figure 4**) [7], substrate mediated excitation dominates.

Cross sections for the loss of α AsH_3 drop as the number of photons used increases; a model involving three different binding sites fits the data (**Fig. 4**) [7]. The fit gives coverages of the three sites (0.14±0.1, 0.26±0.1, and 0.16±0.1 ML) and, for 193 nm, the corresponding cross sections are σ_1 = 3.6±0.5 ($x10^{-17}$), σ_2 = 1.9±0.3 ($x10^{-18}$), and σ_3 = 2.0±0.3 ($x10^{-20}$) cm^2. These cross sections for 193 nm, divided by a single wavelength dependent constant, C = 1.0 for 193 nm, 2.3 for 248 nm, and 7.7 for 351 nm, provide a "universal" fit. Slightly different binding sites for α AsH_3 on the reconstructed GaAs(100) surface can account for these three cross sections [7]; a proposed structural model [8], has at least five different Ga sites with potential for binding AsH_3.

5. Summary

There is now good evidence that bond breaking (chemistry) at adsorbate-substrate interfaces can compete with quenching, particularly for cases where, in the gas phase, the excited molecule of interest undergoes prompt bond cleavage. Bond cleavage is inhibited in other cases, e.g., when the adsorbed molecule, has gas phase properties involving long-lived excited states. Yields depend strongly, and often in subtle ways, on the structural details of the adsorbate-substrate interface.

Acknowledgment: This work was supported in part by the National Science Foundation, Grant CHE 8920120 and CHE 9015600, by the Robert A. Welch Foundation and by the U.S. Department of Energy, Office of Basic Energy Sciences.

References

[1]   a. W. Ho, Comments Cond. Mat. Phys. **13**, 293 (1988).
b. X.-L. Zhou, X.-Y. Zhu, and J. M. White, Surf. Sci. Rep. **13**, 73 (1991).
c.   J. C. Polanyi and H. Rieley "Dynamics of Gas-Surface Interactions", ed. by C. T. Rettner and M. N. R. Ashford (Roy Soc. of Chem., London, 1990).

[2]   J. G. Calvert and J. N. Pitts, Photochemistry (Wiley, New York, 1966).

[3]   Z.-J. Sun, S. Gravelle, R. S. Mackay, X.-Y. Zhu and J. M. White (in progress).

[4]   M. Wolf, X-Y. Zhu and J. M. White (to be published in DIET V proceedings).

[5]   J. Kiss, D. Alberas, and J. M. White (in progress).

[6]   Z.-J. Sun, Z.-L. Zhou and J. M. White (in progress).

[7]   X.-Y. Zhu, M. Wolf and J. M. White, J. Chem. Phys. (in press).

[8]   D. K. Biegelsen, R. D. Bringans, J. E. Northrup, and L.-E. Swartz, Phys. Rev. B. **41**, 5701 (1990).

Part III

Laser-Induced Ion Desorption

Desorption of Ions and Neutrals with >6 eV Laser Light*

Y. Murata, K. Fukutani, A. Peremans, and K. Mase

Institute for Solid State Physics, The University of Tokyo,
7-22-1 Roppongi, Minato-ku, Tokyo 106, Japan

Abstract. Photostimulated desorption of NO and CO chemisorbed on a Pt(001) surface at 80 K has been studied with an ArF excimer laser ($h\nu$=6.41 eV) irradiation onto the surface. Desorption of both neutrals and ions was observed. Neutral species were state-selectively detected by the resonance-enhanced multiphoton ionization technique. The yields of NO, NO^+, CO and CO^+ are proportional to the first, third, third, and ~1.8th powers of laser fluence, respectively. Desorption mechanisms for these species are discussed.

1. Introduction

Surface science studies have begun to shift from static problems to dynamical phenomena. The photo-induced process in the gas-surface interaction system is one of the typical dynamical phenomena and a subject of great interest and important both fundamentally and practically. In particular, photostimulated desorption (PSD) induced by valence electron excitation has attracted an interest of many researchers in both scientific and technological fields, because the non-thermal process is one of the most elementary surface reactions and is expected to show a high reaction selectivity. There have been, however, only a few studies on laser-stimulated desorption induced by valence electron excitation in molecules chemisorbed on metal surfaces [1-5]. Most of PSD studies have been confined to adsorbates physisorbed on insulator surfaces, because deexcitation of excited molecules often proceeds much more rapidly on metals.

We observed ultraviolet(UV)-laser-stimulated desorption induced by valence electron excitation of molecules chemisorbed on a metal surface. Positive ions and neutrals were detected from NO- or CO-saturated Pt(001) surface at 80 K under ArF excimer laser irradiation. Desorbed neutrals were state-selectively detected by the resonance-enhanced multiphoton ionization (REMPI) technique.

2. Experiment

The present apparatus is composed of ultra-high vacuum chamber equipped with a LEED-AES system, an ArF excimer laser (λ=193 nm, pulse duration: 11 ns, repetition rate: 10 Hz) for the pump laser, a tunable pulsed Coumarin 460 dye laser (line width: 0.07 cm^{-1}, pulse duration: 6 ns) pumped by the third harmonic of a Nd:YAG laser for the probe laser, and a positive-ion measurement system [3]. The ArF excimer laser was linearly polarized and the polarization was changed with a $\lambda/2$ wavelength plate. The dye laser was frequency-doubled

*This work is supported by a Grant-in-Aid on Priority-Area Research on "Photo-Excited Process" supported by the Ministry of Education, Science, and Culture, Japan.

in a β-barium borate crystal for $(1+1)$- and $(2+1)$-REMPI detection of desorbed NO and CO neutrals, respectively, and irradiated at a distance of 1.5-3.5 mm from the sample surface. The probe laser for REMPI was fired after pump laser irradiation as a function of the time-interval; therefore, the velocity distribution of the desorbed neutrals was measured as a time-of-flight (TOF) spectrum.

The NO^+ and CO^+ ions generated by the lasers were accelerated to the inlet of a flight tube biased at a negative voltage and detected by a microchannel plate assembly. The signals were averaged and stored as a TOF spectrum in a transient recorder. The two signals corresponding to the NO and NO^+ PSD or the CO and CO^+ PSD were clearly discriminated by the TOF spectrum, because the light path and the timing are different between the pump and probe lasers. The velocity distributions of the desorbed ions can be obtained from the TOF spectrum measured without acceleration. The Pt(001) sample was mounted on a liquid nitrogen cooled holder. The surface was exposed to NO or CO gas at 80 K to the saturation coverage, prior to each measurement.

3. Results

Figures 1 and 2 show the laser fluence dependence of the desorption yield from NO and CO adsorbed on Pt(001), respectively. Desorbed NO neutrals probed in the ground vibrational state (v=0), the rotational states of J=15/2-21/2, and the spin-orbit state of $\Omega=1/2$, and desorbed CO neutrals probed at the Q head maximum position of the $B^1\Sigma^+(v=0) \leftarrow X^1\Sigma^+(v=0)$ transition. For NO neutrals, the desorption yield is proportional to pump laser fluence, as shown in Fig. 1. So, the NO molecules are desorbed via a one-photon process. On the other hand, NO^+ species generated by pump ArF laser irradiation is proportional to the third power of laser fluence over a wide laser fluence range, as shown in the inset of Fig. 1. The formation process of the ions is considered that the first step is desorption of neutral NO with one-photon absorption and then the desorbed NO molecule is ionized in the vicinity of the surface via the two-photon non-resonant ionization process [6]. This means that the desorption mechanisms for NO and NO^+ are the same and the desorption cross section takes a reasonable value.

As shown in Fig. 2, we observed a strong supralinear dependence of both CO and CO^+ PSD yields versus pump laser fluence. The third and ~1.8th dependences of laser fluence could be inferred for CO and CO^+, respectively,

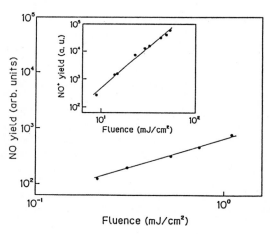

Fig. 1. Laser fluence dependence of the PSD yields of NO and NO^+ (inset).

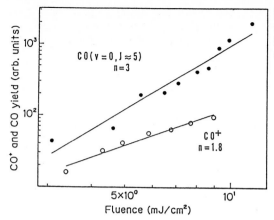

Fig. 2. Laser fluence dependence of the PSD yields of CO (closed circles) and CO^+ (open circles).

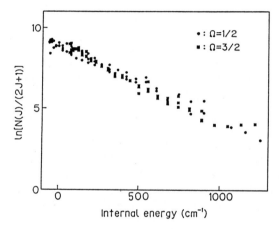

Fig. 3. Internal energy distributions of NO desorbed in the v=0 state and the $\Omega=1/2$, 3/2 spin-orbit state.

the exponent accuracy being limited by the yield measurement range of only one decade. It is considered that CO and CO^+ are desorbed via three-photon and two-photon processes, respectively. Thus, it is concluded that the desorption mechanisms for both species are different. Since the CO and CO^+ signals are comparable in intensity, the PSD yield of CO^+ must be several order of magnitude lower than that of CO because of the low efficiency of the REMPI process in CO neutrals ($\ll 10^{-4}$).

Figure 3 shows the internal-state distribution of the molecules in both the $^2\Pi_{1/2}$ and $^2\Pi_{3/2}$ spin-orbit states. NO molecules in the v=0 state with the translational energy of 0.05-0.13 eV were detected. Each internal state distribution exhibits a nearly Boltzmann form except in the high energy region, and is appropriately represented by a rotational temperature (T_r), 300 K. No distinct difference for the populations of two spin-orbit states was observed. Taking account of the Franck-Condon factors and assuming the Boltzmann distribution for vibrational excitation, the vibrational temperature (T_v) of 1200 K was obtained.

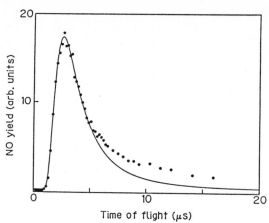

Fig. 4. TOF spectrum of NO (v=0, J=15/2-21/2, Ω=1/2). The solid curve shows the Maxwell-Boltzmann fit with T_t=650 K.

Figure 4 shows a TOF spectrum of desorbed neutral NO (v=0, J=15/2-21/2, Ω=1/2), which exhibits a peak at the flight time of 2.7 μs and a broad shoulder at around 10 μs. Ignoring the latter low energy component, this TOF spectrum can be fitted by the Maxwell-Boltzmann distribution as illustrated by the solid curve and the translational temperature (T_t) is 650 K. The state-specific detection for the first excited vibrational state (v=1) was also carried out. The result of the internal state distribution is similar to that for v=0, and T_r=340 K.

These temperature values are considerably higher than the sample temperature 80 K. The expected temperature rise of the sample is lower than 10 K at pump laser fluence used in this experiment (<2 mJ/cm^2) [7]. This result definitely shows that the desorption is a non-thermal process. The desorption cross section (σ_d) in the v=0 state is estimated from the decay of the desorption yield versus the photon flux and is given by 3.4×10^{-20} cm^2.

Since the internal state distribution could not be obtained for the CO neutrals with the (2+1)-REMPI scheme, the simulated spectrum was compared with observed one. It is found to be rotationally cold (~150 K) and vibrationally excited (~800 K) [8]. The translational energy at the maximum of the TOF spectrum is strongly dependent on the vibrational state and the molecules in the v=1 state propagate much faster than the ones in the v=0 state [8].

4. Discussion

In order to determine whether excitation in the initial step of PSD occurs at the adsorbate or at the substrate, we measured the incidence angle dependence of the desorption yield using a polarized pump laser. The relative yield was measured at 25° and 81° incidence from the surface normal in both p- and s-polarizations of the pump laser. NO molecules were probed in the v=0, J=15/2-21/2, and Ω=1/2 states. The absorption coefficient of a Pt substrate and the mean square electric field outside the substrate are calculated following the Fresnel equations [9] using the refractive index of Pt. The ratios of the observed yield at 25° to that at 81° for both polarizations are in good agreement with those of the calculated absorption coefficients [10].

This result suggests the valence electron excitation of the substrate in the initial step of NO PSD. The electron is excited indirectly via the conduction band or directly to an unoccupied state of adsorbates and a negative ionic state is formed. That is, desorption occurs via this intermediate state following the Antoniewicz mechanism [11]. This consideration is supported by the following experimental results. We found that the internal energy distribution is independent of the pump laser polarization [12] and another significant result that the translational, rotational, and vibrational temperatures are independent of photon energies of 3.5-6.4 eV obtained by ArF, KrF, and XeF excimer lasers [10]. The latter result shows that the desorption of NO follows the same mechanism at these excitation energies. The initial step of the electronic excitation should be the same. This model is consistent from the electronic structure of the Pt bulk and adsorbate NO. In photoelectron-spectroscopic studies, the Pt valence band was found to have ~8 eV width below the Fermi level (E_F), while the unoccupied adsorbate-induced electronic level ($2\pi_a$) was found to lie at 1.5 eV above E_F with the energy spread of ~1 eV by inverse photoemission study on Pt(111)-NO [13].

Next, we will compare the present results with the desorption study on Pt(111)-NO. Two desorption works have been reported by PSD [1] and electron-stimulated desorption (ESD) [14]; substrate excitation followed by negative ion state formation drives desorption below 3.5 eV in PSD, whereas desorption is induced by direct adsorbate excitation followed by positive ionic state formation above 6 eV in ESD. The energy distributions of desorbed NO from Pt(001) are definitely different from those on Pt(111) in PSD even though the same mechanism is operative. In particular, the rotational energy distribution exhibits a Boltzmann form and inversion of the populations in the two spin-orbit states was not observed in the present study. In constrast, T_t and T_r, but not T_v are similar to those observed in ESD in spite of the different desorption mechanism. The potential energy surface of the ground and excited states is considered to vary depending on the crystal orientation of the surface, in fact, a high adsorption state selectivity was observed on Pt(001) [12].

Another significant difference is the desorption cross section; $\sigma_d \sim 10^{-20}$ cm^2 for Pt(001) and $\sigma_d \sim 10^{-22}$ cm^2 for Pt(111) [1]. The total σ_d is determined by two factors: the cross section of the ionic state formation and the desorption probability (P_d) at the ionic state, where P_d is governed by lifetime (τ). According to the theoretical work by Gadzuk et al. [15] who described the desorption via a negative ionic state, the lifetime of the intermediate state is expressed by the population of the v=1 state (β) as $\tau \sim \sqrt{\beta}$. Folowing this expression, τ of the present study becomes 1.5 times as long as that on Pt(111). Since the τ different by a factor of 1.5 can explain the variation of P_d by several orders [15], the larger cross section in the present study is qualitatively consistent with the study for Pt(111) [1].

CO and CO$^+$ desorption mechanisms are clearly different from those of NO and NO$^+$. We propose that neutral species desorption proceeds via three photon ionization of the adsorbate followed by reneutralization without recapture, while CO$^+$ desorption results from a two photon electronic excitation of the adsorbate followed by its ionization due to the transfer of the electron to the metal and desorption before neutralization. Since three-photon energy exceeds the CO ionization potential (14 eV), the direct adsorbate photoionization is likely the first excitation step of the neutral CO desorption process. This mechanism has indeed already been suggested in an ESD experiment of CO on Ru(001) [16]. It is difficult to consider desorption via multiphoton excitation involving hole-electron excitation in the substrate except for using ultrashort pulses [4]. It is remarkable to observe that CO$^+$ is desorbed via only the two-photon process, as compared with the three-photon process for CO PSD. The initial step of CO^+ PSD must be adsorbate valence electron excitation and ionization follows as a result of the excited electron transfer to the substrate, as proposed by Burns et al. [17]. The CO^+ desorption can proceed from this ion-resonance before reneutralization. An inverse photoemission spectrum of Pt(011)-CO shows that the level separation between the 5σ and $2\pi_a$ states is ~12 eV [18], which is close to the two-photon energy of the pump laser.

References

1. S.A. Buntin, J.L. Richter, D.S. King, and R.R. Cavanagh, J. Chem. Phys. **91**, 6429 (1989).
2. R. Schwartzwald, A. Modl, and T.J. Chuang, Surf. Sci. **242**, 437 (1991).
3. K. Mase, S. Mizuno, Y. Achiba, and Y. Murata, Surf. Sci. **242**, 444 (1991).
4. J.A. Prybyla, T.F. Heinz, J.A. Misewich, M.M.T. Loy, and J.H. Glownia, Phys. Rev. Lett. **64**, 1537 (1990).
5. E. Hasselbrink, S. Jakubith, S. Nettesheim, M. Wolf, A. Cassuto, and G. Ertl, J. Chem. Phys. **92**, 3154 (1990).
6. K. Mase, S. Mizuno, M. Yamada, I. Doi, T. Katsumi, S. Watanabe, Y. Achiba, and Y. Murata, J. Chem. Phys. **91**, 590 (1989).
7. J.F. Ready, *Effect of High-Power Laser Radiation* (Academic, New York, 1971).
8. A. Peremans, K. Fukutani, K. Mase, and Y. Murata, Phys. Rev. B, to be published .
9. R.G. Greenler, Surf. Sci. **69**, 647 (1977).
10. K. Fukutani, A. Peremans, K. Mase, and Y. Murata, Phys. Rev. B, to be published.
11. P.R. Antoniewicz, Phys. Rev. B **21**, 3811 (1980).
12. K. Mase, K. Fukutani, and Y. Murata, J. Chem. Phys. **96**, 5523 (1992).
13. V. Dose, Surf. Sci. Rep. **5**, 337 (1980).
14. A.R. Burns, E.B. Stechel, and D.R. Jennison, Phys. Rev. Lett. **58**, 250 (1987).
15. J.W. Gadzuk, L.J. Richter, S.A. Buntin, D.S. King, and R.R. Cavanagh, Surf. Sci. **235**, 317 (1990).
16. P. Feulner, R. Treichler, and D. Menzel, Phys. Rev. B **24**, 7427 (1981).
17. A.R. Burns, E.B. Stechel, and D.R. Jennison, Phys. Rev. Lett. **58**, 250 (1987).
18. S. Ferrer, K.H. Frank, and B. Reifl, Surf. Sci. **162**, 264 (1985).

Laser-Induced Desorption of Positive Ions from Wide Band Gap Insulators

O. Kreitschitz[1], W. Husinsky[1], G. Betz[1], and N.H. Tolk[2]

[1]Technische Universität Wien, Institut für Allgemeine Physik,
Wiedner Hauptstraße 8–10, A-1040 Wien, Austria
[2]Department Physics and Astronomy, Vanderbilt University,
Nashville, TN 37235, USA

1. Introduction

The intensity dependence of the yield and energies of positive ions desorbed from SrF_2, CaF_2 and MgO under 193 nm and 308 nm excimer laser irradiation has been investigated by the Time-of-Flight method. Absorption of visible or UV photons in large band gap materials can take place by multiphoton excitation [1] of valence band electrons, defects or particle inclusions [2]. The creation of an electron-hole plasma in the interaction volume of the laser with the target [3,4] is the most typical mechanism of non-thermal laser annealing, leading to ablation. Holes may contribute to positive ion desorption by the formation of a positive ion in an excited and antibonding state on the surface with subsequent desorption [5] or two holes can be localized on the surface and lead to desorption [6]. At a material and wavelength dependent laser intensity, the flux of desorbing particles can become so high that a plasma will be formed at the surface. It is known from other experiments that ions emitted under laser irradiation can gain high kinetic energies [11,12], while measurements on CaF_2 showed that emitted neutral species have nearly pure thermal energy distributions [13]. The origin of these energies are plasmoid extension effects first introduced by Bykovskii [14] and experimentally found by Akhsakhalyan [12]. In the present work we have addressed these questions analyzing the ion emission for two different laser wavelengths by measuring their thresholds, ion yield dependence with laser intensity and the energies of the emitted ions .

2. Experimental setup

We used excimer laser radiation at 308 nm (XeCl) and 193 nm (ArF) with a pulse width of 17 ns. The angle of incidence of the laser beam was 30° with respect to the target normal. A focus spot diameter of approx. 300 μm was obtained. The single crystals were heated for several days to 300°C and all experiments were performed at this temperature. The base pressure in the vacuum chamber was 7×10^{-10} mbar.

The experimental Time-of-Flight apparatus [15,16] (Fig. 1) enables us to determine the yield of all ion species simultaneously using geometry A (reflectron mode). This mode was used for the intensity dependence measurements using single laser shots. The laser intensity has been varied using a dielectric attenuator and could be determined at every single shot. Using geometry B (Fig. 1) we calculated the energy distributions of the emitted positive ions from the measured flight time.
In this case up to 100 laser shots were made for one velocity distribution, as shown for SrF_2 in Fig. 2a. Different extraction voltages (drift potentials) between 100 and 1000 eV have been used, but no differences in the calculated ion energies have been observed. The distance from the drift tube to the target was 2 cm. Ions were detected by microchannel plates,

Springer Series in Surface Sciences, Vol. 31
Desorption Induced by Electron Transitions DIET V
Editors: A.R. Burns · E.B. Stechel · D.R. Jennison   © Springer-Verlag Berlin Heidelberg 1993

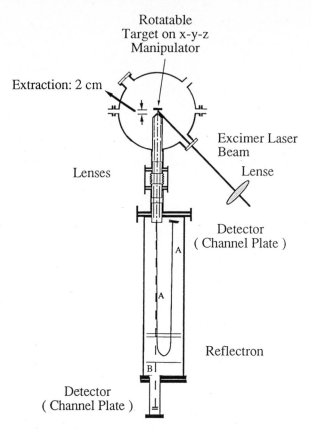

Rotatable
Target on x-y-z
Manipulator

Extraction: 2 cm

Excimer Laser
Beam

Lenses

Lense

Detector
(Channel Plate)

A

A

Reflectron

B

Detector
(Channel Plate)

Fig.1: Experimental setup: geometry A (reflectron mode) was used for the intensity dependence measurements and for determination of the emitted positive ion species, while geometry B is suited to determine the energy distributions of the particles. The distance between the target and the extraction conus is 2 cm and the drift region has a length of 1.34 m.

which were sensitive not only for particles but also for photons with energies of more than 6 eV (~200 nm). Because the dedector was gated with a time delay of 60 ns no reflected laser light could be recorded, but only energetic photons emitted from excited atoms/ions or from the target surface.

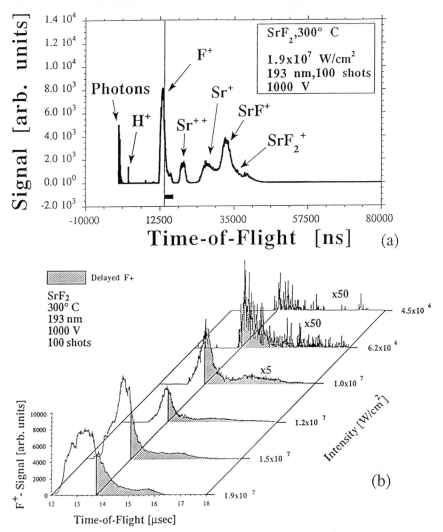

Fig.2: a) TOF spectrum (geometry B) shows all the observed positive ion species at 193 nm for a SrF₂ target.. The F⁺ peak is divided by a solid line, which represents the flight time of the F⁺ ions with zero initial energy. F⁺ ions emitted with non-zero energies are to the left of the line. The delayed emission yield of F⁺ ions is marked by the filled rectangle.

b) Part of Fig. 2a. Only the F region in the TOF spectrum is shown for different laser intensities. The delayed emission part is shaded.

3. Results

3.1. Yield versus laser intensity

For both laser wavelengths the dominantly observed ion species were the metal ions (Sr^+, Ca^+, Mg^+) and F^+ or O^+, respectively. In addition typically in a narrow intensity range around the onset of ion desorption also molecular ions (SrF^+, CaF^+, MgO^+) were observed. Minority species were doubly ionized metal atoms and larger molecular ions like SrF_2^+. The latter were only observed in measurements done in geometry B, indicating that they were metastable ions, which decayed before reaching the detector.

In all cases a distinct threshold for ion emission was observed, above which the ion yield typically increased by orders of magnitude (Fig. 3 for SrF_2 at 193 nm) in a narrow intensity range above which the total ion yield remained nearly constant. We attribute this to the formation of a plasma at the surface due to dielectric breakdown of the material. The laser intensities for which the total ion yield begins to stay constant have been identified with the values for dielectric breakdown (see below) and are listed in Table 1. As can be seen for example from Fig. 3 and Table 1 the ion yield increases by 3 to 4 orders of magnitude between 1 and 4.10^6 W/cm^2. At the plasma threshold the F^+, Sr^+ and SrF^+ yield increase strongly by orders of magnitude and then show no significant increase at higher laser intensities, with the exception of SrF^+, which can be only observed in the threshold region and then vanishes. This behavior is typical for all cases studied, except that for MgO at 193 nm where the O^+ yield starts only at much higher intensities than the plasma threshold.

Analysis of this near threshold region has shown no correlation with a multiphoton process in neither case. In the case of SrF_2 and 308 nm careful studies have revealed that the Sr^+ yield and the SrF^+ yield increase like I^6 at the plasma onset, where I is the laser intensity, if a series of single shot measurements is performed with continually increasing laser intensity, but that the yield scales with I^{11}, if the measurements are done in order of decreasing intensity. As in this case, with a band gap of 11.25 eV for SrF_2, a 3 photon process would be sufficient to create a valence band (vb) electron and because of the high powers (6, 11) and corresponding low cross sections involved, we do not attribute this increase to a multiphoton

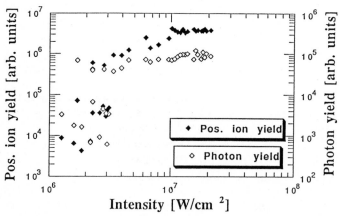

Fig 3: Total positive ion yield and photon yield for a SrF_2 target bombarded with single shot 193 nm laser pulses as a function of the laser intensity.

process. No apparent I^n dependence of the ion yields has been observed for 193 nm in all cases. Typically the onset of ion emission for laser irradiation at 193 nm shifts to lower intensities by more than one order of magnitude as compared with measurements at 308 nm (see Table 1).

Table 1: Values for the bandgap and for the dielectric breakdown in MW/cm^2 at 308 and 193 nm

	SrF_2	CaF_2	MgO
308 nm	31	35	30
193 nm	4	2	1.5
bandgap(eV)	11.25	12.2	7.8

The only exception in all our measurements with respect to the threshold behavior was SrF_2 at 308 nm, where F ion emission starts at much lower intensities and increases linearly with laser intensity until a certain intensity is reached, where the F ion intensity starts to increase dramatically. We will not discuss this point any further in this paper. For a discussion see Ref [15].

Fig. 3 shows also the measured photon yield, as could be measured in geometry B, in addition to the ion yield. We have generally observed that at the same laser intensity where ion emission starts we also start obtaining a photon signal. As discussed in the experimental section this light is not due to reflected laser light and must be from energetic photons to be measured by the detector. This coincidence we take as further evidence that ion emission is connected with the formation of a plasma at and above the surface in all cases studied. With increasing laser intensity, the light peak becomes broader, which can be attributed to a longer life time of the plasma. At laser intensities well above the plasma threshold, the light signal indicates a plasma lifetime of up to 2 μs at both wavelengths.

Another general observation is again depicted for the case of 193 nm and SrF_2 in Fig. 4. As already mentioned above, emission of SrF^+ is only observed in the threshold region. As soon as the total ion yield stabilizes above the plasma onset no more molecular ions are detected. Also in this case the F^+ emission starts at the same intensity, where the molecule emission disappears. We take this as evidence that molecules will dissociate in the expanding plasma formed. Furthermore, comparing Fig. 3 and 4 one can observe, that while the total ion yield is remarkably constant within a factor of 2 this is not the case for the components Sr^+ and F^+. Their emitted yields undergo large fluctuations showing a pronounced anticorrelation, which results in the almost constant total yield. We have analyzed this behavior in terms of consecutive laser shots, to see if a laser shot does precondition the surface for the next laser shot. This seems to be indeed the case, as is seen in Fig. 5, where the yields of F^+ and Sr^+ are plotted versus numbered, successive, single laser shots at intensities well above the plasma threshold. This interesting behavior of the positive ion yields has been observed for both wavelengths and all target materials but was generally more pronounced for 193 nm. A similar behavior has been previously reported by Reif et al [17] for BaF_2 and attributed to possible layer by layer removal. We are not in a situation of giving an explanation, but we would like to point out that changes seem to

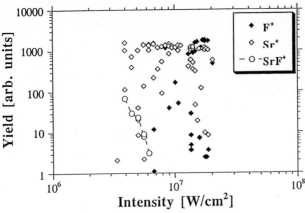

Fig 4: Yield of Sr$^+$ and F$^+$ ions as a function of the laser intensity for 193 nm laser pulses on SrF$_2$

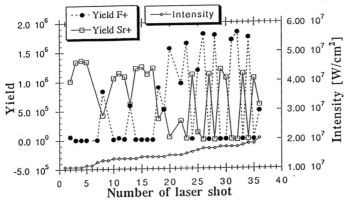

Fig.5: The Sr$^+$ and F$^+$ yields have been recorded at each, successive laser shot (193 nm) at intensities well above the plasma threshold for a SrF$_2$ target. The yields of the single ion yields fluctuate by orders of magnitudes but the sum of both yields remains nearly constant (Fig. 3).

occur more rapidly with increasing laser intensity as seen from Fig. 5. Maybe a metallic overlayer of Sr at the surface develops and is removed by the next laser shot and so on. The target temperature used (300°C) is too low to enable thermal evaporation of metallic layers formed, a mechanism which enables typically in ESD of alkali and alkali earth halides to maintain continuous desorption. But the temperature increase due to laser light absorption by a metallic overlayer could give a sufficient temperature increase to make evaporation possible.

In the case of SrF$_2$ and 193nm we also performed topographical investigations using a *Scanning Electron Microscope* (SEM) after prolonged irradiation, which revealed that surface structures exhibit only non-thermal characteristics. The edges of the craters are very sharp, which is an indication of the occurrence of non-thermal processes during irradiation. Cracks induced by thermal stress can rarely be observed.

3.2. Energy distributions

Energy distributions of the desorbed positive ions have been determined as a function of the intensity for 308 nm and 193 nm using Time-of-Flight geometry B (Fig.1). For the energy spectra typically up to 100 laser shots at the same intensity were added up. In Fig. 2a, a TOF spectrum recorded at 193 nm in the plasma regime is shown. The solid line, which separates the F^+ peak into two parts, represents the flight time of F^+ ions without initial kinetic energy. At the right side of the line the delayed emission of F^+ ions is marked by a filled rectangle while the left side represents energetic F^+ ions. This part of the spectrum for the F^+ peak is shown enlarged in Fig. 2b for different laser intensities. As can be seen, with increasing laser intensity the ratio of energetic particle emission to delayed emission becomes larger.

At intensities above the plasma treshold ions become more energetic with increasing laser intensity (Fig. 2b and 6). Fig. 6 also depicts the light emission, which we take as indication for plasma formation. The maximum kinetic energies of F^+ and Sr^+ reach a plateau for laser intensities well above the plasma onset at laser intensities much higher than the saturation value for light emission. Near the plasma threshold intensity Sr^+ exhibits maximum kinetic energies of 10 - 40 eV. At high laser intensities, the maximum kinetic energies of F^+ can reach values of more than 200 eV and emitted Sr^+ can even gain energies of more than 400 eV.

Similar results have been observed for all targets and both wavelength with ion energies of typically around 10 -40 eV near the plasma threshold and maximum kinetic energies up to 600 eV for laser intensities well above the threshold.

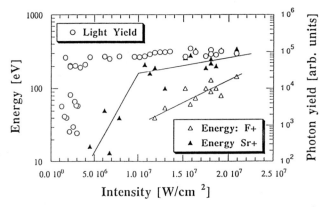

Fig.6: The plot shows the intensity dependence of the maximum kinetic energies of Sr^+ and F^+ at 193 nm laser irradiation for a SrF_2 target. With increasing laser intensity the maximum kinetic energies of the ion species increase and reach a plateau.

4. Discussion

The band gap of SrF_2 is 11.3 eV and for CaF_2 is 12.2 eV. Non-resonant multiphoton absorption of 3 photons at 308 nm (4.0 eV) and 2 photons at 193 nm (6.4 eV) by one electron are necessary to transfer an electron of the vb into the cb. The probability of such a multiphoton absorption event results in a proportionality to I^n, where I represents the primary laser intensity and n is the number of photons absorbed. Itoh et al.[6] proposed a desorption model for ions and neutrals from crystals under laser irradiation. According to the model of Itoh an electron-hole plasma is created and if the density of holes is high enough, the Coulomb repulsion between two holes will be screened. If this two-hole localization takes place at the surface, relaxation of the lattice leads to desorption of neutral or ionic particles from the crystal. Wu proposed an alternative ion desorption mechanism from Si and Ge surfaces [5]. The two models mentioned above consider the electron-hole plasma as a direct precursor of the desorption event. At 308 nm, the SrF^+ signal vanishes at the exact laser intensity where the F^+ emission increases sharply, which was taken as an indication that a plasma at the surface is formed and dissociation of SrF^+ takes place. Both, the Sr^+ and SrF^+ yields at 308 nm behave like I^6. Following the interpretation of Matthias [7], two holes are necessary to induce the desorption of one positive ion rather than the occurrence of a 6 photon absorption process, which is much less probable. However as already mentioned above this apparent dependence was only observed if we performed measurements with increasing laser intensity. For a series, where the laser intensity was continuously decreased, an I^{11} was observed and for CaF_2 and MgO no such correlations could be found.

Furthermore for all targets and at both wavelengths, light emission could be observed for every laser shot, where also ion emission occurs, which is a strong indication of plasma formation. Dielectric breakdown in an ionic crystal induced by the strong laser field is responsible for macroscopic damage of the crystal leading to considerable removal of material.

Dielectric breakdown can be induced by avalanche ionization [8-10] or polaron heating [18]. In the avalanche model, acceleration of electrons by the laser field (inverse Bremsstrahlung [19]) followed by inelastic collisions with vb electrons results in additional conduction electrons, which are again accelerated by the laser field [9,10].

The avalanche ionization model is based only on the energetic conduction band (cb) electrons which start the avalanche, while in the polaron model all cb electrons can contribute to dielectric breakdown. The polaron mechanism should be more sensitive to the density of electron-hole pairs than the avalanche mechanism. In the avalanche model the decisive process is the acceleration of cb electrons to the ionization limit (=band gap energy) via inverse Bremsstrahlung [22]. The absorption coefficient of inverse Bremsstrahlung scales as λ^2 (λ is the laser wavelength) [20]. Therefore the mean kinetic energy of the electrons should be lower at 193 nm than at 308 nm. From Tab.1 it can be seen that dielectric breakdown at 193 nm occurs at considerably lower laser intensity than at 308 nm. This is not in agreement with the avalanche model.

Multiphoton processes create electron hole pairs, which will lead to dielectric breakdown and plasma formation. Furthermore, after a plasma is formed and expands into the vacuum interaction of the laser beam with the plasma, especially the neutral atoms will modify the observed emitted ion yields, especially for 193 nm. For example in the case of Sr anc Ca the low ionization energy (5.7 eV and 6.11 eV) makes postionization of desorbed neutral Sr and Ca by laser photons in the gas phase via single photon absorption possible. The different number of photons needed for postionization of neutral Sr in the gas phase could be an indication, why the increase of the positive ion yields near the threshold intensity at 308 nm is not so steep compared with the increase at 193 nm. As known from measurements on CaF_2, the overwhelming amount of desorbed particles near the threshold intensity is neutral [13].
The high kinetic energies observed for the emitted ions are a further indication that a plasma is formed, which determines the ion energies. The

plasma life time can amount to 2 μs and is proberbly responsible for the delayed emission, which occurs at both wavelengths for all observed positive ion species. In Refs. 11 and 12 such high energies of ions have also been obtained. These high energies cannot ·be transfered to the particles during the desorption process. Bykovskii et al. [14] introduced a model based on the acceleration of ions due to -a self-consistent field in the plasmoid. According to this model, ions can gain kinetic energies of more than 2 keV. Summarizing the results of the measured maximum kinetic energies, there is strong evidence that the model of Bykovskii can explain the high kinetic energies of the emitted ions. If high kinetic energies of positive ions can be measured then a plasma exists. We can use the occurrence of high kinetic energies of the positive ions as an additional indication for plasma formation. The kinetic energies of the ions do not indicate any information about the kinetic energies obtained during a desorption process.

In conclusion the following indications of plasma formation at the surface have been observed:

    a) The sharp increase of the positive ion yield at a threshold
    b) The high kinetic energies of the positive ions
    c) The dissociation of molecular ions in the plasma
    d) The photon yield emitted by the plasma.

The plasma thresholds show that no positive ion desorption occurs below this threshold laser intensity with exception of F^+ emission at 308 nm. Furthermore we conclude that multiphoton absorption processes can influence the threshold laser intensity for dielectric breakdown. Furthermore processes like single or multiphoton ionization of plasma particles by the laser beam in the gas phase have to be taken into account. The experimental results do not agree with the avalanche ionization model for dielectric breakdown. The high kinetic energies of positive ions can be explained by a plasmoid extension effect. The kinetic energies gained during the actual desorption process still remain unknown.

5. Acknowledgements

The authors gratefully acknowledge financial support by the Österr. Fonds z. Förderung der wiss. Forschung (project No.: 6797 and 6386). We also thank Dr. Bangert for giving us the opportunity of using the SEM and the National Science Foundation (Project No. N000 14-86-C-2546).

References

[1]   N. Vaidya, A.H. Günther, S.S. Mitra, J. Opt. Soc. Am. B2, 294 (1985)
[2]   A.J. Glass and A.H. Guenther,Appl.Opt.12,637(1973)
[3]   J.A. Van Vechten, R. Tsu, F.W. Saris and D. Hoonhout, Phys. Lett. 74A, 417 (1979)
[4]   J.A. Van Vechten, J. Phys. C4, 15 (1980)
[5]   Z. Wu, Phys. Lett. A 131, 486 (1988)
[6]   N. Itoh, and T. Nakayama, Phys. Lett. 92A, 471 (1982)
[7]   E. Matthias and T.A. Green, Desorption Induced by Electronic Transitions, DIET IV, (edts.: G. Betz and P. Varga), Springer Verlag Berlin 112 (1990)
[8]   M. Sparks, D.L. Mills, R. Warren, T. Holstein, A.A. Maradudin, L.J. Sham, E. Loh, Jr., and D.F. King, Phys. Rev. B24, 3519 (1981)
[9]   A.S. Epifanov, IEEE J. of Quantum Electron. QE-17, 2018 (1981)
[10] A.S. Epifanov and S.V. Garnov, IEEE J.Quantum Electron. 17,2023 (1981)
[11] H. Cronberg, W. Muydermann, H.B. Nielsen, E. Matthias and N.H. Tolk, Desorption Induced by Electronic Transitions, DIET IV, (edts.: G. Betz and P. Varga), Springer Verlag (1990) 157
[12] A.D. Akhsakhalyan, Yu. A. Bityurin, S.V. Gaponov, A.A. Gudkov and V.I. Luchin, Sov. Phys. Tech. Phys 27, 969 (1982) .
[13] R.C. Estler, E.C. Apel and N.S. Nogar, J. Opt. Soc. Am. B4, 281 (1987)

[14] Yu.A. Bykovskii, N.N. Degtyarenko, V.F. Elsin, V.E. Kondrashov and E.E. Lovetskii, Sov. Phys. Tech. Phys 18, 1597 (1974)
[15] O. Kreitschitz, W. Husinsky and G. Betz, submitted to Surface Science
[16] W. Husinsky, P. Wurz, A. Traunfellner and G. Betz, Fresnius`Journ. Analyt. Chem. 341 (1991) 12
[17] J. Reif, H. Fallgren, H.B. Nielsen and E. Matthias, Appl. Phys. Lett.49 930 (1986)
[18] A. Schmid, P. Bräunlich and P. Kelly, Phys. Rev. B 16, 4569 (1977)
[19] Yu. P. Raizer, Sov. Phys. USPEKHI 8, 650 (1966)
[20] C.R. Phipps, Jr., T.P. Turner, R.F. Harrison, G.W. York, W.Z. Osborne,G.K. Anderson, X.F. Corlis, L.C. Haynes, H.S. Steele, K.C. Spicochi and T.R. King, J. Appl. Phys. 64, 1083 (1988)

Laser-Induced Desorption from Sapphire Surfaces

M.A. Schildbach and A.V. Hamza

University of California, Lawrence Livermore National Laboratory,
Livermore, CA 94551, USA

Abstract. Laser-induced desorption of energetic (~7eV) aluminum ions was observed from clean and water-covered sapphire ($1\bar{1}02$) surfaces using time-of-flight mass spectrometry with laser wavelengths of 1064, 355, and 266 nm. In sharp contrast, O^+ (H^+ and OH^+) ions were observed in electron-induced desorption measurements with 300 eV electrons from the bare (water-covered) ($1\bar{1}02$) surface. Sapphire surfaces were characterized with low energy electron diffraction, reflection electron energy loss spectroscopy, and Auger electron spectroscopy.

1. Introduction

Control and understanding of laser ablation processes requires identifying the initial mechanism of energy deposition in the material surface. In the case of wide bandgap solid surfaces identification of these mechanisms is a challenging problem since these solids are nominally transparent to a wide range of photon energies. Interaction of photons with wide bandgap material surfaces is often considered in terms of absorption by electronic states in the bandgap due to the surface, defects and impurities[1].

The initial mechanism of laser-induced desorption from sapphire (α-Al_2O_3) ($11\bar{2}0$)-(12X4) is electronic[2], as evidenced by the high kinetic energy of the desorbing aluminum. Schildbach and Hamza[2] have proposed that an exciton-mediated process accounts for the preferential desorption of aluminum particles. The bulk exciton should have a projection on other sapphire surfaces. We report here on the laser-induced desorption of aluminum from the clean and water-covered sapphire ($1\bar{1}02$) surface at fluences below the ablation threshold and compare photon- and electron-induced processes.

2. Experiment

The apparatus has been described previously in detail[2]. Briefly, the ultra-high vacuum chamber (base pressure $< 2 \times 10^{-10}$ torr) was equipped with reverse-view low energy electron diffraction (LEED) optics, a double pass cylindrical mirror analyzer for Auger electron spectroscopy (AES) and reflection electron energy loss spectroscopy (ELS), and a differentially pumped quadrupole mass spectrometer (QMS). Flight distances from the sample to the electron multiplier of the QMS were 46.0 cm. Samples were mechanically polished single crystal sapphire oriented within ± 2° of the ($1\bar{1}02$) face. They were heated in vacuum to over 1300 K to produce the equilibrium reconstructed surface evidenced by the (2X1) LEED pattern[3]. AES revealed the surface was clean and undamaged since the 67 eV peak characteristic of aluminum bound to aluminum was absent. Prolonged exposure to the electron beam (10 minutes at 50 $\mu A/cm^2$) could induce the 67 eV peak. ELS from the undamaged (2X1) surface revealed a surface loss feature at 4.1 eV, which is consistent with the measurements on other surfaces[2,4,5]. Vacuum-cleaned triply distilled water was used for producing water overlayers at room temperature. Temperature programmed desorption (TPD) showed that water desorbed from the surface between 320 and 700 K and saturation coverage was 1.4 monolayer (based on one water molecule per (1x1) unit cell).

3. Results and Discussion

Figure 1 shows the energetic nature of the laser-induced desorption of aluminum ions from sapphire ($1\bar{1}02$)-(2X1). The observed kinetic energy matches that for laser-induced desorption of aluminum from the ($11\bar{2}0$)-(12X4) surface.

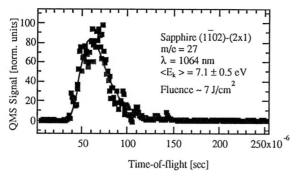

Figure 1. Positive aluminum ion time-of-flight distribution from sapphire ($1\bar{1}02$)-(2X1). Filled squares represent the quadrupole mass spectrometer signal at m/e=27. The solid line represents the non-linear least squares fit to the data.

The nature of the aluminum desorption was unchanged even on the water-covered ($1\bar{1}02$) surface. The average kinetic energy of the desorbing ions was independent of the laser fluence below the ablation threshold (though the intensity increased with fluence until the exposed area appeared to become depleted) and independent of the number of laser pulses (though the intensity decreases as a function of the number of pulses). In addition, the time-of-flight distributions were independent of the laser wavelength at 1064, 355, and 266 nm. The desorbing species is predominately positive aluminum ions, although the desorption of aluminum neutrals can not be ruled out. However, to within the experimental uncertainty no difference in the yield with the ionizer on versus ionizer off was discernible with the QMS.

Zirconium was sputter deposited on the sapphire ($1\bar{1}02$)-(2X1) surface and the surface annealed to 1300K. Coverage of Zr was small (<0.01 ML) as measured by

AES. The LEED pattern remained (2X1); however, the intensity of the half-order beams increased to nearly that of the first order beams. Laser-induced desorption of Zr$^+$ was observed with high average kinetic energy (~5 eV) with 1064 nm light.

The independence of the high kinetic energies of the desorbing ions from wavelength and fluence and the preference for surface cation desorption in the laser-induced process indicates an electronic mechanism involving an excitation localized on the surface cation. An aluminum-localized exciton with 1 eV binding energy is known from ELS measurements[6] and x-ray absorption measurements[7]. Trapping of the bulk exciton by surface aluminum adatoms or other impurity cations on the surface which would subsequently decay at the surface could account for the wavelength independence and the surface cation preference of the desorption mechanism. The decay of the ~8eV of electron energy may also account for the high average kinetic energies observed.

Figure 2 shows the electron-stimulated desorption of H$^+$, OH$^+$ and O$^+$ from water-covered (1$\bar{1}$02)-(2X1) surface as a function of exposure time to a 300 eV electron beam. Water exposure was sufficient to produce saturation at room temperature (1.4 ML).

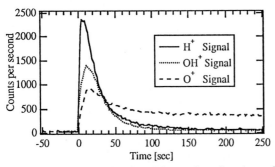

Figure 2. Electron-stimulated desorption signal as a function of irradiation time after a 10 Langmuir exposure of H$_2$O on sapphire (1$\bar{1}$02)-(2X1). The electron beam energy was 300 eV and the primary beam current density was 16 μA/cm^2.

In sharp contrast to the laser-induced desorption no aluminum particles were observed to desorb with electron exposure. Since the electron beam has sufficient energy to excite the aluminum core levels at ~80 (2p) and ~125 eV (2s), we suggest that the electron induced desorption involves an interatomic auger decay process as formulated by Knotek and Feibelman[8]. The H$^+$ and OH$^+$ signals decay away in approximately 150 seconds at 16 μA/cm^2, which corresponds to the removal of the water overlayer. The LEED pattern is (1x1) after the electron beam exposure, suggesting that residual oxygen after H$^+$ desorption has caused a relaxation of the atomic positions from that of the clean equilibrium surface. Electron beam exposure of the clean surface at 16 μA/cm^2 for 10 minutes does not remove the (2X1) reconstruction. The O$^+$ signal persists after electron-stimulated desorption of the water overlayer consistent with the observed O$^+$ signal from the clean surface and the eventual appearance of the 67 eV AES peak on the clean surface.

4. Conclusions

While the mechanisms for laser- and electron-induced desorption are both electronic on sapphire surfaces, the cross sections for the core level based excitations caused by electron irradiation are higher, and this results in the vastly different desorption mechanisms for electron- and laser-induced processes. The preferential desorption of surface cations in the laser desorption case suggests the excitation is localized on the aluminum atoms in the near surface region and may involve surface trapping and decay of sapphire excitons.

Acknowledgments

The work was performed under the auspices of the Division of Material Science, Office of Basic Energy Sciences, USDOE at Lawrence Livermore National Laboratory under contract number W-7405-ENG-48.

References

[1] L. L. Chase and L .K. Smith, in Laser-Induced Damage in Optical Materials, edited by H. E. Bennett, A. H. Gunther, D. Milam, B. E. Newnham, and M. J. Soileau (National Institute of Standards and Technology, Washington, DC, 1987) p. 165.

[2] M. A. Schildbach and A. V. Hamza, Phys. Rev. B **45**, 6197 (1992).

[3] C. C. Chang, J. Vac. Sci. and Technol. **8**, 500 (1971).

[4] W. J. Gignac, R. S. Williams, and S. P. Kowalczyk, Phys. Rev. B **32**, 1237 (1985).

[5] S. Ciraci and I. P. Batra, Phys. Rev. B **28**, 982 (1983).

[6] J. Olivier and R. Poirier, Surf. Sci. **105**, 347 (1981).

[7] I. A. Brytov and Yu. N. Romashchenko, Fiz. Tver. Tela (Leningrad) **20**, 664 (1978) [Sov. Phys. Solid State **20**, 384 (1978)].

[8] M. L. Knotek and P. J. Feibelman, Phys. Rev. Lett. **40**, 964 (1978).

UV Pulsed Laser Desorption from Ag/Al/Sapphire Films: Changes in the Ion Desorption Process for Excitations near the Ag d-Band Threshold*

H.-S. Kim[1], L. Wiedeman, and H. Helvajian[2]

Mechanics & Materials Technology Center, Aerospace Corporation,
Los Angeles, CA 90009, USA
[1]National Academy of Sciences/NRC Postdoctoral Fellow
[2]Person to whom correspondence should be addressed
*Aerospace Sponsored Research

Abstract. Energetic ion species desorption from Al(200nm)/sapphire, Ag(10nm)/Al(200nm)/sapphire and Al(111) targets has been measured for 244 and 320 nm laser irradiation. The product kinetic energy distribution (KED) for samples irradiated with 244 nm light, for Ag^+ and the impurity species K^+ resemble the KED of Al^+ desorbed from Al(111) and Al/sapphire ($<KE> \approx 3-4$ eV). However, for resonance excitation (320 nm) of the Ag adlayer, all product KEDs shift to lower energy ($<KE> \leq 1$ eV). These results imply that the excitation which leads to ion KE is not due to a local excitation process.

1. Introduction

Recent experimental results indicate a new atomic/molecular ion desorption process which originates from the interaction of a low-fluence pulsed UV laser with a surface [1–3]. The desorption process does not appear to be the result of a thermal or plasma excitation. The product kinetic energy distributions are independent of the laser fluence and suggest an electronic excitation induced desorption process. Additional results show that the laser excitation process can eject electronegative species (i.e., F^+, H^+ and O^+) analogous to that observed in ESD/PSD (electron/photon stimulated desorption) [4]. Furthermore, there is experimental evidence that both the desorption yield and the product population distribution are laser wavelength dependent [3,5]. This phenomenon is observed for both metallic (e.g., Ag, Al(111), W(100)) and ceramic (e.g., $Bi_2Sr_2CaCu_2O_8$) substrates and explicitly measured at laser fluences near the single particle desorption threshold.

In this investigation we present experimental results on the UV pulsed laser desorption (UVPLD) from metallic thin film targets i.e. 200 nm Al on sapphire (Al/S) and 10 nm Ag on 200 nm Al on a sapphire (Ag/Al/S). More specifically, the experiment measures the consequence of resonance surface excitation (i.e., plasmon and interband) in UVPLD. Using linear time-of-flight (TOF) mass spectroscopy, we have measured the photoejected ion KED at the laser wavelengths 320 and 244 nm. The 320 nm wavelength is nearly resonant with the surface plasmon excitation in the silver overlayer[6]. We compare the photoejected ion KEDs from the two films and from crystalline Al(111).

The thin films were prepared by the resistive evaporation process and kept in a dry box prior to transferring into the UHV chamber. Both the single layer (Al/S) and the multilayer (Ag/Al/S) film thickness is calculated based on a calibrated deposition rate. The films were used without annealing. The single crystal Al (111) target is commercially purchased and used without sputter cleaning.

2. Experiment

The experiment is conducted in a UHV apparatus and employs a 1.4 m TOF mass spectrometer. The two laser wavelengths are generated by frequency up converting the output from a Nd-Yag pumped dye laser. The laser beam is apertured to minimize spatial inhomogeneities. The incident angle is 45° relative to the target normal and the laser fluence is kept at the threshold for removing a fraction of a monolayer (30 mJ/cm^2 @ 244 nm). The photoejected product nascent KED is derived by calculating the species TOF based on a calibrated model of our apparatus (we estimate a ±1.5 eV error for a mass to charge ratio between 20 to 108 amu). Details of both the calibration and data analysis may be found in Refs. [1–3]. Critical to measuring the nascent KED is acquiring the data by the single shot data acquisition technique. There is a natural proclivity to operate at higher fluences due to the strong (I^{10}) fluence dependence in the ejected species yield [1–3]. However, at higher yields, above surface chemistry (i.e., ion-ion interactions) tends to alter the translational distributions thereby masking the nascent process.

3. Results and Discussion

Figure 1 shows a contour plot made by collating 200 single shot TOF traces as a function of the measured incident laser energy. The TOF signal is that of the Al$^+$ ion photoejected from the Al(111) target for 244 nm laser excitation. The data in the figure show that the Al$^+$ TOF arrival times do not change with increasing laser fluence. The result delineates that the measured KEDs shown in Figs. 2 and 3 are nascent and not affected by space charge or interatomic collisional processes. Figure 2 shows the Al$^+$ ion KEDs measured using the Al(111) and Al/S targets for the 244nm laser wavelength. Figure 3 presents the data from the Ag/Al/S film. The photoejected Ag$^+$ ion KEDs are shown for the two laser wavelengths. In both Figs. 2 and 3, each trace is the sum of 200 laser shots digitized at 10 nsec intervals. The upper axis in both figures defines the photoejected KE for a particular isotope.

Several findings are apparent in the data (Figs. 2, 3). First, Fig. 2 shows that, within our error, the Al$^+$ KED from the Al(111) and the Al/S targets is the same.

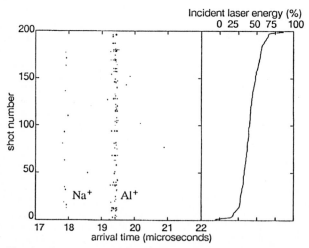

Fig. 1. Contour plot of Al$^+$ as a function of laser energy.

Kinetic Energy of Mass 27 ions (eV)

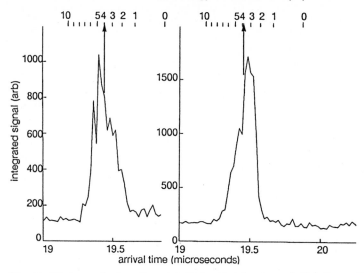

Fig. 2. High resolution time–of–flight distribution of Al^+ from Al(111) [left] and Al/S [right].

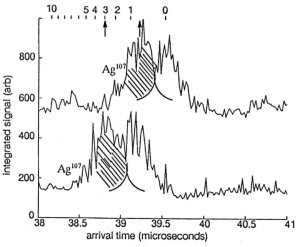

Fig. 3. High resolution TOF distribution of Ag^+ at 244 (top) and 320 nm (bottom). Shaded area represents ^{107}Ag isotope.

This result suggests that the surface crystallinity does not influence the ion KED in the UVPLD process. Second, in Fig. 3 the Ag^+ KED shows a wavelength dependence. The data shows that for 244 nm laser excitation, the Ag^+ ions have a most probable KE, $<KE>$, of $\approx$3–4 eV. However, there is a noticeable shift in the $<KE>$ for 320 nm laser excitation. In both experiments the laser fluence is

97

maintained at the threshold for Ag^+ species desorption. The Fig. 3 result comes somewhat as a surprise since the UVPLD process does not usually show such dependence [1–5]. Third, for the 244nm excitation the Ag^+ KED matches that of the Al^+ (Fig. 2). Both products, from different surfaces, have the same < KE > (Figs 2, 3). Not presented but corroborating this result are the KEDs measured for the desorbed K^+ impurity ions. The alkali ion KED always mimics the majority desorption-product [1–2, 4]. In effect, the observed < KE > shifts are also measured for K^+. Similarly, the K^+ ion KED from the aluminum targets also agree with that shown in Fig. 2.

We perceive that the UVPLD mechanism is a DIET process. The results however cannot easily be explained by the current ESD/PSD theories. The UVPLD process seems to originate via a nonlinear excitation by the laser electric field. As previously recorded [1], both inter and intraband excitations show no change in the product KED. However, direct excitation of the Ag (adlayer) plasmon (320 nm) significantly alters the KE available to the desorbate. The measured KED is not wholly from thermal origin, but resembles that following a charge transfer desorption (CTD) process. One could speculate that plasmon excitation enhances the CTD process, while other excitations lead to a higher KE product. There is some experimental evidence for a plasmon induced desorption of neutral Ag from silver at 532 nm [7]. The Ag neutral KED measured at the incident plasmon angle ($\approx$44.3° @532 nm) agrees with our KED measurement of Ag^+ ion measured at the plasmon wavelength (@ $\approx$320 nm). These results suggest that a plasmon excitation may influence ion species desorption in UVPLD. Further experiments are warranted.

References

[1] H. Helvajian, and R. Welle, J. Chem. Phys. **91**, 2616 (1989).

[2] H-S Kim, and H. Helvajian, J. Phys. Chem. **95**, 6623 (1991).

[3] L. Wiedeman, and H. Helvajian, J. Appl. Phys. **70**, 4513 (1991).

[4] H-S Kim, H. Helvajian, "Laser Ablation Mechanisms and Applications," J. C. Miller, and R. F. Haglund, Jr., Eds. (Springer-Verlag, NY, 1991) pg. 87.

[5] L. Wiedeman, H-S Kim, and H. Helvajian, Mat. Res. Soc. Symp. Proc. **201**, 575 (1991).

[6] H. Ehrenreich, and H. R. Phillip, Phys. Rev. **128**, 1622 (1962).

[7] E. T. Arakawa, I. Lee, and T. A. Callcott, "Laser Ablation Mechanisms and Applications," J. C. Miller, and R. F. Haglund, Jr., Eds. (Springer-Verlag, NY, 1991) pg. 82.

Part IV

High Energy Processes

Core-Induced Photodissociation of Surface Molecules Investigated by Ionic Desorption and Correlated Electronic Spectra

D. Menzel

Physik-Department E 20, Technische Universität München,
W-8046 Garching, Germany

The generalized MGR and similar DIET mechanisms stress the importance of survival vs. delocalization of the primary excitation for desorption to occur. A frequent assumption is that dissociation happens on comparatively slow time scales (10^{-13} seconds or more) which leads to the requirement of long localization times. Increasing evidence accumulates, from studies on both molecular and surface systems, that there are also "ultrafast" processes, i.e. dissociation processes on time scales of 10 femtoseconds or even less. In such cases there may be no need for localization to make DIET sucessful; DIET happens simply because the fastest process always wins. Corroboration for such a mechanism, which in principle can be expected for any sufficiently repulsive excitation, is easiest in the case of core excitations because 1) the core life time constitutes an internal time mark, and 2) the ready accessibility of highly resolved excitation and decay spectra contributes detailed information on the electronic evolution as well. While it appears that in all the examples known so far, these ultrafast dissociation processes are dominated by intramolecular properties (as would be expected), their investigation by DIET from condensed and adsorbed layers is valuable because in these layers the existence of competing processes which quench slower dissociation channels allow a more clearcut decision on the time scale, and because additional information on the selective influences of interactions and delocalizations in these layers can be derived. Examples for ultrafast dissociation before (and possibly after) core decay will be given using comparisons of ionic product yields with core absorption spectra and selective decay spectra. The first vibrationally resolved core-induced desorption data (for condensed N_2) are also shown, in which strong influences of the surrounding medium and possibly even evidence for dynamic effects are found.

1. Introduction

The basic assumption of most mechanisms proposed to date to explain DIET processes, starting with the earliest one [1], is that of a two-step process: the primary excitation which can be treated locally is followed by an evolution in which competition exists between the bond breaking and the delocalization of the excitation from the primary location. For the intrinsic time scale of bond breaking, times of the order of 0.1 ps have usually been assumed, in analogy to vibrational motion. As a consequence, the adjustable parameter determining the outcome of the mentioned competition is the localization time of the excitation which is determined by its coupling to the continuum in the substrate and by the complexity of the excitation. Indeed, the particular desorption efficiency of core excitations and, even more pronounced, of coupled core-valence excitations (core shake-ups) of adsorbates on metal surfaces [2], can be well understood in terms of the increased localization of such strongly correlated excitations and/or of their Auger decay products, caused by the strong Coulomb interaction existing in them [3]. The discussion has therefore often focussed on the factors influencing localization.

Springer Series in Surface Sciences, Vol. 31
Desorption Induced by Electron Transitions DIET V
Editors: A.R. Burns · E.B. Stechel · D.R. Jennison   © Springer-Verlag Berlin Heidelberg 1993

It is clear, of course, that the competition can also be influenced from the other side, i.e. by the time scale of bond breaking, and in the extreme it could be envisaged that very fast dissociation processes would become independent of the delocalization events because they outrun all such competition. The existence of steeply repulsive potentials for dissociative excited states, which can lead to dissociation times of a few femtoseconds or less, appears probable; nevertheless, such "ultrafast" processes in molecular processes have only been discussed in recent years for molecular [4] and surface [5,6] species. One consequence of influencing the DIET competition from this side is that for such a particular process there should be no fundamental difference any more between the isolated molecule and that adsorbed on a metal surface - except that in the former case all other, slower processes are effective as well, while these would be quenched on surfaces, leading to a selective enhancement of the corresponding primary excitation in the desorption yield. So, while the general question of ultrafast processes is as relevant for molecular photodissociation as for DIET, it may be easier to detect them in the latter.

The direct proof for the existence of such a channel is obviously not easy, even in the era of very fast techniques in the time domain. So far, the decision is more easily accessible via indirect information from spectroscopic data. A particular advantage exists if core-induced processes are investigated: since the life times of core holes are comparatively well known, these constitute an internal tick mark against which other processes can be compared. Indeed, the first clear example of a very fast photodissociation process of a free molecule, that of HBr, used the deexcitation spectrum of the free Br atom which was observed after excitation of the Br 3d core in HBr as proof that an appreciable part of the molecules had dissociated into H + Br before core decay [4], which takes about 7 fs [7]. This investigation triggered our interest in such processes, now for molecular photodissociation processes at surfaces. Obviously, the best chance to pinpoint them was again for long-lived core holes and fast-moving fragments, i.e. hydrogen. We believe that we have accumulated a number of examples for ultrafast DIET processes at the surface of condensed layers and in part also for the corresponding adsorbates. We use a variety of arguments, in addition to the knowledge of the core decay times, to corroborate the time scales of dissociation, and believe that the existence of such ultrafast processes is certain now. We stress that in our usage the term "ultrafast" refers to the time scale of dissociation and is quite loosely defined. In the case of core excitations, any event which is faster than electronic core decay is certainly ultrafast. On the other hand, if the core excited state is stable and the repulsive state leading to dissociation is a product of core decay, the dissociation itself may still be ultrafast in our sense, but this would be much more difficult to prove. The fundamental importance of such processes for the theoretical understanding is that for them the electronic and the nuclear evolutions may proceed on comparable time scales, so that a Franck-Condon treatment may break down. Because of our interest in such situations, we shall discuss both repulsive and bound core excitations and their evolutions.

In the following, a summary of earlier data will be supplemented with recent results and the various arguments used and their consequences will be compared and discussed. While some open questions remain, our findings should be of interest to both molecular and surface investigators.

2. Core-induced dissociation of hydrogenic molecules in surfaces: water, ammonia, and benzene

As first clue for possible ultrafast processes we have used the finding of a strong enhancement of dissociation for specific core-to-bound excitations, compared to most other bound and continuum core excitations. The desorption of H^+/D^+ ions served as indicator of dissociation (the ambigui-

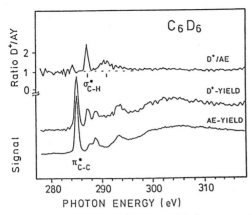

Fig. 1: Comparison of the D^+ fragment ion signal and the Auger electron yield from a condensed layer of C_6D_6, caused by photon irradiation in the C1s region. The enhancement of ions in the C1s → σ^*_{C-H} resonance and its shake-up is most obvious in the ratio plotted on top. In this case the backgrounds prior to the core threshold have not been subtracted; if this is done, the enhancement factor of the σ^*_{C-H} excitation becomes about 4. After ref. 9.

ties which can be introduced by this will be discussed below). The first system showing a clear effect of this type was water [6]: For H_2O or D_2O molecules, both in the surface of thick ice layers or in adsorbate layers on a metal surface, the $1a_1$ (O1s) → $4a_1$ and, to somewhat lesser extent, the $1a_1$ → $2b_2$ core-to-bound excitations showed a very strong enhancement of H^+/D^+ signal compared to other core-to-bound or to-continuum excitations. A similar effect was seen for condensed NH_3 [8] and benzene (fig. 1, [9]). In all these cases the selective excitations are such that the excited electron in the final state resides in an orbital which is expected to be strongly antibonding for the O-H, N-H, and C-H bond, respectively. This is particularly obvious for benzene where many other excitations to antibonding states exist which, however, are antibonding with respect to the C-C bonds and therefore have no effect on the H^+ signal, as expected [9]. No enhancement was found in the case of condensed CH_4 [8]. While we originally [8] discussed the possibility that this may be due to the missing hydrogen bonds, it is now clear that this is a molecular property (see also below). In a simple (Z+1)-view, the strong enhancement of dissociation for these specific core-to-bound excitations could then be explained by the strongly repulsive nature of the respective FH_2, OH_3 and NH entities at the OH_2, NH_3, and CH (in C_6H_6) distances, and consequent ultrafast dissociation.

However, this argument is not sufficient. Since the primary excitation leads to an overall neutral state, the direct ultrafast process should predominantly lead to two <u>neutral</u> fragments, and specifically to H fragments rather than the H^+ ion monitored. To be sure, there might well be a much higher abundance of neutral H fragments in our cases, but these would be much harder to detect; while plans for their analysis exist, this has not yet been accomplished so far. To explain the observed <u>ionic</u> fragments from a <u>neutral</u> excitation, we have argued [6,8] that, while most of the energy transfer to the H atom occurs in very fast and direct fashion, core decay happens when there is still overlap of the H electron cloud with the O (N, C) core, i.e. within a few tenths of an Å from the ground state equilibrium position. This also makes sense in view of the possible steepness of the repulsive potentials (a few eV/Å) and the known core life times (O1s: ~ 4-6 fs; N1s: ~ 7-11 fs; C1s: ~ 10-13 fs [7]): on a

103

potential of steepness 5 eV/Å, an H atom will move about 0.6 Å in 5 fs and collect 3 eV of energy. Further repulsion can then follow on the potential curve of the product of electronic decay.

The difficulty of this view is, of course, that ionic fragments are explainable equally well if dissociation is assumed to happen in the core decay states only, i.e. if core decay precedes dissociation (which, as mentioned, could still be very fast). In this case the $1a_1^{-1}4a_1^{+1}$ state (for the H_2O and NH_3 cases; correspondingly for the others) will lead to $v_1^{-1}v_2^{-1}4a_1^{+1}$ final states by spectator decay (where v_1 and v_2 are valence orbitals filled in the ground state), and all final states with strongly bonding v_1 and/or v_2 will be strongly antibonding for the molecule, with very repulsive potential energy curves. Part of the consequent fast dissociation will then leave the charge on the H fragment, leading asymptotically to the H^+ ion. Without further information then, we cannot conclude that dissociation is faster than core decay for the fragment ions from the observed enhancement alone, although direct dissociation is indeed safely concluded, and it could still be very fast. In fact it may well be that by detecting the (possibly few) H^+ ions instead of the H atom fragments, we select from the life time distribution of core decay the fastest part, and dissociation indeed proceeds as just sketched. However, for a clear identification of ultrafast processes in the sense defined above we do need further evidence.

Different pieces of information can be used for this. In the case of water, an interesting information was the observation that the primary excitation appears to break the ground state symmetry. This was concluded from the dependences of H^+ signal on the angles of polarization and emission [6]. These data indicated that the selectively dissociating excitation does not possess its maximum intensity for the E-vector of the light in the molecular (C_{2v}) symmetry axis, but when pointing along the OH axis of the bond to be broken, corresponding to a $\sigma_c \rightarrow \sigma^*$ excitation in $C_{\infty v}$. The complementary behavior was observed for the $2b_2$ excitation. This could mean that already the primary excitation leads to the dissociating state. Alternatively, however, the asymmetric hydrogen bonding situation at the surface could induce this symmetry breaking [6]. Indeed, very recent gas phase dissociation measurements [10] showed that core-induced dissociation leaves the C_{2v} molecular symmetry intact, so that our observations must be explained by surface symmetry breaking in the ground state.

While this argument did not hold up, several others do. We summarize them in order of increasing strength. The weakest is the fact that the selectivity of these excitations to the ionic fragment was significantly higher for D_2O compared to H_2O [6]. This is understandable in terms of the argument given above for the production of ionic fragments from the primary neutral excitation: the slower D atom has a better chance to become ionized by core decay within the regime of residual overlap, leading to a higher ionic/neutral fragment ratio. Secondly, the enhancement persisted for adsorbed molecules on a metal surface, even though the total signal was strongly decreased. This shows that quenching processes into the metal do not change this selectivity and suggests that dissociation proceeds on a time scale comparable to that of transfer of the excitation into the metal (quenching times for simple excitations such as those relevant here are expected to be of the order of some femtoseconds). This again implies ultrafast dissociation, either in the core-excited or in the core decay state. We therefore suggested ultrafast core-induced dissociation for these molecules which should also exist in the free molecules [6,8].

The third and most direct indication for dissociation in the core-excited state (and thus also for ultrafast dissociation) would be the finding in the core absorption spectra that even very high resolution cannot resolve vibrational structure. Because of the broadening of molecular absorption spectra by coupling to a surface, such measurements for surface species could be undecisive (see, however, sect. 4. below). While so far no gas phase core-induced dissociation measurements have been done (which could

utilize the powerful coincidence methods available for isolated molecules),
very recent highly resolved photoabsorption measurements for the free
molecules [11] have clearly corroborated our earlier suggestion [6]. Using
resolution of the photon beam which would have amply been sufficient to
resolve the vibrational structure of the core-to-bound excitations, this
work has shown that the $1a_1 \rightarrow 4a_1$ and $\rightarrow 2b_2$ of H_2O and the corresponding
excitations of NH_3 consist of broad structures typical for transitions into
repulsive states, in agreement with our postulate (in the case of NH_3 the
existence of strong vibrational coupling could also be responsible for the
failure to detect vibrational structure [11]). In contrast, the corresponding excitations for CH_4 do show vibrational fine structure as expected for
a transition into a bound final state [11], in agreement with our finding
that dissociation is not enhanced for this molecule [8].

Another possibility of additional evidence for ultrafast processes is
that used by Morin and Nenner [4] in their pioneering work mentioned above
and very recently also applied by other authors [12], namely the recording
of decay (autoionization) spectra. If the observed decay spectrum for a
certain core-to-bound excitation is that expected for the mixture of
spectator and participant decay (with the former usually dominating) of the
intact molecule, then core decay must have happened before dissociation; if
not, then the opposite conclusion is indicated. To be sure, in view of the
signal/noise problems with selectively excited decay spectra this indicator
will only be conclusive in a positive sense: a channel not obvious in the
decay spectra may still possess considerable strength. For our measurements
with H_2O and NH_3 another, experimental, problem was more important. In
these cases the resolution of the photon beam was not sufficient to
separate off the selective channels, so that the observed decay spectra
were always dominated by the strong background of unselective channels, and
all decay spectra looked similar. The situation was better for the benzene
work, where the photon resolution was sufficient to separate channels. In
this case we indeed observed [9] that the strongest selective dissociation
channel was coupled to a decay spectrum which was very different from all
the others (which could be explained by spectator or normal Auger spectra
of the intact molecule) and was not explainable by decay of a molecular
benzene species. While so far we have not been able to positively identify
this spectrum, this fact clearly shows that by the time core decay happens
something dramatic has happened to the molecule [13]. Unfortunately the
best variant of this method which is very successful in isolated molecules,
i.e. the correlation of a particular primary excitation to a particular
decay spectrum by coincidence methods, is very difficult if not impossible
with condensed and adsorbed species, due to the very high background of
unwanted coincidences with unspecific secondary electrons.

In all we conclude that there are several cases for which dissociation
clearly occurs before core decay and thus is necessarily ultrafast.

3. Some recent results on H_2S

In view of the situation described so far, it would be very desirable to
find an example where the time scales could be varied between the extremes
of dissociation before and after core decay. In an attempt to find such a
situation, we carried out dissociation measurements on condensed and
adsorbed H_2S [13]. In this case three types of core levels with very
different decay times are accessible: The shallow S2p level has a life time
of about 7-10 fs, the slightly deeper S2s level is, due to Coster-Kronig
decay, very short-lived (~0.4 fs), and the deep S1s level lies in between
(~2 fs[7]). Between the extremes, a variation by a factor 20 exists which
covers the times that may be expected to embrace those required for
dissociation. To be sure, a clear answer could easily be obscured by Auger
cascades. Not only does the Coster-Kronig decay of S2s lead mainly to S2p
holes, but even for primary S1s holes a considerable number of final decay

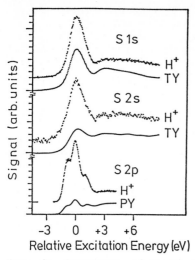

S 1s

H⁺

TY

S 2s

H⁺

TY

S 2p

H⁺

PY

Relative Excitation Energy (eV)

-3   0   +3   +6

Fig. 2: Comparison of the H⁺ fragment ion signal and the total (TY) or partial (PY) electron yields from a condensed layer of H_2S, caused by photon irradiation in the S1s (~2467 eV), S2s (~229 eV), and S2p (~165 eV) regions, respectively. An enhancement of ion signal around threshold is obvious in all three cases.

states will contain the same holes. So the answer of experiment might not be as conclusive as expected at first sight. Unfortunately this is exactly what happens.

As can be seen from fig.2, core-to-bound excitations starting with all three types of core levels lead to considerable enhancement of dissociation. The utilization of variations of polarization and emission angles, independently possible with our new machine which can be turned as a whole around the synchrotron beam under ultrahigh vacuum [14], allowed the assignment of the final states and the resolution of the contributions of the various initial states for the 2p spectra [13]. However, the enhancement of dissociation is quite similar in all cases: the maximum ratios of ion to absorption signal (normalized to 1 at continuum energies 10-30 eV above threshold) are about 1.6 for S1s, 2.7 for S2s, and 4.4 for S2p. We believe the reason to be that the $(2p^{-1}6a_1)$-excited molecules dissociate on a time scale of 5-10 fs, the 2p hole life time, similarly to $(1a_1^{-1}4a_1)$-H_2O. The (smaller) enhancements for the shorter lived 1s and 2s holes then can be attributed to their 2p-containing decay channels which constitute only a fraction of all channels. The observation that these enhancements prevail in monolayer H_2S on S/Ru(001) only indicates that the overall process is ultrafast, but cannot distinguish between dissociation before or after core decay. That the $2p^{-1}6a_1$ state is indeed repulsive in the isolated molecule can be concluded from the missing vibrational resolution of this peak in highly resolved absorption spectra [15] (our data for the condensate, fig.1 and ref. 13, cannot resolve them either, despite the sufficient photon resolution. But this could be due to coupling in the hydrogen-bonded substrate, as suggested by the fact that higher excited states are vibrationally resolved in the gas phase but not in our spectra). For the free molecule, this is also shown by the decay spectra which can be ascribed to the neutral HS* fragment [12] (for the condensate, the recording of decay spectra for the condensate with sufficiently narrow primary excitation has so far been impossible). The predominance of this product again suggests that the ionic decay channel investigated by us may well be a minority decay channel, making the planned investigation of the neutral H fragments very desirable. In any case, whether our H^+ fragments stem from direct dissociation of the core excited state with "late" ionization by core decay, or whether we select the fastest core decays, with dissociation of the strongly repulsive 2h1e final states, the processes must be direct and very fast.

4. Dissociation after core decay: Vibrationally resolved photodissociation via ionic photodesorption.

The most convincing argument for the existence of an ultrafast channel leading to ionic fragments,which we used for H_2S (2p), H_2O and NH_3, rested on the nonexistence of vibrational structure in the absorption and/or H^+ yield curves. The fact that both these types of signal indicate the repulsive nature of the core-excited state proves that both see essentially the same processes and excludes the possibility that the ionic channel might be indicative of the (few) excited species which survive as intact molecules beyond core decay, so that the $v_{1\;2}{}^{-2}a_1{}^*$ final states arising from spectator decay would be responsible for dissociation (the ionic fragments would then selectively detect the short-lived core holes within their distribution of life times). As indicated above, this proof of dissociativeness of the primary core-excited state is a sufficient, but not a necessary condition for an ultrafast dissociation process. If we go to the other extreme, core excitations with vibrational structure indicative of bound core-excited states, any dissociation must happen after core decay. As discussed above, these dissociation events in the decay products (mainly 2-hole 1-electron states, but also more highly excited states) may still be ultrafast, but the proof for this is more difficult. However, from Auger spectroscopy it is known that dynamic effects are to be expected if the core hole life time and the vibrational frequency are not too different. Such effects should be even more important for dissociation. It appeared interesting, therefore, to investigate highly resolved photodesorption also for clearly stable core-excited states.

The N1s → π_{1g} excitation of N_2 has been shown to possess vibrational structure in photoabsorption, both in the free molecule [16] and in the condensate [17]. To our knowledge, no data have been published of vibrationally resolved core-induced photodissociation. We have carried out such measurements on various types of condensed N_2 layers, directly comparing desorbing ion yields to photoabsorption, as indicated by partial electron yields (PY). Fig. 3 gives a comparison of the N^+ yield (the strongest ion signal) to the PY, for a thin condensed layer (about 4 monolayers thick). Both yields show clear vibrational structure, indicating that both for the majority channel(s) monitored by the decay electrons and in the channel monitored by the ionic fragment, the relevant core-excited state is stable against dissociation, so that ionic dissociation can only proceed from the electronic decay products ($v_{1\;2}{}^{-2}\pi^*$ states). At first glance, the two spectra are quite similar, and also are similar to the photoabsorption measurements of the free and condensed molecules [16,17]. Close comparison shows that there are some small but reproducible differences, the main one being a slight increase of the ratio of v=1 to v=0 in the N^+ yield, as compared to the absorption curve. Such an effect would be indicative of dynamical effects. Because of its smallness, we refrain from detailed dicussion here. Much stronger are differences of the vibrational distributions (for both signals) for various differently prepared nitrogen layers (thick N_2 layers, N_2 in Ar matrix, N_2 monolayer on a Xe spacer layer) [17], which indicate that there are influences of the surroundings ("cage effects") on the vibrational excitation in core absorption. Also, differences of the vibrational structures in these two signals from those of N^{2+}, $N_2{}^+$ and larger cluster ions indicate different production mechanisms for these species. Interesting desorption effects are

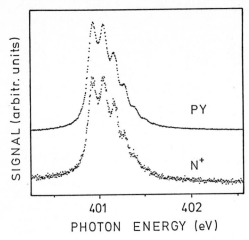

Fig. 3: Vibrationally resolved N1s → π* resonance of condensed N_2, as observed in partial electron yield and N^+ fragment ions, recorded simultaneously, from a thin N_2-film (about 5 monolayers). PY spectra for 1 ML N_2 on a Xe-spacer on Ru(001) were identical, while PY spectra of a thick (~50 ML) N_2 layer contain bulk effects.

also found in the Rydberg region. In all cases data for variation of the angles of polarization and detection show clear alignment effects [17,18].

It is difficult at present to make a definite statement about the time scales in this case. The fact that alignment has been found says that the molecules do not rotate before dissociation, but this does not imply a very short time scale. If the differences of vibrational occupations between absorption and dissociation are real, a coupling of electronic and nuclear coordinates would become more probable, which might be expected from the times involved (the core life time is about a quarter of the vibrational time). But our observations may still be explainable by residual environment effects. Highly resolved dissociation data for the isolated molecule would be very desirable to decide this. Also, good theoretical treatments are desparately needed to completely understand these fast processes.

5. Conclusions

Using older as well as very recent data, we have shown that the time scales of dissociation processes in molecules can be extremely fast. Applied to the surface situation, this may mean that dissociation can outrun all quenching processes, eliminating the need for localization to make DIET processes efficient. But even for somewhat slower processes which would be quenched in a monolayer adsorbate and allow core decay essentially before dissociation, electronic and nuclear time scales can be comparable and may mix , which would make a coupled treatment mandatory. These observations which have largely been made on condensates, but also on adsorbates, have importance not only for DIET but also for molecular photodissociation. This continues the cross-fertilization between the investigation of gas phase and surface processes in which not infrequently surface investigations have preceded corresponding investigations on free molecules.

6. Acknowledgements

The work surveyed here rests on the contributions of many excellent coworkers over the years, in particular of Peter Feulner, Wilfried Wurth, Don Coulman, Gerd Rocker and Roland Scheuerer, with whom it was a pleasure to collaborate and who not only did the experimental work, but also were very important in the development and shaping of the concepts presented here.

This work has been supported by the German Bundesministerium für Forschung und Technologie through grant 05 466 CAB and by the Deutsche Forschungsgemeinschaft through SFBs 128 and 338.

References

1 D. Menzel and R. Gomer, J. Chem. Phys. $\underline{41}$, 3311 (1964);
   P.A. Redhead, Can. J. Phys. $\underline{42}$, 886 (1964).
2 R. Franchy and D. Menzel, Phys. Rev. Letters $\underline{43}$, 865 (1979);
   R. Treichler, W. Riedl, W. Wurth, P. Feulner and D. Menzel, Phys. Rev. Lett. $\underline{54}$, 462 (1985).
3 M. Cini, Solid State Commun. $\underline{24}$, 681 (1977);
   G.A. Sawatzki, Phys. Rev. Letters $\underline{39}$, 504 (1977).
4 P. Morin and I. Nenner, Phys. Rev. Lett. $\underline{56}$, 1913 (1986).
5 R. Treichler, W. Wurth, W. Riedl, P. Feulner and D. Menzel, Chem. Physics $\underline{153}$, 259 (1991).
6 D. Coulman, A. Puschmann, W. Wurth, H.-P. Steinrück, and D. Menzel, Chem. Phys. Letters $\underline{148}$, 371 (1988);
   D. Coulman, A. Puschmann, U. Höfer, H.-P. Steinrück, W. Wurth, P. Feulner and D. Menzel, J. Chem. Phys. $\underline{93}$, 58 (1990).
7 F.C. Brown, Solid State Phys. $\underline{29}$, 1 (1974);
   J. McGuire, Phys. Rev. $\underline{185}$,1 (1969).
8 G. Rocker, D. Coulman, P. Feulner, R. Scheuerer, Zhu Lin, and D. Menzel, in DIET-IV, G. Betz, P. Varga, eds.; Springer, Berlin 1990, p. 261;
   D. Menzel, G. Rocker, D. Coulman, P. Feulner, and W. Wurth, Physica Scripta $\underline{41}$, 588 (1990).
9 D. Menzel, G. Rocker, H.-P. Steinrück, D. Coulman, P.A. Heimann, W. Huber, P. Zebisch, and D.R. Lloyd, J. Chem. Phys. $\underline{96}$, 1724 (1992).
10 D.M. Hanson, this conference
11 K.J. Randall, J. Schirmer, A. Barth, J. Feldhaus, A.M. Bradshaw, Y. Ma, C.T. Chen and F. Sette, presented at "Synchrotron Radiation and Dynamic Phenomena", 48th Internat. Meeting, Soc. Francaise Chimie, Grenoble, Sept. 1991, Abstract book p.63; and to be published.
12 H. Aksela, S. Aksela, O.-P. Sairanen, A. Kivimäki; G.M. Bancroft, and K.H. Tan; Physica Scripta $\underline{T41}$, 122 (1992), and Phys. Rev. A (in press).
13 R. Scheuerer, F. Pellowski, M. Scheuer, W. Wurth, P. Feulner and D. Menzel, to be published.
14 R. Scheuerer, K. Eberle and P. Feulner, to be published.
15 T.D. Thomas, M. Coville, R. Thissen and P. Morin, Synch. Rad. News $\underline{5}$, 9 (1992).
16 C. T. Chen, Y. Ma, F. Sette, Phys. Rev. $\underline{A40}$, 6737 (1989);
   D. Arvanitis, H. Rabus, M. Domke, A. Puschmann, G. Comelli, H. Petersen, L. Tröger, T. Lederer, G. Kaindl and K. Baberschke, Appl. Phys. $\underline{A49}$, 393 (1989)
17 P. Feulner, R. Scheuerer, M. Scheuer, G. Remmers, W. Wurth, and D. Menzel, Appl. Phys. (in press);
   R. Scheuerer, P. Wiethoff, W. Wurth, P. Feulner and D. Menzel, in preparation.
18 P. Feulner, these proceedings.

The Relevance of Gas-Phase Core-Electron Excitation and Decay Phenomena to Molecular Adsorbate Structure and Desorption

D.M. Hanson

Department of Chemistry, State University of New York,
Stony Brook, NY 11794-3400, USA

The results of recent experiments with two new techniques involving the excitation of core electrons and the decay of core holes in isolated molecules are reviewed. The techniques involve measurements of fragment-ion angular distributions and energy-resolved electron-ion-ion coincidences. The results pertain to the photoabsorption anisotropy and core-hole excited-state symmetries, the assignment of resonances, the nature of the resulting valence-hole states, multiple electron ionization, and core-excitation induced fragmentation. Results for H_2O, SiF_4, CF_4, N_2O and acetone are reported.

1. Introduction

Experiments on molecules in the gas phase or in molecular beams can provide information that is critical to understanding the electronic and chemical structure of molecular adsorbates and chemical reactions such as desorption. This paper focuses on the excitation of core electrons and the decay of core holes in molecules. Results pertaining to the photoabsorption anisotropy and excited state symmetries, to the electronic decay of the core hole, and to multiple electron ionization are presented.

2. Photofragmentation Anisotropy and Excited State Symmetry

Since synchrotron radiation is polarized, an anisotropic orientational distribution of molecules can be produced in the gas phase by core electron excitation. This distribution is determined by the orientation of the transition dipole moment with respect to the electric vector of the radiation. Information about this alignment and subsequent time evolution is contained in the angular distributions of fragmentation products resulting from the excitation. Such angular distribution data can reveal details of the fragmentation dynamics, serve to identify the symmetry of the core hole excited state, and provide an important characterization, namely the photoabsorption anisotropy, of how the electronic wavefunction varies through a shape resonance. The photoabsorption anisotropy can be used to determine the orientation of a molecular adsorbate on a single crystal surface if the excited state symmetry is known.

Results for water are reported here, and those for nitrogen and oxygen were reported previously. [1] The fragmentation anisotropy parameter β can be determined, assuming a dipole-allowed transition, from ion yield spectra collected simultaneously at two angles relative to the polarization axis of the radiation [1a]. If the fragmentation is fast relative to the rotational period and if the axial recoil approximation is valid, then the value for β is determined by the angle between the transition dipole moment and the bond that ruptures (or equivalently, the fragment velocity vector). [2]

Springer Series in Surface Sciences, Vol. 31
Desorption Induced by Electron Transitions DIET V
Editors: A.R. Burns · E.B. Stechel · D.R. Jennison   © Springer-Verlag Berlin Heidelberg 1993

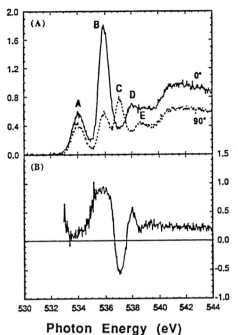

Fig. 1. Top: total ion yield spectra of H_2O detected at $0°$ and $90°$ with respect to the soft x-ray polarization axis. The ion yields are plotted on a common relative intensity scale, and the background has been subtracted. Bottom: the fragmentation anisotropy spectrum giving the anisotropy parameter β as a function of photon energy.

For the case of H_2O, five resonances are observed and labeled A through E in Fig. 1. The experimental values of β then can be used to identify these resonances as having symmetries $^1A_1$, $^1B_2$, and a mixture of $^1A_1$ and $^1B_1$, respectively for the first three. At higher energies, transitions overlap and β is determined by the mixture of all three transitions.

3. Electronic Decay of the Core Hole

The decay of a core hole excited state is dominated by autoionization or Auger processes for the light elements. By these means, ions with vacancies in valence molecular orbitals are produced, and fragmentation occurs with high probability. The nature of the valence-shell vacancies determines which bonds in the molecule rupture.

Studies of the Auger spectra of solids and electron and photon stimulated desorption from surfaces has led to the concept of localized valence-hole states. The essential idea is that Auger decay of a core hole produces two valence holes associated with a single atom because the core hole was localized on that atom. The separation of the two holes is retarded because a large amount of potential energy must be dissipated or converted into kinetic energy, and the two holes are localized at the site of the initial core hole because the rate for two-hole hopping also is slow in many systems since it depends on the ratio of the hopping matrix element to the hole-hole repulsion. Such localization has been called Coulomb localization [3].

Quantum mechanical calculations have supported the relevance of this concept for molecules [4], and such localization has been used to explain some molecular Auger spectra [5]. The relevance of Coulomb localization is important because it may lead

111

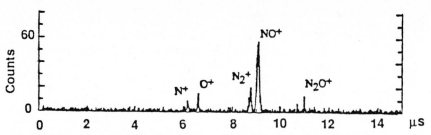

Fig. 2. The electron-ion coincident time-of-flight mass spectrum associated with the electronic decay channel that leaves N_2O in a 2-hole, 1-electron state ($2\pi^{-2}\,{}^1\Delta\,3\pi^1$) following excitation of a 1s electron on the terminal nitrogen to the 3π molecular orbital. The high yield of NO^+ shows that the N-N bond is ruptured preferentially. This result was predicted previously by the molecular orbital depopulation model. [9] This prediction was based on the fact that the 2π orbital is N-N bonding and N-O antibonding. Two holes in this orbital then greatly weaken the N-N bond and may actually strengthen the N-O bond.

to atomic-site specific chemistry, which may be useful in technology, for example, in the modification of surfaces.

The concept of valence hole localization has been used to explain the Auger spectra of CF_4 and SiF_4. The presence of six rather than three main peaks in the Si(LVV) spectrum and the similarity of the F(KVV) spectrum to that of Ne were attributed [6] to contributions from localized two-hole states. The contribution from the localized states is larger for SiF_4 than for CF_4. This difference is attributed to the larger size, greater F-F distance, and higher bond ionicity of SiF_4; all of which decrease the interactions that lead to valence hole delocalization.

The fragmentation patterns observed [7] following ionization of a fluorine 1s electron in these compounds is consistent with this analysis. In particular, the two-hole ion F^{2+} is detected directly for SiF_4, but not for CF_4, showing that a localized valence hole state persists for a time after Auger decay at least comparable to the time for bond rupture to occur. The observation of F^{2+} for SiF_4 and not for CF_4 implies that in some molecules a localized picture of bonding (the valence bond model) may be more appropriate for describing the electronic and chemical relaxation accompanying decay of the core hole than a molecular orbital model, which appears relevant in other cases. To emphasize this point, we identify the former as a valence bond depopulation mechanism and the latter as a molecular orbital depopulation mechanism for the electronic decay of the core hole and concomitant fragmentation.

In nitrous oxide, the molecular orbital depopulation mechanism appears to be active. Auger and resonance Auger spectra can be explained [8] in terms of one electron molecular orbitals as can the fragmentation patterns [9] and the correlation between bond rupture and electronic decay as determined by the electron - ion coincidence data shown in Fig. 2.

A major problem for polyatomic molecules with multiple functional groups is the assignment of resonances associated with the excitation of core electrons of different atoms. The resonances can be characterized and this assignment facilitated by use of secondary processes, i.e. the electronic decay of the core hole and associated fragmentation. A recent study of acetone illustrates this principle. At high resolution, two prominent absorption resonances are observed. Fig 3 shows the resonance Auger

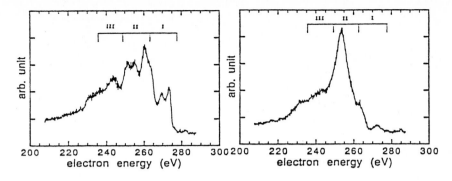

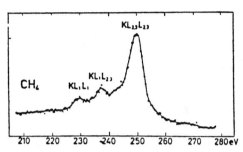

Fig. 3. Left: Resonance Auger spectrum of acetone obtained by exciting at the lower energy resonance, 286.8 eV. Right: Resonance Auger spectrum obtained by exiting at the higher energy resonance, 288.4 eV. Bottom: Auger electron spectrum for methane.

spectrum obtained by exciting at 286.8 eV and at 288.4 eV. The similarity of one to the Auger spectrum of methane, also shown in Fig. 3, associates the higher energy resonance with the methyl carbon in acetone and the lower energy resonance with the carbonyl carbon.

4. Multiple Electron Ionization

Studies of electron and photon stimulated desorption of ions from solid surfaces led to the recognition that multiple electron excitation and ionization often are necessary conditions for ion desorption to occur. Multiple electron ionization often produces electrons with very low kinetic energy. Such processes can be studied in some detail in the gas phase because the secondary electron background is absent. Fig. 4 shows a coincidence map of ionic fragments from H_2O associated with the production of near zero kinetic energy electrons by exciting the resonance structure at 537 eV. Even though the primary excitation leaves the molecule in a neutral core-hole state and the major decay channels produce a singly-charged ion, double ionization yielding low-energy electrons also is significant to produce a doubly-charged ion, which fragments into two ions. In the coincidence map, the flight time of the first ion detected in a time-of-flight mass spectrometer is plotted on the abscissa, and the flight time of the second ion is plotted on the ordinate. The timing is triggered by the detection of a near zero energy electron with the electron energy analyzer. Each point in the coincidence map represents one ionization event. Such experiments involving Auger electron [10] and valence electron [11] coincidences were reported previously. The technique was called single-event ERAEMICO mass spectroscopy (energy-resolved, Auger-electron, multiple-ion coincidence). Two fragmentation processes are revealed.

113

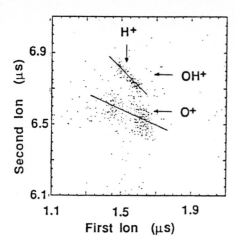

Fig. 4. Coincidence map for H_2O produced by excitation of the resonance structure at 537 eV. Each point represents a single ionization event producing a near zero energy electron and two ions.

One involves coincidences between H^+ and OH^+ and necessarily represents fragmentation and charge separation to produce two cations. The characteristic slope of -1 for the loci of coincidence points is determined by conservation of energy and momentum. The second involves coincidences between H^+ and O^+. These ions could be produced in coincidence by sequential or simultaneous rupture of the two bonds in H_2O. The loci of coincidence points has a slope differing from -1, differing from a ratio of parent and daughter ion masses, and the cluster of points is more diffuse. These characteristics identify simultaneous fragmentation as producing the two ions and a hydrogen atom. The distribution of points on the coincidence map can be analyzed to give the kinetic energy released in the fragmentation and the momentum distributions of the fragments.

5. Summary

Two experimental techniques in addition to conventional time-of-flight mass spectroscopy and electron energy spectroscopy were used in these studies of gas phase molecules in a molecular beam . One involves measurements of fragment-ion angular distributions. Such data allow the photoabsorption anisotropy and excited state symmetries to be determined. Results for H_2O were given. The other involves single-event, energy-resolved, electron-multiple-ion coincidence mass spectroscopy as applied to the study of multiple ionization in H_2O. Additional results documenting valence bond and molecular orbital mechanisms in core hole decay for SiF_4 and N_2O, respectively, were reviewed. The use of secondary processes to characterize excitation resonances was illustrated by the case of acetone.

Acknowledgements

The collaboration of recent coworkers K. Lee, C.I. Ma, D.Y. Kim, and D. Lapiano-Smith on this research is acknowledged with much appreciation. Support from the National Science Foundation under Grant CHE-8921729 also is acknowledged. Part of the work was conducted at the National Synchrotron Light Source, which is supported by the Department of Energy, Division of Material Science and Division of Chemical Science.

References

1. (a) K. Lee, D.Y. Kim, C.I. Ma, D.A. Lapiano-Smith, and D.M. Hanson, J. Chem. Phys. **93** (1990) 7936; (b) A. Yagishita, H. Maezawa, M. Ukai, and E. Shigemasa, Phys. Rev. Lett. **62** (1989) 36; (c) N. Saito and H. Suzuki, Phys. Rev. Lett. **61** (1988) 2740.
2. G.E. Busch and K.R. Wilson, J. Chem. Phys. **56** (1972) 3655.
3. (a) D.E. Ramaker, C.T. White, and J.S. Murday, Phys. Lett. **89**A (1982) 211; (b) H.H. Madden, D.R. Jennison, M.M. Traum, G. Margaritondo, and N.G. Stoffel, Phys. Rev. **B26** (1982) 896; (c) G.A. Sawatzky, Phys. Rev. Lett. **39** (1977) 504;.(d) M. Cini, Solid State Commun. **20** (1976) 605.
4. (a) P. Weightman, T.D. Thomas, and D.R. Jennison, J. Chem. Phys. **78** (1983) 1652; (b) T.D. Thomas and P. Weightman, Chem. Phys. Lett. **81** (1981) 325.
5. (a) R.R. Rye and J.E. Houston, Acc. Chem. Res. **17** (1984) 41; (b) D.R. Jennison, J.A. Kelber, and R.R. Rye, Phys. Rev. **B25** (1982) 1384.
6. R.R. Rye and J.E. Houston, J. Chem. Phys. **78** (1983) 4321.
7. D.A. Lapiano-Smith, C.I. Ma, K.T. Wu, and D.M. Hanson, J. Chem. Phys. **90** (1989) 2162.
8. (a) F.P. Larkins, J. Chem. Phys. **86** (1987) 3239; (b) F.P. Larkins, W. Eberhardt, I.W. Lyo, R. Murphy, and E.W. Plummer, J. Chem. Phys. **88** (1988) 2948.
9. J. Murakami, M.C. Nelson, S.L. Anderson, and D.M. Hanson, J. Chem. Phys. **85** (1986) 5755.
10. D.M. Hanson, C.I. Ma, K. Lee, D. Lapiano-Smith, and D.Y. Kim, J. Chem. Phys. **93** (1990) 9200.
11. (a) L.J. Frasinski, M. Stankiewicz, K.J. Randall, P.A. Hatherly, and K. Codling, J. Phys. B: At. Mol. Phys. **19** (1986) L819; (b) F.S. Wort, R.N. Royds, and J.H.D. Eland, J. Electron Spectrosc. Relat. Phenom. **41** (1986) 297.

Dissociation in Fast Molecule–Surface Collisions

A.W. Kleyn

FOM Institute for Atomic and Molecular Physics,
Kruislaan 407, NL-1098 SJ Amsterdam, The Netherlands

At high translational energies both collision induced dissociation, due to direct momentum transfer, and electronic dissociation are energetically allowed for fast molecules. In the case of negative ion formation, e.g. the formation of O_2^- in O_2 surface scattering, most of the dissociation observed can be attributed to mechanical dissociation. The dissociation has been shown to scale mainly with the normal energy and shows a strong orientation dependence.

In case of the formation of positive hydrogen ions in molecule surface collisions we have found that the degree of dissociation scales with the total energy times the total scattering angle squared. This points to the importance of dissociation induced by a collision with a single substrate atom.

Introduction

In low energy (thermal - 2 eV) molecule surface collisions dissociative adsorption occurs in some cases instantaneously, in a single collision so to speak. In many cases, however, it is an activated process in which one or more precursors are involved [1]. In all cases the nature of the dissociation is electronic or chemical by nature [2]. This is self-evident, since the translational energies in these collisions are well below the dissociation energy of the gas-phase molecule. It would be very interesting to increase the translational energy of the molecule until it exceeds the dissociation energy. However, this is very difficult experimentally and has only been done for heavy molecules, such as I_2 by Amirav and co-workers [3]. For most other systems a jump in energy needs to be made, going from nozzle beam technology to ion beam technology. Many scattering studies of low energy ion beams have been performed, also for systems in which dissociative chemisorption does occur, see e.g. [4]. Although ion beams are used, the situation might be similar to that of scattering of a fast neutral, because often complete neutralization (>99.9%) occurs in the ion surface collision. In some experiments grazing incidence is used to keep the normal energy $E_n = E_i \cos^2(\theta_i)$ low. Here E_i is the incident beam energy and θ_i the angle of incidence measured from the surface normal. For an uncorrugated surface E_n is the quantity of relevance, rather than E_i. Scaling with E_n is indeed observed for dissociative scattering of O_2 molecules. By contrast, the much smaller H_2 molecule does not exhibit normal energy scaling when dissociatively scattered from an Ag(111) surface. Dissociative collisions of fast H_2 molecules are very reminiscent of a gas phase H_2-Ag collision. Both O_2 and H_2 scattering and dissociation will be discussed in this report.

Scattering of O_2 molecules

Reijnen et al. have performed scattering experiments in which a 100-5000 eV O_2^+ beam is reflected from an Ag(111) surface at specular scattering for $\theta_i = 85^\circ$ [5]. Time-of-flight (TOF) analysis has been applied to the scattered neutrals and negative ions. Positive ions have not been observed. O atoms and O_2 molecules can be seen in the TOF spectra. The atoms are formed in dissociative neutralisation yielding a broad TOF distribution; ground state O_2 molecules are formed by an Auger process, leading to a much narrower TOF distribution which is superimposed on that of the atoms [6]. In some cases a separation between O-atom and O_2-molecule yield can be easily made. Also O_2^- ions are clearly observed at most experimental energies. From the yield of O_2^- with respect to O_2 we can see that the probability for charge transfer is at least a few percent. From the analysis of the data it has been concluded that the probability to form O_2^- close to the classical turning point along the trajectory is higher, but that most of the O_2^- reneutralizes in the exit channel or autodetaches when vibrationally hot [5]. By con-

Springer Series in Surface Sciences, Vol. 31
Desorption Induced by Electron Transitions DIET V
Editors: A.R. Burns · E.B. Stechel · D.R. Jennison   © Springer-Verlag Berlin Heidelberg 1993

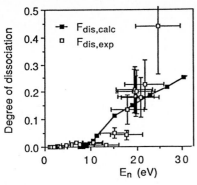

Figure 1   Degree of dissociation of O_2^- $F_{dis,exp}$ for the tof experiment
and of O_2 $F_{dis,calc}$ for the classical trajectory calculations as a function
of $E_n = E_i \cos^2(\Theta_i)$. The line through the computed points was drawn to
guide the eye. From ref. 5.

trast to the reasonably efficient O_2^- formation, O^- ions are almost absent in the spectrum at low
energies (<2000 eV), although O-atoms are abundantly made by dissociative neutralization.
Apparently the molecules can pick up another electron whereas the atoms cannot, as judged
from the absence of O^- ions at low energies. By contrast the probability for forming O_2^+ is very
low and positive ions are not detectable at grazing angles.

Increasing E_i leads to a sudden threshold-like appearance of dissociation of O_2^-, i.e. O^- formation. This is clearly shown in fig. 1 [5]. The x-axis shows $E_n = E_i \cos^2(\Theta_i)$, the cosine factor
being 0.0076 in this case. The threshold is consistent with the onset of collision induced dissociation in trajectory calculations, also shown in the figure. In this context collision induced dissociation is used when the dissociation is driven by direct momentum transfer to the atoms in
the molecule during their collision with the surface, not by a change in the intermolecular potential due to an electronic effect. It can be seen that collision induced dissociation occurs when
E_n is about four times the dissociation energy of O_2^- and E_i is about 500 times this value.
Clearly E_n is the important factor in the dissociation and not E_i. The normal component of the
velocity is the only active component. This can very nicely be seen in calculations of the average internal excitation of the O_2 molecule and the excitation of the solid as a function of E_i,
while keeping E_n constant [7]. From the calculations by Van den Hoek et al. it is clear that experiments at grazing angles of incidence can mimic experiments with low total energy but more
normal angles of incidence very well [5,7]. This is experimentally very convenient.

From the good agreement between theory and experiment it was concluded that collision induced dissociation is the primary mechanism in the dissociation of O_2^-. To avoid any interference from charge transfer to O-atoms formed by dissociative neutralization, it is preferable to
perform experiments using neutral beams. Such experiments have been performed for negative
ion formation when scattering O_2 or NO from Ag(111) and Pt(111) by Reijnen et al. [8]. It
turns out that the results obtained for incident neutrals and incident positive ions are very similar
[8,9]. This similarity demonstrates that dissociative neutralization yielding initially O-atoms
does not lead to additional negative ion formation, and that all negative ions formed are due to
O_2 molecules either in the beam or formed by Auger neutralization.

The dissociation for the case of O_2/Ag(111) has been attributed to collision induced dissociation. Calculations using pairwise additive potentials derived from HFS-LCAO pairpotentials in
fact overestimate the degree of dissociation [7,10]. However, in those calculations the charge
transfer is entirely neglected and this may lead to corrections to the degree of dissociation. One
would expect that similar calculations for O_2/Pt(111) would give rise to a smaller degree of dissociation, because the potential for the bigger Pt-atom would lead to an effectively smaller corrugation and a smaller anisotropy of the potential with respect to molecular orientation, both
leading to less dissociation. This is what one would get for a 'universal' potential like the
Ziegler-Biersack-Littmark potential [11]. However, the degree of dissociation of O_2^- observed
in case of Pt(111) is much higher. At first this has been attributed by Reijnen et al. to electronic
dissociation [8]. However, a HFS/LCAO calculation of the O-Pt pair potential by Kirchner et

117

al. shows that the potential for O-Pt is actually steeper than for O-Ag [12]. Due to relativistic contraction of the wave functions of the Platinum core electrons, the valence orbitals also shrink. This makes Pt-atoms effectively smaller and increases the corrugation of the molecule surface potential and the anisotropy of the potential with respect to molecular orientation. Both effects increase the degree of dissociation.

The calculations by Kirchner et al. showed that the strategy to construct a potential developed by Van den Hoek et al. does not always work. Van den Hoek et al. have fitted for several systems a Born-Mayer potential to the HFS/LCAO pair potential above say 10 eV, which leads to a useful extrapolation at lower energies. This was evident in both experiments on O_2 scattering in the 50-200 eV range as for hyperthermal scattering (1-2 eV) [13]. For O_2-Pt this extrapolation did not give the right repulsion at lower energies and the presence of a well has to be included in the fit. To what extend these effects are specific to the O-Pt system is not clear at present.

Dissociation mechanism for O_2 scattering

The classical trajectory calculations also give more detailed information about the dissociation mechanism. Calculations by Van den Hoek et al. showed that the impact points for dissociative and non-dissociative scattering for O_2/Ag(111) are almost identical [10]. Therefore, the effect of the impact parameter is only slight, except at acute grazing angles of incidence. By contrast the orientation dependence on the degree of dissociation is large. This has been shown by Van den Hoek and Kleyn, who present the average orientation of molecules undergoing dissociation [7]. It is clear from their work that for grazing incidence molecules with their internuclear axis

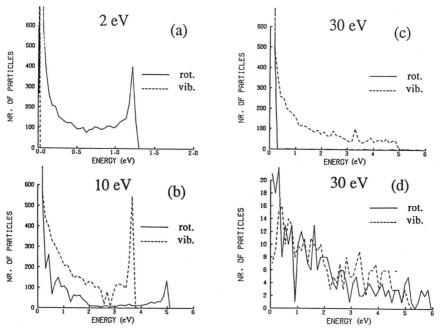

Figure 2   Distributions of rotational and vibrational excitation of O_2 molecules scattered from Ag(111). The distributions have been obtained in classical trajectory calculations. In panels (a) - (c) the molecules have been scattered from a flat rigid surface for different energies at normal incidence as given in the figure. In panel (d) the molecules have been scattered from the corrugated Ag(111) surface at normal incidence and E_i=30 eV. From ref. 7.

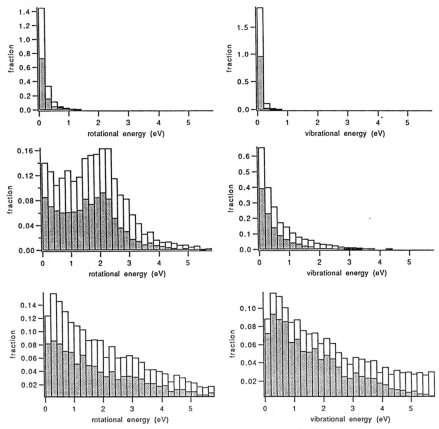

Figure 3   Distribution between rotational (left hand side) and vibrational energy (right hand side) as calculated for O_2 scattered from Pt(111) at E_i=70eV and Θ_i=70°. The shaded areas represent the finally non-dissociating molecules and the unshaded areas the dissociating molecules. The unshaded distribution is plotted on top of the shaded one. The distributions are evaluated along the trajectory at three positions as discussed in the text. From ref. 12.

along the surface normal have a much higher dissociation probability than when the axis is parallel to the surface plane. This orientation dependence has also been seen in the analysis of rotational excitation of oriented NO molecules scattered from a Ag(111) surface, see e.g. [14]. From this similarity one can deduce that the initial excitation in a collision leading to dissociation is also rotational excitation. This can be concluded in figure 2, where the probability of rotational and vibrational excitation is plotted for several incident energies at normal incidence [7]. At low incidence energies mainly rotational excitation occurs and a beautiful rotational rainbow is seen [15]. At higher energies the amount of rotational excitation decreases and vibrational excitation sets in. Even a vibrational rainbow is seen. The decrease of the rotational excitation is due to chattering; i.e., the molecule hits the surface with both ends and most of the rotational excitation introduced in the collision with the first end of the molecule is removed when the second end of the molecule hits the surface [16]. In this second collision efficient rotational to vibrational energy transfer can occur. This model has been proposed by Gerber and Elber in a series of papers [17]. The prevalence of rotational excitation in molecule surface conditions has also been found in an analytical model by Bitenski and Parilis [18,19].

The transfer of rotational energy to vibrational energy can be seen very nicely for a set of classical trajectories for O_2 scattering from Pt(111) at $E_i=70$ eV and $\Theta_i=70°$. In figure 3 the rotational and vibrational state distributions are shown at three positions along the trajectory [12]. The first position is when the molecule has lost 5% of its initial translational energy. The second position is when the O-atom nearer to the surface reverses its normal velocity. The third position is when the other O-atom reverses its velocity. Until position 2 is reached the molecule gains about twice as much rotational excitation as vibrational excitation: the first hit leads mainly to rotational excitation. For the studies discussed in this report so far the importance of the normal velocity and normal energy is predominant.

Fast hydrogen scattering

The discussion so far has been centred on O_2 molecules. However, the interaction between hydrogen and surfaces also deserves some special interest. Hydrogen is a model system, which is rather easily accessible for theoretical studies of the potential energy surface. Quantum effects are important for hydrogen, which gives the system an additional richness compared to systems like O_2, the nuclear motion of which can be described mainly by classical mechanics [20]. (Nevertheless, the presence of an isotope effect in dissociative chemisorption of O_2 on Ag(111) points to the importance of tunneling even for 'heavy' molecules [21]). Another interest in hydrogen surface interactions comes from fusion research. Intense beams of D-atoms (1 MeV, many Amps equivalent) are needed for heating and current drive in next generation tokamaks like ITER, the International Experimental Reactor being developed by Europe, Japan, Russia and the USA. These beams are produced by neutralization of D⁻ ion beams and this has triggered a large interest in the construction of hydrogen negative ion beams [22]. These beams in turn can be made by negative surface ionization [23]. At FOM we have studied these processes extensively and it has triggered studies on the interaction of energetic neutral H_2 and solid surfaces. Most recently, Van Slooten et al. have studied scattering of H_2 from Ag(111) [24,25].

In the experiments only charged products could be detected. Both H_n^+ and H^- ions have been detected, roughly in equal amounts. A crude measurement indicates that less than 1% of the scattered particles is charged. Energy spectra of scattered positive ions are shown in fig. 4. Results are shown at two incident energies, several final scattering angles and for both incident neutral molecules and molecular ions. Two peaks can be seen in the spectra for incident neutrals. These are attributed to H^+ and H_2^+. The amount of H_2^+ is far less in case of incident molecular ions than for incident neutrals. The positions of the peaks is approximately the same for incident ions and neutrals. The proton peak is clearly shifted to the left of the binary collision value minus the dissociation energy of the molecule or molecular ion, which is approximately half the beam energy. The latter value is found for H^- formation, which indicates that negative ion formation is a resonant process. Clearly positive ion formation is a non-resonant process irrespective of the charge state of the incident particles. This points to resonant neutralization followed by re-ionization for incident positive ions and collisional ionization of ground state molecules for incident neutrals [26]. The degree of dissociation is much higher for incident positive ions. We attribute this to the predominance of dissociative neutralization of H_2^+, followed by re-ionization of the H-atoms formed. This channel is absent for incident neutrals. Dissociative neutralization has been studied extensively, see e.g. [27].

The angular distributions shown by Van Slooten et al. in their paper show that the angular distributions for the protons approximately peak around the specular angle, whereas the molecules are preferably found along the surface at very grazing exit angles [25]. This can also be seen in fig. 4.

That the formation of molecular ions in collisions of H_2 and Ag(111) occurs at all is in strong contrast to what is observed for the reverse process, neutralization of positive ions. Here it has been observed that molecular ions yield dominantly neutral atoms due to dissociative neutralization [5,27]. Dissociation is predominant, but it is due to resonant neutralization, a channel which is absent for the scattering of neutrals and formation of positive ions. Nevertheless, survival of positive ions in surface scattering has been seen Eckstein et al. and later by several other groups at ion energies which are at least an order of magnitude larger that the molecular dissociation energy [28-30]. These surviving molecules have been attributed to very special trajectories in which each atom of the molecular ion is scattered from a different surface atom, but the relative velocity after scattering is sufficiently small to prevent dissociation. This has been demonstrated theoretically by Parilis and coworkers [18,19,31] and Jakas and Harrison [32].

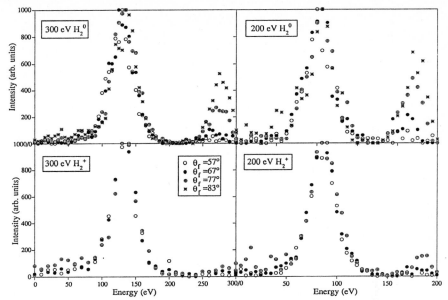

Figure 4 Intensity distributions for different outgoing angles as a function of final energy for product positive ions in the scattering of hydrogen from Ag(111) along the [11$\bar{2}$] direction for Θ_i=80°: (a) 300 eV H_2^0, (b) 300 eV H_2^+, (c) 200 eV H_2^0, (d) 200 eV H_2^+. The peak in the middle of the spectrum is attributed to H^+, the one to the right of a spectrum to H_2^+. No correction for the energy dependent transmission of the energy analyzer has been applied. From ref. 25.

The molecular survival in the present experiment [24,25] is observed under different conditions than in the earlier work mentioned [28-30]. E_i is much smaller. Due to the difference the collision dynamics might be different. Therefore, the scattered intensity I and the degree of dissociation has been studied as a function of the azimuthal orientation of the crystal ϕ, Θ_i, Θ_f and E_i. It was found that the intensity strongly depends on ϕ but that the degree of dissociation F_{dis}= I(H$^+$)/(I(H$^+$+H$_2^+$) does not depend on ϕ. F_{dis} does show a strong dependence on Θ_i, Θ_f and E_i. At first it was tried to correlate the observed F_{dis} with E_i or E_n, but no correlation was observed at all. However, it was found that F_{dis} scaled very well with $E_i\Theta^2$, where Θ is the total laboratory scattering angle ($\Theta = 180^o$-Θ_i -Θ_f). This is shown in figure 5. Note that the same scaling is found both for H_2 and D_2.

Collision induced dissociation in molecule surface scattering is expected to differ from collision induced dissociation in molecule atom scattering in the gas phase. In the surface case the breaking of the bond is normally viewed as being due to the cooperative contribution of the surface atoms while in gas phase the dissociation has to proceed via just one collision. A typical surface case is found for the O_2-Ag(111) system as discussed above. From the range of the interaction potential it is clear that for O_2-Ag(111) scattering the dissociation process involves the interaction with many surface atoms. Some feeling for the potential range governing the interaction with the molecule is given by the potential for Ag-O [7]. This potential reaches a value of 50 eV at an interatomic distance of 1.01 Å, a value of 500 V is reached at 0.56 Å. The distances for the corresponding Ag-H potential are considerably smaller: 0.63 Å for 50 eV and 0.28 Å for 500 eV [33]. This smaller range, the observed scaling of F_{dis} with total scattering angle, and the absence of crystal orientation dependence are reminiscent of gas phase collisions. Thus, in the case of H_2 scattering it is likely that the dissociation process can be described by independent H_2-Ag atom collisions. In a first paper Van Slooten et al. used a gas phase model based on atom-molecule collisions [24]. In this model developed in the group of Mahan [34] the collision between a fast molecule BC and a stationary atom A is treated as a collision involving three hard

121

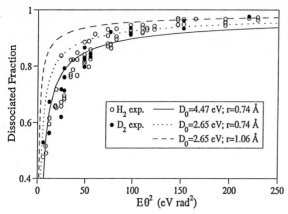

Figure 5　Dissociated fraction $F_{dis}=[I(H^+)/(I(H^+)+I(H_2^+))]$ as a function of the product of beam energy and total scattering angle squared $(E\Theta^2)$ for H_2 and D_2 scattering from Ag(111). In the figure also the calculated curves are drawn: (——) bond length H_2, $D_0=4.47$ eV; ($\cdots$) bond length H_2, $D_0=2.65$ eV; (- - -)bond length H_2^+, $D_0=2.65$ eV. From ref. 25.

spheres. One of the atoms in the molecule, say B hits the atom A and recoils, while in the collision C is not affected. This so-called spectator model has often been applied to inelastic scattering processes [35]. In the high velocity limit, where the collision is vibrationally and rotationally sudden, Mahan and coworkers showed for the angle region studied here that the energy transferred to the internal modes of BC can be written as $E_{int}=mE\Theta^2$ where m is defined as $m=m_A m_C/(m_B(m_A+m_B+m_C))$. This factor is very close to one for both H_2 and D_2 colliding with Ag. The scaling with $E\Theta^2$ is indeed the one we found experimentally for F_{dis}. The threshold for dissociation, which is smooth in figure 8, would appear as a step function in such a model and modifications of the model are needed. After a simple amendment [24] Van Slooten et al. have extended the model to treat BC as a rigid rotor, so that the A-B atom collision is rotationally sudden. With only one fitting parameter, there was good agreement with the data [24]. Next, Van Slooten et al. derived an expression for the dissociated fraction in terms of the rotational excitation of a hard ellipsoid [25]. Korsch and Schinke studied the rotationally inelastic scattering of a hard ellipsoid within the infinite order sudden approximation [36]. For small ellipsoid eccentricity they derived a formula for the classical total differential cross section which has served as a starting point for Van Slooten et al. These authors finally obtained:

$$F_{dis} = \frac{\sqrt{1-\sqrt{\frac{\mu D_0}{M}}\frac{r}{(A\text{-}B)}\frac{1}{\sqrt{E\Theta^2}}(1+\frac{1}{6}\varepsilon^2(2+\sqrt{\frac{\mu D_0}{M}}\frac{r}{(A\text{-}B)}\frac{1}{\sqrt{E\Theta^2}}))}}{(1+\frac{1}{3}\varepsilon^2)} \qquad (1)$$

where ε denotes the ellipsoid eccentricity defined as $\varepsilon=(A^2-B^2)^{1/2}/A$ (A and B being the main axes of the ellipse); r denotes the bond length and μ the reduced mass of the molecule. For a collinear collision the turning point is reached where the atom-atom potential equals D_0; for a perpendicular collision it is reached at $D_0/2$ [36]. D_0 is the dissociation energy. These distances are directly extracted from the potential and lead to a value of 0.68 for ε [33]. Here we would like to point out that equation 1 following from the model just described contains no fitting parameters in contrast to the model used earlier [24]. In fig. 5 the solid line represents the calculated curve according to equation 1. The agreement between theory and experiment is surprisingly good. In the calculation the dissociation energy of H_2 ($D_0=4.47$ eV) is used. This is an upper limit because the beam is not in the ground state. However, the calculated curve for slightly excited H_2 almost coincides with the curve drawn in fig. 5. A similarly good agreement is obtained when taking the dissociation energy of H_2^+ (D_0=2.65 eV), but keeping the molecular bond length fixed to that of H_2. Thus for low $E\Theta^2$ values the dissociation can be described solely as the consequence of rotational excitation. However, for high $E\Theta^2$ values the calculated curve underestimates F_{dis}. This might be due to vibrational excitation, which is ignored in the model but certainly plays a role at high $E\Theta^2$ values.

Recently, classical trajectory calculations have been carried out to investigate if the H_2 molecules indeed only collide with a single surface atom, or that the nature of the collision is such that it can be described using only a single collision. It appears that for $\Theta_i<75^o$ indeed collisions with a single surface atom are predominant. In addition, the scaling with $E\Theta^2$ has also been found in the classical trajectory studies [33].

Summarizing the studies of Van Slooten et al.: in the experiments using incident H_2^+ the observed dissociation of the product ions is mainly caused by the dissociative neutralization into the repulsive $b^3\Sigma_u^+$ state. The dissociation of the product ions in neutral H_2 scattering from Ag(111) can be described in terms of the rotational excitation following from the collision between the incident species and a single Ag atom. The ionization step and the dissociation step in the experiments seem to be decoupled, which is verified by classical trajectory calculations.

Conclusion

Concluding this overview of dissociation in fast molecule-surface interactions it is good to state first that it is far from complete and is centred around some experiments, that have been carried out in the authors group at the FOM-Institute of Atomic and Molecular Physics. In our studies two mechanisms for dissociation in fast neutral-molecule surface collisions have been encountered. One is an impulsive collision between a (O_2) molecule and a flat structureless surface. The other is an impulsive collision between a fast (H_2) molecule with essentially a single surface atom. In both cases the initial orientation of the molecule is determining the internal excitation of the molecule. Often rotational excitation is the first step towards dissociation.

The work described in this progress report is the result of the work and collaboration of and stimulating discussions with Dag Andersson, Eric Gislason, Paul Van den Hoek, Eric Kirchner, Joop Los, Paul Reijnen, Udo Van Slooten, and Ken Snowdon. Udo Van Slooten is thanked for his careful reading of the manuscript. This work is part of the research program of FOM and is financially supported by NWO and by EURATOM.

References

[1] Dynamics of Gas-Surface Interactions, Eds. C.T. Rettner and M.N.R. Ashfold, Advances in Gas-phase Photochemistry and Kinetics, Volume III (Royal Society of Chemistry, Cambridge, UK, 1991).

[12] A.W. Kleyn, J. Los and E.A. Gislason, Phys. Rep. 90 (1982) 1; J.W. Gadzuk and J.K. Norskov, J. Chem. Phys 81 (1984) 2828; J.W. Gadzuk, Comm. At. Mol. Phys 16 (1985) 219; J.W. Gadzuk and S. Holloway, Phys. Scr. 32 (1985) 413; A.W. Kleyn in: XVI ICPEAC, New York 1989, AIP Conf. Proc. 205 (1990), edited by. A. Dalgarno et al. p. 451.

[3] A. Amirav, Comm. At. Mol. Phys. 24 (1990) 187.

[4] S.R. Kasi, H. Kang, C.S. Sass and J.W. Rabalais, Surf. Sci. Rep. 10 (1990) 1; H. Akazawa and Y. Murata, Phys. Rev. Lett. 61 (1988) 1218; Phys. Rev. B39 (1989) 3449; Nucl. Instr. Meth. B33 (1989) 442; J. Chem. Phys. 92 (1990) 5551; 5560.

[5] P.H.F. Reijnen, P.J. Van den Hoek, A.W. Kleyn, U. Imke and K.J. Snowdon, Surf. Sci. 221 (1989) 427.

[6] K.J. Snowdon, B. Willerding and W. Heiland, Nucl. Instr. Meth. B14 (1986) 467.

[7] P.J. Van den Hoek and A.W. Kleyn, J. Chem. Phys. 91 (1989) 4318.

[8] P.H.F. Reijnen, U. Van Slooten and A.W. Kleyn, J. Chem. Phys. 94 (1991) 695.

[9] Pan Haochang, T.C.M. Horn and A.W. Kleyn, Phys. Rev. Lett. 57 (1986) 3035; P.H.F. Reijnen and A.W. Kleyn, Chem. Phys. 139 (1989) 489; Pan Haochang, P.H.F. Reijnen, T.C.M. Horn and A.W. Kleyn, Rad. Eff. Def. in Sol. 109 (1989) 41.

[10] P.J. Van den Hoek, T.C.M. Horn and A.W. Kleyn, Surf. Sci 198 (1988) L335.

[11] J.F. Ziegler, J.P. Biersack and U. Littmarck, The stopping and ranges of ions in matter, Ed. J.F. Ziegler (Pergamon, New York, 1985), Vol. 1.

[12] E.J.J. Kirchner, E.J. Baerends, U. Van Slooten and A.W. Kleyn, J. Chem. Phys. (1992) in press.

[13] P.J. Van den Hoek, E.J. Baerends and A.W. Kleyn, Comm. At. Mol. Phys. 23 (1989) 93.

[14] M.G. Tenner, E.W. Kuipers, A.W. Kleyn and S. Stolte, Surf. Sci. 242 (1991) 376; F.H.Geuzebroek, A.E. Wiskerke, M.G. Tenner, A.W. Kleyn, S. Stolte and A. Namiki, J. Phys. Chem. 95 (1991) 8409.

[15] R. Schinke and J.M. Bowman in: Molecular Collision Dynamics, edited by J.M.Bowman (Springer, Berlin, 1983) p. 61; A.W. Kleyn and T.C.M. Horn, Phys. Repts 199 (1991) 191.

[16] J.C. Polanyi and R.J. Wolf, Ber. Bunsenges. Phys. Chem. 86 (1982) 356; J. Chem. Phys. 82 (1985) 1555.

[17] R.B. Gerber and R. Elber, Chem. Phys. Lett. 102 (1983) 466; R.B. Gerber and A. Amirav, J. Phys. Chem. 90 (1986) 4483.

[18] I.S. Bitensky and E.S. Parilis, Nucl. Instr. Meth. B2 (1984) 364.

[19] I.S. Bitensky and E.S. Parilis, Surf. Sci. Lett. 161 (1985) L565.

[20] S. Holloway, in:Dynamics of Gas-Surface Interactions, Eds. C.T. Rettner and M.N.R. Ashfold, Advances in Gas-phase Photochemistry and Kinetics, Volume III (Royal Society of Chemistry, Cambridge, UK, 1991), p. 88.

[21] P.H.F. Reijnen, A. Raukema, U. Van Slooten and A.W. Kleyn, J. Chem. Phys. 94 (1991) 2368; Surf. Sci. 253 (1991) 24.

[22] M. Bacal and D.A. Skinner, Comm. At. Mol. Phys. 23 (1990) 283; A.W. Kleyn, in: Proc. of the 5th International Symposium on Production and Neutralization of Negative Ions and Beams, Brookhaven National Labaratory, 1989, AIP Conf. Proc. 210 (1989), edited by A. Hershcovitz, p.3.

[23] C.F.A Van Os, R.M.A. Heeren and P.W. Van Amersfoort, Appl. Phys. Lett. 51 (1987) 1495; R.M.A. Heeren, D. Ćirić, H.J. Hopman and A.W. Kleyn, Appl. Phys. Lett. 59 (1991) 158; R.M.A. Heeren, D. Ćirić, H.J. Hopman and A.W. Kleyn, Nucl. Instr. Meth. B 69 (1992) 389; and references cited therein.

[24] U. Van Slooten, D. Andersson, A.W. Kleyn and E.A. Gislason, Chem. Phys. Lett. 185 (1991) 440.

[25] U. Van Slooten, D.R. Andersson, A.W. Kleyn and E.A. Gislason, Surf. Sci. (1992) 274 (1992) 1.

[26] M. Aono and R. Souda, Nucl. Instr. Meth. B 27 (1987) 55; R. Souda, M. Aono, C. Oshima, S. Otani and Y. Ishiwaza, Surf. Sci 179 (1987) 199; M. Tsukada, S. Tsuneyuki and N. Shima, Surf. Sci. 164 (1985) L811.

[27] B. Willerding, W. Heiland and K.J. Snowdon, Phys. Rev. Lett. 53 (1984) 2031; U. Imke, K.J. Snowdon and W. Heiland, Phys. Rev. B34 (1986) 41; 48; W. Tappe, A. Niehof, K.Schmidt and W. Heiland, Europh. Lett. 15 (1991) 405. J.H. Rechtien, W. Mix and K.J. Snowdon, Surf. Sci. 259 (1991) 26; J.H. Rechtien, R. Harder, G. Hermann, C. Röthig and K.J. Snowdon, submitted to Surf. Sci.

[38] W. Eckstein, H. Verbeek and S.Datz, Appl. Phys. Lett. 27 (1975) 527.

[39] W. Heiland and E. Taglauer, Nucl. Instr. Meth. 194 (1982) 667.

[30] L.L Balashova, Sh.N. Garin, A.I. Dodonov, E.S. Mashkova and V.A. Molchanov, Surf. Sci. 119 (1982) L378.

[31] I.S. Bitensky, E.S. Parilis and I.A. Wojchiechowski, Nucl. Instr. Meth B47 (1990) 243.

[32] M.M. Jakas and D.E. Harrison Jr., Surf. Sci. 149 (1985) 500.

[33] U. Van Slooten, E.J.J. Kirchner, and A.W. Kleyn, subm. to Surf. Sci..

[34] M.H. Cheng, M.H. Chiang, E.A. Gislason, B.H. Mahan, C.W. Tsao and A.S. Werner, J. Chem. Phys. 52 (1970) 6150.

[35] N. Andersen, M. Vedder, A. Russek and E. Pollack, Phys. Rev. A 21 (1980) 782; P. Sigmund, J. Phys. B 11 (1978) L145; 14 (1978) L321 (1981); F. Budenholzer, M. Chang and P. Lü, J. Phys. Chem. 89 (1985) 199.

[36] H.J. Korsch and R.Schinke, J. Chem. Phys. 75 (1981) 3850.

Orientation Effects in the Scattering of H_2 and H_2^+ from Cu(111)

R. Harder, G. Herrmann, J.-H. Rechtien, and K.J. Snowdon

Fachbereich Physik, Universität Osnabrück,
Postfach 44 69, W-4500 Osnabrück, Germany

Abstract. We report measurements of the molecular axis orientation dependence of dissociative scattering of 2.98 keV H_2^+ and H_2 molecular beams from Cu(111) under glancing incidence conditions (perpendicular energies from 0.2-4 eV). The results show that neutralization of H_2^+ at Cu(111) occurs predominantly to the $X^1\Sigma_g^+$ state, and that the neutralization step is orientation dependent.

1. Introduction

We have recently described a technique for the direct measurement of the molecular axis orientation dependence of dissociative scattering of molecular beams from surfaces, and have used it to investigate the dissociative neutralization of H_2^+ at a Cu(111) surface [1-3]. The spectra representing the molecular axis orientation dependence of the distribution of kinetic energy released in the dissociation event were shown to exhibit two distinct structures. We tentatively correlated these structures with scattering on two distinct regions of the potential energy hypersurface (PES) describing the interaction. Furthermore, the relative contribution of each structure to the total spectrum exhibits a threshold-like behaviour as a function of the perpendicular energy of the beam to the surface in the range 1-2 eV. The perpendicular energy at which this threshold appears was found to be dependent on the molecular axis orientation. We correlated this threshold with a barrier on the PES, and the observed dependence on orientation with an orientation dependent barrier height. The proposed PES thus has a topography similar to that of the H_2/Cu adiabatic PES calculated by Harris and Andersson [4].

Central to our interpretation of the data is the assumption that charge transfer occurs to the H_2 singlet ground state on the incident trajectory, before the molecule reaches the supposed barrier on the PES. An alternative explanation for the existence of two distinct structures in the observed spectra could be that both the H_2 singlet and triplet states ($X^1\Sigma_g^+$ and $b^3\Sigma_u^+$) are accessed on the incident trajectory, and that the two structures observed simply represent dissociation via these two charge transfer channels. This latter possibility can be checked by comparing the spectra obtained using H_2^+ and H_2 incident beams under otherwise identical conditions. We report such a comparison in this paper.

2. Experiment

The experiments were performed under UHV conditions using 2.98 keV H_2^+ and H_2 beams incident at grazing angles θ_i to a carefully prepared Cu(111) surface. The neutral H_2 beam was generated by charge exchange by passing the H_2^+ beam through a gas cell containing H_2. The gas cell was operated under single collision conditions. The detection technique has been described in detail elsewhere [1-3]. In principle, we measure the final laboratory-frame velocities of the dissociation products and from them calculate the relative velocity vector $\vec{v} = \vec{v}_2 - \vec{v}_1$ of the atoms. The direction of $\vec{v}$ gives us the final orientation θ of the molecular axis to the surface normal (after all interaction between the molecular fragments and the surface has ceased). Its magnitude defines the kinetic energy $\epsilon = \frac{1}{2}\mu v^2$ released in the dissociation event (μ is the reduced mass of the molecule). We plot the data in the differential form $d^2 N(\epsilon, \theta)/d\epsilon d\omega$, where $d\omega$ represents a solid angle element whose size is independent of ϵ and θ. For simplicity we call these $N(\epsilon, \theta)$ distributions.

3. Results

We have measured the $N(\epsilon, \theta)$ distributions (representing coincidences between neutral hydrogen atoms) obtained using 2.98 keV H_2^+ and H_2 beams at incidence angles $\theta_i = 0.47, 0.88, 1.05, 1.17, 1.28, 1.48, 1.81$ and $2.09°$ corresponding to incident beam perpendicular energies of $E_{\perp i} = 0.2, 0.7, 1.0, 1.25, 1.5, 2.0, 3.0$ and 4.0 eV. A subset of this data is shown in Fig. 1.

4. Discussion

The $N(\epsilon, \theta)$ distributions obtained using H_2^+ and H_2 beams are qualitatively similar. The only significant difference is that higher orientations are more strongly favoured in the H_2 beam distributions. This result provides strong evidence that the $b^3\Sigma_u^+$-state plays no significant role in the charge transfer to H_2^+ at the Cu(111) surface. The $N(\epsilon, \theta)$ distributions are composed of two components; namely, a broad "peak" superimposed on a "background" structure whose shape is independent of $E_{\perp i}$ [1-3]. We can (after suitable normalization) subtract the "background" distribution from those $N(\epsilon, \theta)$ distributions containing the second structure ("peak"). We have used the spectra corresponding to $E_{\perp i} = 4$ eV (Fig. 1) for this "background" subtraction. In this way we can generate spectra of the "peak" alone (Fig. 2).

We see a strong influence of the initial charge state of the beam on the orientation distribution of both the "peak" and "background" structures. Provided the orientation distribution of both the H_2^+ and H_2 incident beams is isotropic, the change in the shape of the orientation distribution of both the "peak" (at very small $E_{\perp i}$) and "background" structures reflects the orientational anisotropy of the charge transfer to the $H_2 X^1\Sigma_g^+$ state at the Cu(111) surface.

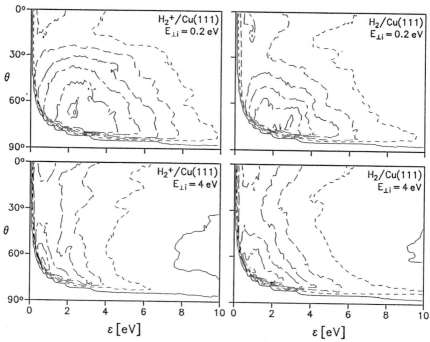

Fig. 1: Distribution of released kinetic energy ϵ as a function of the final orientation θ of the molecular axis to the surface normal for dissociative neutralization of 2.98 keV H_2^+ and for dissociation of 2.98 keV H_2 at a Cu(111) surface for incident beam perpendicular energies $E_{\perp i}$ of 0.2 and 4 eV and incident beam azimuth $\varphi = -20°$ to the [1$\bar{1}$0] direction.

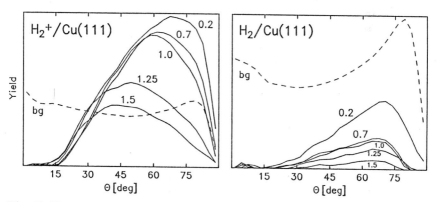

Fig. 2: Dependence of the yield in the "peak" (see text) on orientation angle θ for several values of $E_{\perp i}$ (in eV). The curves labeled "bg" represent the θ-dependences of the "background" structures.

5. Conclusion

We have compared the molecular axis orientation dependence of dissociative scattering of 2.98 keV H_2^+ and H_2 beams for incident beam perpendicular energies from 0.2-4 eV. These measurements conclusively demonstrate that charge transfer to H_2^+ at a Cu(111) surface occurs predominantly to the $X^1\Sigma_g^+$ ground electronic state, and furthermore, that the charge transfer is orientation dependent. The observed orientation dependence is consistent with that expected for a state of this symmetry [5].

6. Acknowledgements

We wish to thank J. Los and W. van der Zande for enlightening discussions. This work was generously supported by the Deutsche Forschungsgemeinschaft.

7. References

[1] J.-H. Rechtien, R. Harder, G. Herrmann, C. Röthig, K. J. Snowdon, Surf. Sci. 269/270 (1992) 213

[2] J.-H. Rechtien, R. Harder, G. Herrmann, K. J. Snowdon, Surf. Sci. (in press)

[3] R. Harder, G. Herrmann, J.-H. Rechtien, K. J. Snowdon, Surf. Sci. (submitted)

[4] J. Harris, S. Andersson, Phys. Rev. Letts. 55 (1985) 1583

[5] G. H. Dunn, Phys. Rev. Letts. 8 (1962) 62

Desorption from LiF(100) by Singly- and Doubly-Charged Hyperthermal He Ions

T. Neidhart, M. Schmid, and P. Varga

Institut für Allgemeine Physik, T.U. Wien,
Wiedner Hauptstraße 8–10, A-1040 Wien, Austria

Sputtering of a LiF(100) surface by singly- and doubly-charged He ions with impact energies between 10 and 500 eV has been performed. The yield of sputtered Li^+ and F^- ions is only slightly higher for doubly-charged ions compared to singly-charged projectiles at impact energies below 100 eV, whereas the F^+ yield is substantially increased if doubly-charged projectiles are used. The experimental data are explained within the framework of a model [1,2] based on calculations by Walkup and Avouris [3].

1 Introduction

Most experimental information on sputtering concerns kinetic processes on metal surfaces [4]. Only few data are available for insulating targets, and information about the conversion of potential energy (ionization energy) of the primary projectile into kinetic energy of sputtered particles is still very scarce.

On the basis of experimental results [5,6], Bitensky and Parilis [7] have developed the model of "Coulomb explosion" for sputtering of insulators. Williams [8] reported on ion-stimulated desorption of F^+, where an interatomic Auger process, similar to the Knotek-Feibelman mechanism [9] in electron stimulated desorption, was proposed to explain the relatively high energy of emitted F^+. In Ref. [10] a relationship between the potential energy of the primary ion and the F^+ yield has been demonstrated for LiF sputtering. With singly-charged He and Ne ions, a constant sputtering yield down to zero impact energy was extrapolated from measurements with ion energies between 200 and 2000 eV, and connected to an inter-atomic Auger transition between the projectile and the F^- ion at the surface. In this process, the F^- is doubly ionized to F^+ and becomes desorbed because of the repulsive Madelung potential.

We have set up a UHV experiment where singly- and doubly-charged ions with kinetic energy from 500 eV down to 10 eV can be used for ion-surface interaction studies [11,12,13]. We use a quadrupole mass spectrometer (QMS) to detect particles sputtered at an angle of 90° by a mass-filtered ion beam. The target was a LiF(100) single crystal heated to 400°C, where LiF is a good ionic conductor and a stoichiometric surface is restored, whereas it is Li-enriched if sputtered at room temperature [14].

In the present work we seek further insight into the sputtering of insulators like LiF due to the conversion of projectile potential energy into kinetic energy of sputtered particles. We show that the formerly [10] extrapolated constant sputtering yield for F^+ down to zero impact energy for ions with sufficiently large potential energy can not be confirmed.

2 Results and discussion

For all projectiles, the dominantly emitted charged species were Li^+ and F^- ions. Furthermore, cluster ions like Li_2^+, LiF^+, Li_2F^+, F_2^-, LiF^- and LiF_2^-, have also be detected. To study the influence of the projectile potential energy, emission of Li^+, F^+ and F^- has been studied systematically.

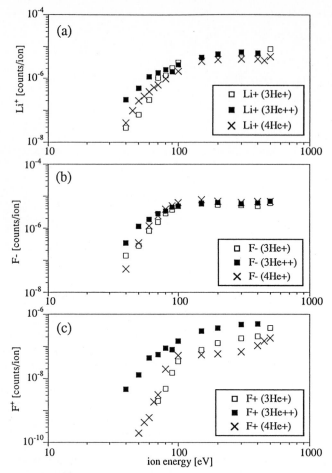

Fig.1. Emission of (a) Li^+, (b) F^- and (c) F^+ induced by impact of $^4He^+$, $^3He^+$ and $^3He^{++}$ on LiF(100) at 400°C target temperature.

In fig. 1a the influence of the primary ion charge state on the Li^+ emission is presented for impact of singly- and doubly-charged He ions with different impact energies. Only below 100 eV a slightly higher Li^+ yield for doubly-charged He^{++} can be observed, and a sputtering threshold energy of about 25 eV can be extrapolated. For singly-charged projectiles, the threshold energy is slightly higher. These data show that the emission of all Li particles is a classical sputter process, dominated mainly by momentum transfer, and the potential energy of the projectile is not of great influence. The difference at low impact energy may be caused by the huge difference in F^+ emission discussed below; but in principle no complete explanation can be given for this phenomenon. The slightly higher Li^+ yield of $^4He^+$ with respect to $^3He^+$ is also a clear indication for the momentum transfer during sputtering.

For F^- emission (fig. 1b), this isotope effect can not be seen because of the larger mass difference between projectile and target atom. The F^- yield is approximately equal to the Li^+ yield, which indicates a stoichiometric surface composition. Again, a slight difference between He^+ and He^{++} is observed at impact energies below 100 eV.

In fig. 1c results for F^+ emission induced by impact of He^+ and of He^{++} are compared. From these results a clear influence of the potential energy on the emission of F^+ is evident. The binding energy of the fluorine 2p valence band is between 11 eV and 16 eV [15]. Therefore, He^+ can ionize F^- to F^+ only by one Auger-neutralization (AN) process, whereas He^{++} is neutralized at first into He^+ by resonance-neutralization (RN) into the $2s^2S$ state, followed by Auger deexcitation and Auger neutralization. This can explain why the yield for He^{++} is higher than for He^+. Nevertheless, both yields decrease monotonically with decreasing impact energy leading to a sputtering threshold below 30 eV for He^{++} and about 50 eV for He^+.

From this data, it is evident that the potential energy of the projectile is of great importance for the production of F^+, but in addition a minimum momentum transfer seems to be necessary for the emission process as stated previously [1,2]. The repulsive energy between the F^+ particle and the surface lattice is obviously not sufficient for the emission.

Walkup and Avouris [3] have performed classical trajectory calculations on the behavior of an alkali halide lattice if one of the negative halide ions of the surface (e.g. F^- in NaF) is suddenly changed into a positive one. In their calculations, it is clearly shown that the initial large repulsive energy of 11 eV between F^+ and the NaF-lattice becomes slightly attractive within 15-30 fs, due to lattice distortions where the Na^+ ions are pushed back and the F^- are attracted. This time is too short for desorption of the F^+, which therefore finally becomes trapped in a stable state at the surface, with a binding energy of 4 eV in a bridge position between two F^- surface atoms.

These results may be applied to low energy ion sputtering as well, because the electron transition processes (AN or doubly RN) take place when the projectile ion is still several Å away from the surface [16,17]. Therefore, the sudden change from F^- into F^+ will be similar to the formation of F^+ by electron impact, with a consequently similar reaction of the lattice. However, one has to consider the subsequent impact of the neutralized particle which can change the rearrangement of the lattice. The low current densities used in the present experiments exclude formation and emission of F^+ by two different subsequently arriving projectiles. Assuming that NaF is not principally different from LiF as already indicated in [1,2], we want to apply the time scale given above to our experiment.

For metal surfaces, the transition rate for the above mentioned electron transfer from the surface valence band to an ionized particle in front of the surface depends exponentially on the surface-particle distance, because it is governed by the overlap of the respective atomic wave functions. The transition rate is independent of the ion impact energy but the distance where the process most probably takes place moves closer to the surface with increasing particle velocity. In a rough approximation [16,17], the transitions start at a distance in front of the surface which is about three times the initial charge state in Ångströms. This means about 3 Å for singly-charged ions and 6 Å for doubly-charged ones. This distance may be smaller for alkali halides where more localized wave functions have to be assumed because of the larger binding energy of the F2p electrons, as compared to valence electrons in metals.

He^{++} with an energy of 100 eV touches the surface within 10 fs (depending on the impact angle), after it has been neutralized. Therefore the impact of He^0 takes place when the lattice is still reacting and the F^+ sees still a repulsive potential. For lower impact energy (longer time between neutralization and impact), the F^+ ion may feel an increasing binding energy and additionally the available kinetic energy of the projectile decreases. As a consequence the emission decreases and reaches its threshold at some 30 eV.

At higher impact energies, the projectile hits the surface already when the F^+ is still in a repulsive state. Finally, the kinetic processes will dominate and the electronic transitions will become more and more unlikely (time scale is too short for the transitions). Therefore a steep increase in the emission yield is observed, and the relative differences between He^+ and He^{++} become smaller with increasing impact energy.

3 Conclusion

Sputtering of a LiF(100) surface with hyperthermal singly- and doubly-charged He ions has been studied. For Li^+ and F^- ejection, only a small influence of the charge state has been observed at impact energies below 100 eV. In contrast to this, it has been shown that sputtering of F^+ strongly depends on the potential energy of the projectile. This can be explained by assuming that in this low energetic impact region, F^+ is formed by an Auger neutralization process between the lattice F^- ion and the He ions, since the potential energy of the projectiles exceeds more than two times the band gap at the surface. From the experimental data it is also evident that for the emission of the F^+ ions a minimum value of momentum transfer from the projectile to the ion to be emitted is necessary, and the repulsive potential between the lattice and the F^+ alone is not able to eject the projectile from the surface.

This work was supported by the Austrian Fonds zur Förderung der wissenschaftlichen Forschung (project P 8969)

References

[1] P.Varga, U.Diebold and D.Wutte, Nucl.Instr.Meth.B 58 (1991) 417
[2] D.Wutte, U.Diebold, M.Schmid and P.Varga, Nucl.Instr.Meth.B 65 (1992) 167
[3] R.E.Walkup and Ph.Avouris, Phys.Rev.Lett.56 (1986) 524
[4] R.Behrisch, ed., Sputtering by Particle Bombardment 1 & 2, Topics Applied Physics Vol .47 & 52 (1981,1983) Springer
[5] Sh.S. Radzhabov, R.R. Rakhimov and D.Abdusalomov, Izv. Akad.Nauk SSSR,Ser.Fiz. 40(1976) 2543
[6] S.N.Morozov, D.D.Gruich and T.U.Arifov, Iz.Akad.Nauk SSR, Ser.Fiz., 43 (1979) 612
[7] I.S.Bitensky, M.N.Murachmedov and E.S.Parilis, Zh.Techn.Fiz.49 (1979) 1042
    I.S.Bitensky and E.S.Parilis, Journ.de Physique C2 (1989) 227
[8] P.Williams, Phys.Rev.B 23 (1981) 6187
[9] M.L.Knotek and P.J.Feibelmann, Phys.Rev.Lett.40 (1978) 964
[10] J.A.Schultz, P.T.Murray, R.Kumar, Hsin-Kuei Hu and J.W. Rabalais, Springer Series in Chemical Physics 24 (1983) 191, "Desorption Induced by Electronic Transitions, DIET I", eds.: N.H.Tolk, M.M.Traum, J.C.Tully and T.E.Madey
[11] U.Diebold, thesis T.U.Vienna 1990, unpublished
[12] U.Diebold and P.Varga, Surface Sci.Lett. 241 (1991) L6
[13] U.Diebold and P.Varga, in Springer Series in Surface science Vol.19 (1990) 193, "Desorption Induced by Electronic Transitions, DIET IV", eds. G.Betz and P.Varga
[14] P.Wurz, C.H.Becker, Surf.Sci.Lett.224 (1989) L559
[15] M.Piacentini and J.Anderegg, Sol.State Comm.38 (1981) 191
[16] P.Varga, Appl.Phys. A 44 (1987) 31
[17] P.Appell, Nucl. Instr. Meth. B 23 (1987) 242

Ion-Induced Electron Excitation at Magnetic Surfaces

C. Rau, N.J. Zheng, M. Rösler*, and M. Lu

Department of Physics and Rice Quantum Institute, Rice University,
Houston, TX 77251, USA
*Permanent address: KAI, W-1000 Berlin, Germany

Abstract. Using spin-polarized electron emission spectroscopy (SPEES), we have studied electronic excitations at clean and oxygen-covered Fe surfaces. Employing small angle ($\alpha=1^{\circ}$) surface interaction of 7-28 keV Ne^{+} ions, we investigated the spin-polarized, angle-resolved energy distribution (ARED) of electrons emitted from these surfaces. For small α, the ARED is significantly different from electron- or ion-induced secondary electron spectra obtained at larger α. We note that 25 keV Ne^{+} ions incident at $\alpha=1^{\circ}$ are specularly reflected and probe the <u>topmost</u> surface layer whereas for $\alpha>2^{\circ}$, the ions penetrate the surface thereby probing <u>bulk</u> layers. At clean Fe surfaces, we find for $\alpha=1^{\circ}$ and 45° average values of the electron spin polarization (ESP) of (33 ± 2)% and (25 ± 2)%. These values show that the surface magnetization (33%) is enhanced by approximately 30% compared to the bulk value (25%). These findings give clear evidence that SPEES is a powerful technique to study <u>layer-dependent</u> magnetic properties.

1. Introduction

Ion and electron scattering experiments at clean and adsorbate-covered surfaces are powerful means to investigate their physical properties, in particular their magnetic, electronic and chemical properties.

Despite many breakthroughs in the interpretation of data [1-5] obtained from particle-surface interaction experiments at magnetic/nonmagnetic surfaces, there is still no fundamental understanding of how these data are linked to the band structure of clean and adsorbate-covered surfaces.

A promising approach to elucidate details of the physics of electronic processes involved in ion-surface interaction includes not only the determination of the angle-resolved energy distribution (ARED) of emitted electrons but also the detection of sign and magnitude of the electron spin polarization (ESP) which can be used as an additional "label" to identify various processes occuring in ion-surface interaction processes [8,6,7,9].

Recently, we reported on SPEES data from magnetic Ni(110) surfaces [6,7]. There is evidence that the ARED and the ESP of the emitted electrons contains most valuable information on the spin-polarized surface electronic structure of Ni(110) and on the physics of various electron excitation processes. occuring during ion-surface interaction.

In this paper, we present information about SPEES experiments at clean and oxygen covered Fe surfaces. Changing α from 1° up to 45° allows us to vary the <u>probing depth</u> of the incident ions from the topmost surface layer to deeper layers. The use of Ne^{+} ions instead of H^{+} ions allows us to connect our data to data obtained using ion neutralization spectroscopy (INS) [1]. Using $\alpha=45^{\circ}$ allows us to link our SPEES data to well-known electron- or ion-induced secondary electron spectra [2-5].

Springer Series in Surface Sciences, Vol. 31
Desorption Induced by Electron Transitions DIET V
Editors: A.R. Burns · E.B. Stechel · D.R. Jennison   © Springer-Verlag Berlin Heidelberg 1993

2. Experimental

Experimental details are reported elsewhere [6,7]. We give only a brief discussion. In SPEES, surface scattering of energetic (5-150 keV) ions is used to study the emission of spin-polarized electrons occuring during particle-surface interaction. Using an einzel lens, we detect electrons emitted along the surface normal (emission cone angle 11°) of a remnantly magnetized ferromagnet. For energy analysis an electrostatic energy analyzer is used. For spin analysis, the energy-analyzed electrons are accelerated to 150 eV and enter a low-energy ESP detector which allows to measure the component of the ESP, P, parallel to the target magnetization. We note that P>0 is related to a predominance of so-called majority-spin electrons (ESP parallel to the total magnetization), and P<0 refers to a predominance of minority-spin electrons (ESP antiparallel to the total magnetization) [10].

3. Results and Discussion

Figure 1 gives for $\alpha=1°$ and Ne^+-ion energies $E_0=7-28$ keV the ARED of electrons emitted normal to the reflecting Fe surfaces as function of the electron energy E above the vacuum level. We note that the distance of closest approach of the ions towards the Fe surface is characterized by the energy component $E_\perp =E_0\sin^2\alpha \approx E\alpha^2 =2.1 -8.5$ eV for $\alpha=1°$ and $E_0 =7-28$ keV. Using appropriate surface potentials [10], it can be shown that the Ne^+ ions are specularly reflected and do not penetrate the Fe surfaces. We note that our ARED for 7 keV Ne^+ ions ($E_\perp =2.1$ eV) closely resembles the ARED (see Fig.2 in Ref.1), obtained in ion neutralization spectroscopy (INS) (two electron Auger-type processes) using slow noble gas ions such as (5 eV Ne^+) at Cu(110) surfaces [1]. Increasing in our experiments the ion beam energy up to 28 keV results in a narrowing of the ARED which may be due to a transition from potential (Auger-type) to predominantly kinetic (direct ion-induced electronic excitation) electron emission [11,2].

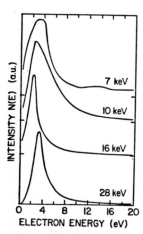

Fig.1: ARED of electrons emitted from Fe surfaces for $\alpha=1°$ and Ne^+-ion energies $E_0=7-28$ keV.

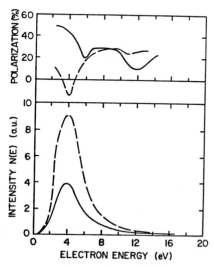

Fig.2: ARED and ESP of electrons emitted from clean (solid line) and and oxygen-covered (dashed line) Fe for 25 keV Ne^+ ions and $\alpha=1°$.

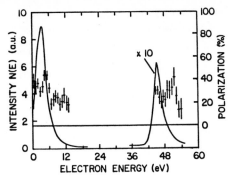

Fig.3: ARED and ESP of electrons emitted from clean Fe surfaces for 25 keV
Ne[+] ions and $\alpha=45°$.

Figure 2 shows for 25 keV Ne[+] ions and $\alpha=1°$ the ARED and the ESP of
electrons emitted from clean and oxygen-covered Fe surfaces. As regards the
ESP of electrons excited at the topmost surface layer of clean Fe, we observe an increase of the average ESP from P=33% (E=10 eV) to P=48% (E=4
eV). These values are far above the bulk magnetization of Fe which amounts
to 28%. For Fe surfaces with one monolayer of O, the average ESP amounts to
32% for E=10 eV and decreases to -14% for E=4 eV implying the existence of
O bands polarized antiparallel to the Fe magnetization.

In the following, we discuss these values in connection with our SPEES
data using $\alpha=45°$ where bulk layers are probed.

Fig.3 gives for 25 keV Ne[+] ions and $\alpha=45°$ the ARED and the ESP of
electrons emitted from clean Fe surfaces. The ARED is similar to that obtained in electron-induced secondary electron emission experiments [3-5].
We observe an increase of the average ESP from P=25% for E=10 eV to P=45%
for E=4 eV. Fig.3 shows a pronounced peak in the ARED at around 45 eV which
can be attributed to emitted spin-polarized MVV Auger electrons [12,7] with
an average ESP of 30% which is close to the bulk magnetization of Fe (28%).

At present, it is of considerable interest of whether the ESP reflects
the <u>layer-dependent</u> net magnetization of Fe. From ion- [5,8,11,7] and
electron-induced spectra [4], there is evidence that the ESP of electrons
emitted at high energies (≈10 eV above vacuum level) scales roughly with
the average net magnetization. We find an increase in the ESP of "high-
energy" electrons from 25% to 33% by changing α from 45° to 1° which corresponds to a reduction in the probing depth from bulk layers to the surface layer implying that for polycrystalline Fe surfaces, the net magnetization increases in going from the bulk to the surface. It is tempting to
correlate this surface enhancement (≈30%) of the ESP to the influence of
theoretically predicted magnetic surface states which cause enhancements of
the magnetization of 32% and 20% for Fe(100) and Fe(110) surfaces [13].

From Figs.2 ($\alpha=1°$) and 3 ($\alpha=45°$) it is obvious that the ESP of electrons
emitted from clean Fe increases with decreasing (≈10 eV → ≈3-4 eV) electron
energy. For $\alpha=1°$, the average ESP increases from 33% to 48% and for $\alpha=45°$,
the average ESP increases from 25% to 40%.

We believe that these enhancements of the ESP are not predodminantly due
to the spin dependence of the electron mean free path [14] which is caused
by an excess of unfilled minority-spin d states over unfilled majority-spin
d states in which excited electrons can be scattered during transport to
the surface. Assuming that electron transport processes are less important
for electrons excited at the surface than for electrons excited at subsurface layers, we correlate these ESP-enhancements for low-energy electrons
to Stoner excitations (electron hole pairs with antiparallel spin orientation) which occur during inelastic electron exchange-scattering processes.

135

[4,5] Further details as regards the fine structure of the ESP spectra, which can be associated to details of the spin-polarized band structure above the vacuum level [15], will be reported elsewhere [16].

We discuss the ESP of electrons emitted from oxygen-covered Fe surfaces (see Fig.2). For $E \approx 10$ eV, the ESP amounts to 32%, which is close to 33% found at clean Fe surfaces, and decreases to -14% ($E \approx 3-4$ eV). These findings reveal the absence of a magnetic dead layer at Fe surfaces covered with one covalently bound O layer [17] and the existence of spin-split bands in the occupied and unoccupied parts of the band structure of oxygen-covered Fe surfaces [12,18,19]. For further details, we refer to Ref. 16.

This research has been supported by the National Science Foundation, by the Welch Foundation and by the Texas Higher Education Coordinating Board.

4. References

1. H.D. Hagstrum, Phys. Rev.150, 495 (1966).
2. D.E. Harrison, Jr., C.E. Carlston, and G.D. Magnuson Phys. Rev. 139, 737 (1965).
3. R.A. Baragiola, E.V. Alonso, and A. Oliva-Florio Phys. Rev. B61, 121 (1979).
4. M. Landolt, in "Polarized Electrons in Surface Physics", ed. by R. Feder (World Sci. Publ. Co., 1985), Chap.9; and Refs. cited therein.
5. J. Kirschner, in "Surface and Interface Characterization by Electron Optical Methods" ed. by A. Howie and U. Valdre (Plenum Publ. Co., 1988), p. 297; and Refs. cited therein.
6. C. Rau and K. Waters, Nucl. Instr. Meth. in Phys. Res. B40, 127 (1989).
7. C. Rau, K. Waters and N. Chen, Phys. Rev. Lett. 64, 1441 (1990).
8. J. Kirschner, K. Koike, and H.P. Oepen, Phys. Rev. Lett. 59, 2099 (1987).
9. M.S. Hammond, F.B. Dunning, and G.K. Walters, Phys. Rev. B45, 3674 (1992).
10. C. Rau, J. Magn. Magn. Mater 30, 141 (1982).
11. J. Kirschner, K. Koike, and H.P. Oepen, Vacuum 41, 818 (1990).
12. M. Landolt, Appl. Phys. a41, 83 (1986); and M. Landolt, R. Allensbach, and M. Taborelli, Surface Science 178 311 (1986).
13. C.L. Fu, A.J. Freeman, and T. Oguchi, Phys. Rev. Lett. 54, 2700 (1985).
14 D. Penn and P. Apell, and S.M. Girvin, Phys. Rev. Lett. 55, 518 (1985).
15. E. Tamura and R. Feder, Phys. Rev. Lett. 57, 759 (1986).
16. C. Rau and N. J. Zheng, to be published.
17. S.R. Chubb and W.E. Pickett, Phys. Rev. Lett. 58, 1248 (187).
18. R. Allensbach, M. Taborelli, and M. Landolt, Phys. Rev. Lett. 55, 2599 (1985).
19. A. Clarke, N.B. Brookes, P.D. Johnson, M. Weinert, B. Sinkovic, and N.V. Smith, Phys. Rev. B41, 9659 (1990).

Part V

Theory

Excited States Lifetimes Near Surfaces

P. Nordlander

Department of Physics and Rice Quantum Institute, Rice University,
Houston, TX 77030-1892, USA

Abstract. Results from calculations of energy shifts and broadening of neutral and negative ion states near clean and impurity covered metal surfaces are presented. The results of these calculations shows that, in general, several different atomic states can be populated close to the surface. It is shown that for the proper description of charge transfer in such situations, it is crucial to include the effects of intra-atomic electron correlation.

1. Introduction

The energies and lifetimes of excited states of atoms and molecules near metal surfaces are controlling factors in many dynamical phenomena at surfaces.[1] From experimental studies of charge transfer processes in atom-surface scattering and stimulated desorption processes, valuable microscopic information about the surface electronic structure can be deduced.[2-4]

In this paper, the results of two recent calculations of level shifts and broadenings of atomic resonances near metal surfaces will be presented. The discussion will be limited to broadening induced by one-electron effects such as resonant tunneling. In the physisorption regime, i.e. $Z>5$ a.u., the relative contribution from two-electron effects is expected to be negligible.

A calculation of the energy shifts and broadenings of the electronic levels of a hydrogen atom near chemisorbed impurities on a metal surface has shown that the presence of impurities on metal surfaces can drastically modify both the energy shifts and the widths of the electronic levels in atoms near an impurity. When electropositive adsorbates, such as alkali atoms, are present on the surface, the energy levels of a neutral hydrogen atom shift downwards in their vicinity.[5] Such a downshift enables resonant tunneling into several different excited atomic states. In the second calculation, the energy shift and broadening of atomic affinity levels near metal surfaces was studied.[6] It was shown that negative ion states also shift downwards near metal surfaces. The widths of the affinity levels were typically found to be relatively large and negative ion states can therefore readily be populated by resonant tunneling near the surface.

When several atomic levels are positioned below the Fermi energy, tunneling may occur into any of these levels. In typical situations, the intra-atomic correlation will prevent more than one of these states to be occupied. The resulting charge state will thus be determined by a combination of tunneling and correlation effects. The temporary formation of negative ion states or excited states at certain positions near the metal surface can prevent the formation of atomic states of lower energy.[7] These effects lead to strong non-adiabatic behavior and cannot be properly understood without the detailed knowledge of how both the excited neutral states and the negative ion states of an atom shift and broaden as a function of atom-surface separation.

In section 2, the calculated level shift and broadenings of atomic levels near metal surfaces is presented. In section 3, the effects of intra-atomic correlation on charge transfer processes will be discussed.

2. Calculations of the energy of atomic resonances near surfaces.

The broadening and shift of atomic levels near the surface is induced by the surface electron potential. The potential energy of an electron at position $\vec{r}$, outside a metal surface can be written as a sum of different contributions:[5]

$$V^{eff} = V_0(z) + \Delta V_A + \Delta V_{AS} \tag{1}$$

The first part of the potential, V_0, describes the bare electron-surface interaction. The surface may be clean, or it may contain chemisorbed impurities which induce strong local modifications of the surface potential. The second term, ΔV_A, is the bare atomic potential and the surface image of the atomic core. For the description of atoms other than hydrogen, a pseudopotential must be used in ΔV_A. The third term, ΔV_{AS}, describes the change in the potential due to the interaction of the negative ion with the surface. In the present calculations, the atom-surface distances will be assumed large ($Z > 5$ a.u.), so that the ΔV_{AS} term can be neglected.[5]

The energies of atomic resonances near surfaces can be calculated by solving the Schrodinger equation for the electrons:[8]

$$[-\frac{1}{2}\nabla^2 + V^{eff}]\psi = \epsilon\psi \tag{2}$$

under resonance boundary conditions:

$$\psi(r) \rightarrow \frac{1}{r} \exp[ik_R r + k_I r]f(\Omega). \tag{3}$$

The energy of the resonance is related to the complex wavenumber k through $\epsilon = -\frac{1}{2}(k_R + ik_I)^2$. The real part of the energy, ϵ_R, describes the energy of the level and the imaginary part, $\frac{\Gamma}{2}$, is the half-width of the resonance.

Resonances can be directly obtained from Eq.(2) by extending the coordinates to the complex plane.[8] The idea here is to introduce a complex variable substitution in the radial coordinate, $r \rightarrow e^{i\theta}r$. If θ is chosen larger than $\arctan\frac{k_I}{k_R}$, the Hamiltonian can be diagonalized using a normalizable basis.

In Fig. 1, the calculated shifts and broadening of the H(n=2) levels as a function of atom-surface separation Z is shown outside a clean Al surface ($r_s=2.07$) and outside a chemisorbed Na impurity. The hybridization with the surface results in a state oriented towards the surface, ψ_2^+, and a state oriented away from the surface, ψ_2^-. The 2p$_{x,y}$ states do not mix. These states are denoted ψ_2^0. For hydrogen outside a clean metal surface the image potential will shift all levels upwards with decreasing Z. For hydrogen outside a chemisorbed sodium impurity, the sodium-induced dipole field is so strong that the image shift is reversed resulting in a downshift of the atomic levels near the surface.

For a correct description of the physical properties of negative ions, it is important that the calculated electron wavefunctions have the right asymptotic behavior.

$$\psi_{lm}(r) \longrightarrow B\sqrt{2\alpha}\frac{\exp(-\alpha r)}{r}Y_{lm}(\Omega) \quad , \tag{4}$$

where the energy of the negative ion state, $\epsilon_a = -\frac{1}{2}\alpha^2$.

The asymptotic parameters α, B and l for different negative ions have been calculated and tabulated.[9] The asymptotic tail of the wavefunction determines

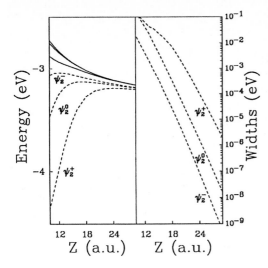

Figure 1: Calculated energy level shifts as function of distance from the surface for the H(n=2) states outside a clean metal surface (solid lines) and outside a Na atom chemisorbed on the surface (dotted lines). The curves for the states ψ_2^+, ψ_2^0 and ψ_2^- are labeled. The units are a.u. and the metal is Al jellium.

many of its physical properties such as the polarizability, cross sections for electron attachment and detachment as well as probabilities for charge transfer in collisions between negative ions and atoms.[9]

From an accurate parameterization of the wavefunction, $\psi(\vec{r})$, of the negative ion state, a pseudo-potential can be derived.[6] In the separable potential method this pseudo-potential has the form:

$$V^A = \lambda |\phi><\phi| \tag{5}$$

The function $< \phi|\vec{r} >$ is assumed normalized. The function $\phi(\vec{r})$ can be obtained from the wavefunction of the negative ion state ψ through

$$|\phi> = \frac{1}{2\lambda < \phi|\psi >}(\nabla^2 - \alpha^2)|\psi > . \tag{6}$$

The constant λ is obtained from

$$\lambda = \frac{< \phi|\nabla^2 - \alpha^2|\psi >}{2 < \phi|\psi >}. \tag{7}$$

In Fig. 2, the calculated energy shifts and broadening for an oxygen negative ion are plotted. It can be seen that the levels shift downward with decreasing atom-surface separation. The downshift follows the bare surface potential, $V_0(z)$, in Eq. (1). The broadening is exponential, which can be expected for a single level interacting with a surface. The oxygen states are of p-character and consequently different orientations with respect to the surface are possible. The m=0 state is oriented towards the surface and the m=±1 states are oriented parallel to the surface. Due to the high symmetry of a jellium surface, the m=± 1 states will remain degenerate. These different states will both shift and broaden differently as a function of atom-surface separation. At large atom-surface separations, the m=0 state lies below the m=1 states. This is due to the larger overlap with the surface electron potential. As the surface is approached, the states cross and the m=1 states have lowest energy. Since the surface potential is smoothly varying, the energy splitting between the different orientations is relatively small. At a distance of Z=5 a.u., the energy difference

141

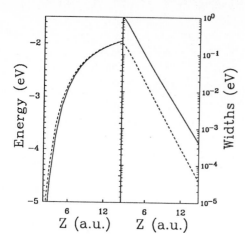

Figure 2. Calculated energy shift (left) and broadening (right) of O^- as a function of distance from an Cu surface ($r_s=2.67$). The solid curve refers to the m=0 state and the dotted curve is the m=$\pm$1 state. The energy unit is eV and the distance Z is measured from the jellium edge.

between the m=0 and m=$\pm$1 is around 0.2eV. The broadening of the atomic levels on the other hand depend strongly on the orientation of the state.[8] This can be seen clearly from the figure where the m=$\pm$1 states are more than one order of magnitude more long lived than the m=0 state.

It can be seen that the widths of the negative oxygen states therefore are very narrow for Z>5 a.u. It can also be seen that the orientation dependence of the width increases with increasing atom-surface separation. This effect has been discussed elsewhere.[10]

3. Effects of intra-atomic correlation on charge transfer processes.

When several atomic levels can be populated by resonant tunneling, it becomes important to include the effects of intra-atomic correlation. These effects are particularly important when levels shift down close to the surface as shown in the previous section. Close to the surface, the formation of shortlived excited states may prevent the formation of lower energy atomic states.[7] A convenient description of the influence of intra-atomic correlation effects in atom-surface scattering can be obtained using the time-dependent multiple-level Anderson model, slave-bosons and non-equilibrium Green's functions.[7] In the finite temperature and low velocity regime the population of the various atomic levels can be calculated by solving a set of coupled master equations. In the limit of strong correlation, for a system of N impurity levels with widths $\Gamma_i(t)$ and energies $\epsilon_i(t)$ interacting with a surface, these equations take the form:

$$\frac{dn_i(t)}{dt} = -\Gamma_i(t)[1 - f(\epsilon_i(t))]n_i(t) + 2\Gamma_i(t)f(\epsilon_i(t))[1 - \sum_i^N n_i(t)] \qquad (8)$$

where $f(\epsilon)$ is the Fermi-Dirac function. This expression clearly shows the importance of intra-atomic correlation. To formation of level $|i>$ is not possible if an excited state already has been formed, i.e., $\sum_i^N n_i(t) = 1$.

As an example of the importance of intra-atomic correlation effects, the charge transfer processes involved during the desorption of a hydrogen atom from a low work function metal surface will be modeled. The atomic states of importance are the H(n=2) states. The negative ion state will not be formed

142

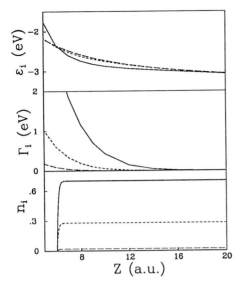

Figure 3: Energy shifts, broadenings and instantaneous populations of the different H(n=2) levels as function of atom-surface separation. The solid line refers to the $|\psi_2^+>$ state, the dotted line is the $|\psi_2^0>$ states and the dashed lines refer to the $|\psi_2^->$ state. The upper figure shows the energy levels $\epsilon_i(Z)$. The middle figure shows the widths $\Gamma_i(Z)$ and the lower figure shows the populations $n_i(Z)$. The kinetic energy of the desorbing H is assumed 1 eV. The work function of the metal is assumed 2 eV.

at distances larger than 2 a.u. from the surface.[6] As discussed in the previous section, the hybridization with the surface results in in a shortlived state oriented towards the surface and a longlived state oriented away from the surface. The lifetimes of the $2p_{x,y}$ states are intermediate. In Fig. 3, the calculated shifts $\epsilon_i(t)$, widths $\Gamma_i(t)$ and instantaneous populations $n_i(t)$ of these levels are plotted during the trajectory. At a distance of 6 a.u., the hydrogen levels shifts below the Fermi energy of the metal. Since the width of the $|\psi_2^+>$ state is very large, this state will be populated fastest. The states $|\psi_2^0>$ are populated at a slower rate. The $|\psi_2^->$ state is so longlived that almost nothing is formed before the total occupation probability of the hydrogen levels is one and the charge transfer stops. A calculation neglecting the intra atomic correlation would have predicted equal population of the different H(n=2) states. Recent experiments studying the relative population of the various H(n=2) states have shown that the most of the desorbing H leaves the surface in the $|\psi_2^+>$ state.[11]

4. Conclusions

The complex scaling technique have been extended to the calculation of the shifts and broadening of atomic affinity levels near clean metal surfaces and neutral excited states near impurities chemisorbed on metal surfaces. Such states are found to sometimes shift downwards with decreasing atom-surface separation. Such a downshift means that in general several different levels can be populated by resonant tunneling. For the proper understanding of charge transfer processes for such systems, it is crucial to include the effects of intra-atomic correlation. This work is supported by the National Science Foundation under grant No. DMR-9117479. Acknowledgment is made to the donors of the Petroleum Research Fund, administered by the American Chemical Society for the partial support of this research.

143

References

1. in *Desorption Induced by Electronic Transitions, Proc. Diet-IV*, edited by G. Betz and P. Varga, Springer-Verlag, Berlin, 1990.

2. P. D. Johnson, A. J. Viescas, P. Nordlander, and J. C. Tully, Phys. Rev. Lett. **64**, 942 (1990).

3. G. A. Kimmel, D. M. Goodstein, Z. H. Levine, and B. H. Cooper, Phys. Rev. **B 43**, 9403 (1991).

4. T. M. Orlando, A. R. Burns, D. R. Jennison, and E. B. Stechel, Phys. Rev. **B45** (1992).

5. P. Nordlander and N. D. Lang, Phys. Rev. **B 44**, 13681 (1991).

6. P. Nordlander, To appear in Phys. Rev. B **July 15** (1992).

7. D. C. Langreth and P. Nordlander, Phys. Rev. **B 43**, 2541 (1991).

8. P. Nordlander and J. C. Tully, Phys. Rev. **B 42**, 5564 (1990).

9. B. M. Smirnov, *Negative Ions*, McGraw-Hill, New York, 1982.

10. J. W. Gadzuk, Surf. Sci. **180**, 225 (1987).

11. K. D. Tsuei, P. D. Johnson, P. Nordlander, and D. C. Langreth, to be published (1992).

ESD from Physisorbed Layers on Metal Surfaces: Theoretical Investigation of Two Desorption Scenarios

Z.W. Gortel

Department of Physics and Theoretical Physics Institute,
University of Alberta, Edmonton, Alberta, Canada T6G 2J1

Abstract. Angle resolved kinetic energy distributions of ESD-desorbed neutrals physisorbed on metal surfaces are theoretically compared for the Antoniewicz and the Wave Packet Squeezing (WPS) scenarios of desorption. Three-dimensional evolution of the system in the electronic meta-stable state is accounted for. It is shown that the shapes of the angle dependent yields are very similar for both models and admit a variety of angular behavior from very narrowly centered around the surface normal to those in which the particles desorb predominantly away from it. In contrast, kinetic energy distributions, particularly in the direction of the surface normal, can be very different in both models and, in some cases of the Antoniewicz scenario, they may exhibit a multi-peak structure without different adsorption states being present.

1. Introduction

A particle adsorbed on the surface of a solid will desorb if it is supplied with enough energy to break its surface bond. In Desorption Induced by Electronic Transitions [1] this energy is initially stored in the electronic part of the solid-adsorbate system and is externally supplied by either the electron beam in Electron Stimulated Desorption (ESD), or by photons in Photon Stimulated Desorption (PSD). A series of electronic transitions may follow which, within some $10^{-16} s$, promote the system to a meta-stable excited electronic state in which the adsorbed particle is bound to the surface by the potential $V_d(\mathbf{r})$ different from $V_0(\mathbf{r})$ - the surface potential in the electronic ground state of the system. The wave function of the adsorbed particle is no longer stationary after the excitation and evolves in time according to the time dependent Schrödinger equation containing $V_d(\mathbf{r})$. Simultaneously, the excited electronic state is unstable due to its interaction with the continuum of delocalized electronic states in the solid, so eventually, after some 10^{-14} - 10^{-13} s, an electronic deexcitation occurs and the particle is returned to the original surface potential $V_0(\mathbf{r})$. If the kinetic energy gained by then by the particle is larger than its binding energy in $V_0(\mathbf{r})$ then it will eventually desorb. The amount and the distribution of the kinetic energies gained in the meta-stable state of the system determines the desorption yield and the kinetic energy distribution of the desorption products.

There exist, at present, two models describing the microscopic processes leading to desorption of neutral particles from physisorbed layers. In the first, due to Antoniewicz [2], it is assumed that an ion is created by the incoming electron beam and that it is then accelerated towards the surface by its Coulomb image. Classical mechanics accurately accounts for the resulting kinetic energy gain and is sufficient to describe the desorption process [3]. In the other desorption scenario, proposed only recently [4] and referred to as the Wave Packet Squeezing (WPS) model, one assumes that due to the electronic relaxation following immediately after the beam-induced ionization of the particle, the latter ends up in an excited state in which it is bound to the surface by a relatively strong chemical surface bond rather than by the image force. It is argued [4,5] that the equilibrium positions of $V_d(\mathbf{r})$ and $V_0(\mathbf{r})$ almost coincide so

Springer Series in Surface Sciences, Vol. 31
Desorption Induced by Electron Transitions DIET V
Editors: A.R. Burns · E.B. Stechel · D.R. Jennison   © Springer-Verlag Berlin Heidelberg 1993

the *classical* kinetic energy gain of the adparticle in the excited state is negligible. Instead, its kinetic energy may still increase due to the increase of its momentum uncertainty brought about by the dynamical squeezing of the wave packet $\psi(\mathbf{r},t)$ describing the particle in the electronically excited state of the system. This is a purely quantum mechanical effect. It was shown recently [4] that, in contrast to the Antoniewicz model, desorption yields and kinetic energy distributions of desorbing neutrals predicted by the WPS model are consistent with *all* existing experimental data for Ar and N$_2$O physisorbed on Ru(001).

The WPS scenario of desorption, despite its success in confrontation with experiments, has a disadvantage in comparison with the Antoniewicz model in the respect that it does not offer as "obvious" candidate for the excited-state potential $V_d(\mathbf{r})$ as the latter model does. It is, therefore, important to confront other predictions of both models which might be different for each of them. The most obvious choice is to look at the distributions of particles desorbing in other directions than that of the surface normal and this is done in the present work. Calculation of angular distributions requires that the existing one-dimensional theories of both scenarios are generalized to all three dimensions by considering also a lateral, i.e. in directions along the surface, evolution of the wave packet $\psi(\mathbf{r},t)$ describing the particle in the meta-stable state. This is important because the uncertainty of the lateral components of momenta of desorbing particles (and, thus, directions in which they desorb) is determined by the lateral width of this wave packet at the instant t at which the electronic deexcitation occurs.

2.   Outline of the theory and results

The quantity of immediate interest is the probability

$$P(\mathbf{q}) = \lambda \int_0^\infty dt\, e^{-\lambda t} |< \varphi_q | \psi(t) >|^2, \tag{1}$$

that the particle in its electronic ground state (i.e. neutral) desorbs with momentum $\hbar\mathbf{q}$. In Eq. (1), $\varphi_q(\mathbf{r})$ describes the detected particle and is the continuum-energy eigen-function in the potential $V_0(\mathbf{r})$. The wave packet $\psi(\mathbf{r},t)$ is a solution of the time dependent Schrödinger equation with the potential energy $V_d(\mathbf{r})$. At $t = 0$ it is the ground state of $V_0(\mathbf{r})$ describing the state of the physisorbed particle before the excitation. λ is the position independent electronic deexcitation rate. The angular and the kinetic-energy distributions can be calculated from $P(\mathbf{q})$ straightforwardly [4].

Series of approximations were used to calculate $\psi(\mathbf{r},t{=}0)$, $\varphi_q(\mathbf{r})$ and to solve the equation for $\psi(\mathbf{r},t)$. The most important one is that the particle on its way to the detector does not experience any surface corrugation but only the average surface potential $V_0(z)$ depending on its distance from the surface. The adsorbed particle in the ground (or excited) electronic state of the system also experiences the average potential $V_0(z)$ (or $V_d(z)$) but in the lateral direction it is, in addition, confined to its adsorption site by the two-dimensional harmonic potential characterized by the angular frequency Ω_0 (or Ω_d). The parameters of the average potentials, which are taken as the Morse potentials, are their depths (V_0 or V_d), inverse range parameters (γ_0 or γ_d) and equilibrium positions (z_0 or z_d).

In the examples presented here we choose parameters pertaining to N$_2$O desorbing from Ru for which $V_0 = 430\ meV$, $\gamma_0 = 1\ \text{Å}^{-1}$ (resulting in the angular frequency $\omega_0 = 1.37 \cdot 10^{13}\ s^{-1}$ at the bottom of $V_0(z)$). The parameters of $V_d(z)$ are, in each model, the same as used in the past [4] in the one-dimensional models (i.e. for $\Omega_0 = \Omega_d$). They afforded the best agreement between the shapes of the experimental and the theoretical kinetic energy distributions in the forward direction. Thus, $V_d = 2.26\ eV$, $\gamma_d = 2.5\ \text{Å}^{-1}$, $z_0 - z_d = 0.07\ \text{Å}$, and $\lambda = 3.9 \cdot 10^{14}\ s^{-1}$ for the WPS model. For the Antoniewicz model $\lambda = 2.4 \cdot 10^{14}\ s^{-1}$ and only the slope $V_d(z)$ at the initial position z_0 is relevant. Setting,

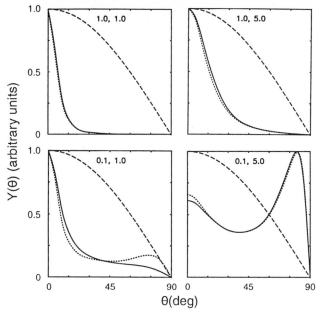

Fig. 1. Angular dependence of desorption yields $Y(\theta)$ (normalized to 1 at their maxima) for the WPS (continuous line) and the Antoniewicz (dotted) scenarios of desorption. Knudsen's $\cos(\theta)$ (long dashes) and values of β_L and r are shown in each panel.

quite arbitrarily, $\gamma_0 = \gamma_d$ and $V_d \exp(2\gamma_0 z_d) = V_0 \exp(2\gamma_0 z_0)$, and choosing $V_d = 3.52\ eV$ we assure that its slope at z_0 matches that of the image potential if $z_0 - z_{im} = 1.5\ \text{Å}$. The only parameters to be varied are those determining the lateral potentials: $\beta_L = \Omega_0/\Omega_d$ and $r = \Omega_0/\omega_0$. Their meaning is clear: for $r > 1$ the initial wave packet is narrower in the lateral directions than it is in the direction of the surface normal, and for $\beta_L < 1$ the time dependent wave packet starts its time evolution from getting narrower in the lateral directions. It is worthwhile to remind here that in the direction of the surface normal the wave packet expands initially in the Antoniewicz model while it starts from narrowing down in the WPS model [4].

In Fig. 1 the angular dependence of the (normalized to 1 at their maxima) desorption yields in both desorption scenarios is presented for a series of values of parameters β_L and r. The angular distributions become wider for larger Ω_0 (i.e. as one goes towards the right side panels in Fig. 1) because for laterally narrower initial wave packet the uncertainty of the lateral components of momenta of desorbing particles is larger. The angular distributions become wider also for smaller β_L (lower panels) because further dynamical lateral squeezing of the wave packet produces even larger lateral momentum uncertainty at an instant of the electronic deexcitation than it was at $t = 0$. It is interesting to note that, in some cases, both effects can result in particles desorbing preferentially in the direction away from the surface normal. Surprisingly, the shapes of angular distributions are almost identical in both scenarios. The absolute values are vastly different: the total desorption yields are 0.85% and 0.017% for the WPS and the Antoniewicz model, respectively (*vs.* few percent, experimentally).

The angle-resolved kinetic energy distributions of desorption products are shown in Figs. 2 (WPS model) and 3 (Antoniewicz model) for a few polar directions θ. We note first that for smaller β_L and/or larger r the energy distributions for $\theta = 60$ degrees (and larger) become very broad. This accounts for the broadening of the angular yields seen

147

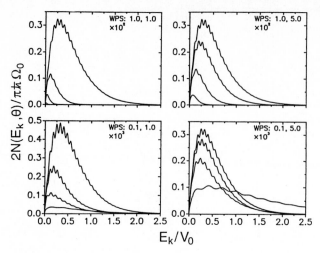

Fig. 2. Angle resolved kinetic-energy distributions $N(E_k,\theta)$ for (counting from the topmost curves) polar angles $\theta = 0$, 15, 30, and 60 degrees in the WPS model. Values of β_L and r and of the numerical factor used to multiply the actually calculated values, are given in each panel.

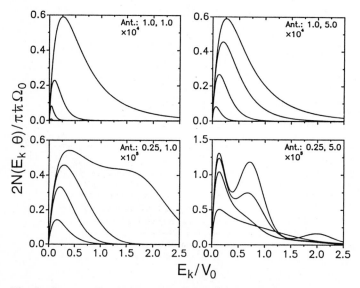

Fig. 3. Same as Fig. 2 but for the Antoniewicz model.

in Fig. 1. The WPS distributions exhibit the Ramsauer-Townsend oscillations discussed previously [4]. The most interesting feature of the presented distributions is, seen in Fig. 3, multi-peak structure of some distributions in the Antoniewicz model, but never observed in the WPS scenario. This may happen when the wave packet has a chance to oscillate few times in the lateral direction within the life time of the excited state (determined by $1/\lambda$): the particles deexcited at instants t at which the lateral wave packet is narrow desorb in a much wider polar cone than those deexcited when the packet is wide. If there is a strong relationship between the kinetic energy of particles desorbing in the direction of the surface normal and the instant at which the electronic deexcitation occurs then there will be a relative deficiency of particles desorbing in this direction with energies corresponding to the deexcitation instants at which the lateral wave packet is narrow. Such a strong energy *vs*. deexcitation time relationship exists for the Antoniewicz model (and for other scenarios which, in the direction normal to the surface, follow essentially classical dynamics) but not for the WPS model. One might speculate that in the MGR model of desorption there should also be a strong coupling between the lateral evolution of the wave packet and the shape of the kinetic energy distributions of desorption products.

This work has been supported by grants from the Natural Sciences and Engineering Council of Canada

References

1. R.D. Ramsier and J.T. Yates, Jr., Surf. Sci. Rep. **12**, 243 (1991) and references wherein.
2. P.R. Antoniewicz, Phys. Rev. B **21**, 3811 (1981).
3. Z.W. Gortel, R. Teshima and H.J. Kreuzer, Phys. Rev. B **37**, 3183 (1988); Z.W. Gortel, Surf. Sci. **231**, 193 (1990).
4. Z.W. Gortel and A. Wierzbicki, Surf. Sci. **239**, L565 (1990); Z.W. Gortel and A. Wierzbicki, Phys. Rev. B **43**, 7487 (1991).
5. D.R. Jennison, E.B. Stechel and A.R. Burns, in: *Desorption Induced by Electronic Transitions, DIET III*, R.H. Stulen and M.L. Knotek (eds.), Springer Series in Surface Sciences **13** (Springer, Berlin, 1988) p. 167.

Computer Simulation of DIET Processes: Molecular Dynamics with Electronic Transitions

J.C. Tully and M. Head-Gordon

AT&T Bell Laboratories, Murray Hill, NJ 07974, USA

Abstract. We describe our progress towards developing self-consistent methods for incorporating the effects of electronic transitions in molecular dynamics simulations. We employ a "surface-hopping" technique to allow switches between potential energy hypersurfaces when localized electronic state changes such as charge transfer occur. To describe dynamics at metal surfaces, we introduce electronic frictions and fluctuating forces to account for energy dissipation and excitation via electron-hole pair transitions. We present preliminary new results for CO on Cu(100) and discuss implications of our results on the possible mechanisms of desorption induced via rapid heating of conduction electrons.

1. Introduction

Molecular dynamics (MD), the numerical integration of the classical mechanical equations of motion for a collection of atoms interacting via an assumed force field, has proved to be a very valuable tool for examining the atomic rearrangements that underlie macroscopic chemical changes. The most important attribute of molecular dynamics is that processes can be examined in full dimensionality without untested assumptions or drastic reduction of the number of degrees of freedom. It would be desirable to apply MD to unravel the various mechanisms that may be operative in desorption induced by electronic transitions (DIET). However, conventional MD is based on the Born-Oppenheimer approximation: electrons respond instantaneously to the slower motions of the atoms, so the motions of the latter are governed by an adiabatic potential energy hypersurface. DIET, as implied by its name, involves transitions between different electronic states; i.e., between different adiabatic potential energy hypersurfaces. Crucial concerns in DIET are the lifetimes of excited states, the presence of avoided crossings between potential energy hypersurfaces, and the rates of energy exchange between adsorbate motion and substrate electrons [1]. Thus DIET is largely about breakdowns of the Born-Oppenheimer approximation, and conventional MD does not apply.

The objective of the research described here is to develop a "Molecular Dynamics with Electronic Transitions" (MDET) that retains the major strength of MD, the classical or classical-like description of atomic motions in full dimensionality, yet accurately and self-consistently incorporates the effects of electronic transitions. The challenge we face is illustrated schematically in Fig. 1. Roughly speaking, adsorbates on metal surfaces can be affected by two kinds of electronic transitions. The first, illustrated in Fig. 1a, is a localized transition that directly modifies the forces exerted on the adsorbate and/or on substrate atoms in its immediate vicinity. An example is charge transfer between the substrate and adsorbate, whereby image forces and other effects can substantially alter the

Springer Series in Surface Sciences, Vol. 31
Desorption Induced by Electron Transitions DIET V
Editors: A.R. Burns · E.B. Stechel · D.R. Jennison   © Springer-Verlag Berlin Heidelberg 1993

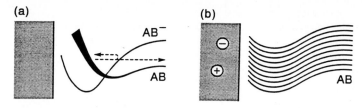

Fig. 1. Schematic illustration of localized, (a), and delocalized, (b), electronic transitions driven by adsorbate motion at metal surfaces.

interaction between adsorbate and substrate. Another example is excitation to or de-excitation from the type of repulsive state invoked in the classic Menzel-Gomer-Redhead (MGR) [2] mechanism of DIET. We have recently developed a "Surface-Hopping" procedure for incorporating such transitions in a molecular dynamics framework [3]. (The word "surface" refers here to potential energy hypersurface, not to surface of the solid.) Surface-Hopping accounts for electronic transitions by instantaneous transitions between one electronic state and another. For many applications, particularly for DIET processes, this is a decisive improvement over alternative procedures for introducing electronic transitions via invoking a single "weighted average" interaction potential with the weights given by the electronic state populations. The advantage of Surface-Hopping is shown by the schematic trajectory of Fig. 1a, which is intended to illustrate an adsorbate attempting to escape from the surface. When the adsorbate has moved to the right of the curve crossing in Fig. 1, if it is in the excited electronic state, i.e., if its motion is governed by the upper of the two potential energy curves (the negative ion state in this illustration), then it will not escape at the indicated energy. On the other hand, if the motion of the adsorbate is governed by the neutral state potential, it will escape. It is impossible to accurately describe both possible outcomes by a single trajectory on any kind of average interaction potential.

As discussed in detail elsewhere [3], the crucial aspect of Surface-Hopping is to develop a self-consistent algorithm for deciding when to switch from one potential energy hypersurface to another. The time-dependent motions of the atoms produce the electronic transitions. The electronic states, in turn, dictate the forces experienced by the atoms. Thus, at least for low-energy processes such as DIET, it is essential that atomic motion and electronic transitions be described self-consistently in a way that conserves total energy. We have proposed an algorithm that meets these requirements. Initial tests against accurate quantum calculations for model problems are very encouraging, even in cases where average potential methods fail completely and quantum coherence effects are critical [3]. Alternative surface-hopping techniques have also been proposed [4]. While the ranges of validity of each of these remain to be determined, preliminary indications are encouraging for all. It appears that Surface-Hopping will provide an accurate and practical way to incorporate discrete electronic transitions into molecular dynamics simulations of a wide variety of processes, including DIET.

In order to implement this scheme, of course, one needs as input the potential energy hypersurfaces corresponding to each electronic state considered, as well as the off-diagonal couplings between states that are responsible for electronic transitions. In view of the present difficultly in obtaining accurate *ab initio* descriptions of even a single potential energy surface, it is clear that this will be a major hindrance to widespread application of surface-hopping to DIET. Nevertheless, there is progress toward this goal, in particular the new techniques for calculating the energies and lifetimes of excited adsorbate states as discussed by Nordlander in this volume [5].

Figure 1b illustrates the second kind of electronic transition we must consider and the one we will discuss in the remainder of this chapter, the coupling of adsorbate motion to the excitation or de-excitation of an electron-hole (e-h) pair of the metal substrate. E-h pair excitations can make an important contribution to the dissipation of adsorbate energy, thereby substantially modifying DIET yields. Conversely, it has been shown recently that the excitation of e-h pairs by photons or electrons can promote desorption, with significant efficiency [6,7]. E-h pairs are delocalized excitations that quickly leave the vicinity of the adsorbate. Thus, to a good approximation the interaction potential governing the adsorbate motion is unchanged by an e-h pair transition. However, the e-h pair transition affects the adsorbate motion through the removal or deposition of energy. We describe in the next Section how we recast the effects of e-h pair transitions into frictional and fluctuating forces that accurately account for energization and dissipation. Furthermore, we give explicit expressions for *ab initio* calculation of the e-h pair couplings required to implement the method. In the following Section we present preliminary applications to a specific system, the carbon monoxide molecule adsorbed on the (100) face of a copper crystal surface. In particular, we present results on the rates of energy flow between substrate electrons and various vibrational degrees of freedom of the adsorbate, and discuss possible consequences of these results on desorption induced by hot substrate electrons.

2. Electronic Friction

A number of workers have proposed that dissipation of energy via e-h pair excitation can be represented in a classical mechanical treatment of atomic motion by a frictional force [8]. We have recently provided an alternative derivation that provides an explicit expression for the friction matrix, including memory and spatial dependence, that can be calculated from first principles [9]. In addition, our formalism introduces a fluctuating force that can also be computed from first principles and that allows, for example, the simulation of desorption induced by hot substrate electrons (see below). We begin by expressing the total electronic wave function $\Psi(t)$ as a linear combination of adiabatic wave functions $\phi_j(R)$ with expansion coefficients $c_j(t)$. The electronic Hamiltonian, $H(R)$, is time-dependent through the motion of the atoms. We can obtain the equations of motion for the coefficients $c_j(t)$ from the time-dependent Schroedinger equation:

$$i\hbar\dot{c}_j = c_j <\phi_j|H|\phi_j> - i\hbar R \sum_i c_i \mu_{ij}(R).$$ (1)

The quantity $\mu_{ij}(R)$, called the "nonadiabatic coupling", is given by:

$$\mu_{ij}(R) = <\phi_i|\nabla_R\phi_j>,$$ (2)

where brackets denote integration over electronic coordinates, **r**, only. Equation (1) gives the amplitude, c_j, of each electronic state as a function of time along any as yet unspecified trajectory.

For this case of nested, parallel potential energy hypersurfaces, we assume that the trajectory is governed by an effective potential, V_{eff}, that is the expectation value of the electronic Hamiltonian:

$$V_{eff} = <\Psi|H|\Psi>. \tag{3}$$

If there were only a few electronic states j, it would be feasible to carry out direct numerical integration of Newton's equations subject to $V_{eff}(\mathbf{R})$, with simultaneous integration of Eq. (1). This procedure has been employed previously in other contexts [10]. The difficulty in the present application to e-h pair excitations is that there are an infinite number of infinitesimally spaced electronic states j, with rapidly varying phases. We have shown [9] that the equations can be simplified by the following three assumptions: First, the potential energy surfaces are parallel. Second, the nonadiabatic coupling, $\mu_{ij}(\mathbf{R})$, is sufficiently small that a weak-coupling limit is valid. Third, the density of states near the Fermi level is smooth. With these approximations, the set of equations (1) and (3) can be replaced by a Langevin equation:

$$\ddot{\mathbf{R}} = - m^{-1}\nabla_R V_{eff}(\mathbf{R}) - \Lambda(\mathbf{R})\dot{x}(t) + \mathbf{F}(t). \tag{4}$$

Thus coupling to e-h pairs can indeed be represented by a position-dependent friction constant $\Lambda(\mathbf{R})$ and random force, $\mathbf{F}(t)$. Equation 4 can describe both dissipation by e-h pair excitation and energization by hot electrons.

3. CO on Cu(100)

As with Surface-Hopping, the major impediment to application of Molecular Dynamics with Electronic Friction is calculation of the required input information, in this case the friction constant matrix $\Lambda(\mathbf{R})$. We have recently developed a practical method for doing this, employing the standard machinery of quantum chemistry [11]. Specifically, we have carried out Hartree-Fock (H-F) calculations of a CO molecule bound to a copper cluster containing from 6 to 14 atoms, for a range of CO orientations and separation distances from the surface. Nonadiabatic coupling matrix elements between the filled and unfilled molecular orbitals of the adsorbate-cluster system were computed from the H-F orbitals. These nonadiabatic couplings cannot be directly used to compute the friction matrix for two reasons. First, H-F is known to give an incorrect representation of the Fermi level density of states for metals. More importantly, for a small cluster the spacings between discrete electronic states are much larger than vibrational spacings, and are thus a poor representation of the continuum of conduction electron states. We have shown, however, that the nonadiabatic couplings in the delocalized molecular orbital representation can be transformed to a local atomic orbital representation, the resulting coupling matrix denoted G. This can then be projected back onto the delocalized conduction electron states of the infinitely large metal using the following expression [11,12].

$$\Lambda = \pi\hbar Tr[\ \rho(E_F)\ \mathbf{G}\ \rho(E_F)\ \mathbf{G}\ ]. \tag{5}$$

The quantity $\rho(E_F)$ is the local density of states at the Fermi level in the same atomic representation as G, and can be adequately approximated in a number of

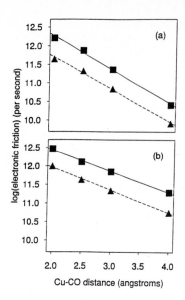

Fig. 2. Diagonal elements of the friction matrix, $\Lambda(R)$, as a function of the distance of the CO center of mass from the Cu(100) surface plane. (a) Squares are for the C-O internal stretch and triangles for the CO - surface stretch. (b) Squares are for the frustrated rotational mode and triangles for the in-plane frustrated translation. In (b), coordinates were chosen to correspond to the asymptotic pure rotational and translational motions, and therefore do not correspond to the normal modes of the bound adsorbate.

ways. This technique has been shown to give quite accurate results that are surprisingly weakly dependent on cluster size [11].

The accuracy of this procedure has been demonstrated by comparing calculated and experimental lifetimes for relaxation of the 4 vibrational modes of CO on Cu(100), C-O internal stretch, CO - surface stretch, CO frustrated rotation, and CO lateral frustrated translation. All are in at least semi-quantitative agreement [11]. Fig. 2 shows new results, the calculated friction constant for these 4 degrees of freedom as a function of the distance of the center of mass of the CO molecule from the surface plane. (The equilibrium CO-Cu separation is 2.5 A.) Note that when the molecule is close to the surface, the internal vibrational and frustrated rotational modes exchange energy on the 1 ps timescale. The normal and lateral frustrated translational modes couple much more weakly to e-h pairs. The origins of this behavior are discussed elsewhere [11].

Now that this friction matrix has been calculated, we can apply Molecular Dynamics with Electronic Friction to DIET processes. For example, we can model the desorption induced by rapid heating of conduction electrons as in the femtosecond laser experiments of Prybyla, et al [7] for CO on Cu(111). We can run trajectories for a period of time on a cold equilibrated surface, and then rapidly heat and cool the electrons, following the same temperature profile as achieved experimentally. We control the electron temperature through the magnitude of the random force $F(t)$ of Eq. 4. These simulations are straightforward when the electron heating and duration are sufficient to give high desorption yields. When yields are low, however, a large number of trajectories will produce only a few desorption events, resulting in poor statistics, especially if details such as translational and internal energy distributions are desired. In order to make these calculations tractable, an efficient method for treating the dynamics of non-equilibrium infrequent events will be employed [13].

These calculations have not yet been completed. Even in the absence of MDET simulation results, however, we can draw some conclusions concerning the expected yields and internal energies of desorbed CO from Cu(100), using the calculated frictions displayed in Fig. 2. For a "back of the envelope" estimate of the yield, we can assume that each vibrational mode of the adsorbate is independent and remains Boltzmann at a time-dependent effective temperature, T_{eff} , that approaches the instantaneous electron temperature, T_{el} , at a rate of

$$T_{eff} = T_{el}[1 - exp(-\tfrac{1}{2}\Lambda t)] .$$ (6)

Λ is the diagonal element of the friction constant corresponding to the designated vibrational mode, taken from Fig. 2. Note that the lifetime of the vibrational mode is equal to $2/\Lambda$. For CO at its equilibrium separation from the surface, $2/\Lambda$ is 2.7, 10.0, 1.6 and 5.1 ps, respectively, for internal vibration, molecule-surface vibration, frustrated rotation and frustrated lateral translation. These lifetimes can be obtained directly from Fig. 2 for a CO - surface distance of 2.5 A. Taking the 10.0 ps. lifetime of the adsorbate - surface vibration, and assuming an Arrhenius desorption law with prefactor 10^{13} s. and activation energy equal to the binding energy, 0.75 eV, it is then simple to estimate the hot electron induced desorption yield. For a Gaussian heating pulse of 1 ps full-width at half-maximum and a maximum temperature of 3500K, this estimate of the desorption yield is $4x10^{-12}$. This is far lower than the 6% yield observed experimentally under similar conditions for the CO - Cu(111) system [7]. Thus direct excitation of the molecule-surface bond by hot electrons does not appear to be responsible for the observed desorption. However, the lifetime of the frustrated rotational mode is much shorter than that of the molecule - surface stretch. If we assume that rotation is equally effective in promoting desorption, then the estimate becomes much more reasonable. A similar role of frustrated rotational motion in DIET has been proposed by Burns, et al. [14]. The estimated desorption yield with the same parameters as assumed above, but with a 1.6 ps. vibrational lifetime, is 1.2%, in rough accord with experiment. Thus the direct excitation of rotational motion by e-h pairs does appear to be capable of producing desorption in yields comparable to the experimental findings. Furthermore, desorption by this mechanism will be prompt and the desorbed molecules will be vibrationally excited, with vibrational temperature of order 900 K for the conditions used in this estimate. This is also in qualitative accord with experiment [6,7]. We should emphasize that these estimates are very uncertain, and should be taken only as a crude indication of the viability of this mechanism. The neglect of coupling between degrees of freedom, the neglect of phonons, the oversimplified role of frustrated rotation, and uncertainties in the assumed parameters certainly preclude drawing any quantitative conclusions. Molecular Dynamics with Electronic Friction simulations, in full dimensionality, are currently in progress.

Misewich, Heinz and Newns [15] have proposed an alternative mechanism for desorption induced by hot electrons which they call DIMET (Desorption Induced by Multiple Electronic Transitions). They suggest that energetic electrons cause frequent transitions to the negative ion resonance state of the adsorbate. This state has a very short lifetime, of order 1 fs, so in a single excitation event evolution of the atomic motion will be insufficient to produce desorption with the observed yield. However, multiple excitations and de-excitations can greatly increase the desorption yield [15]. With knowledge of the energy and lifetime of the excited state, this mechanism could be modelled by the Surface-Hopping method described in Sec. III. The mechanism appears qualitatively quite different from the e-h pair induced thermal mechanism we propose in this Section.

It is interesting to compare these mechanisms. An electron capture into the negative-ion state, followed by a very rapid de-excitation to the ground state can be considered a specific mechanism for excitation or de- excitation of an e-h pair. Thus it may be possible to describe the net effect with appropriate friction and fluctuating forces. As discussed elsewhere [11], the underlying reason for the strong e-h coupling we calculate for both the internal C-O stretch and the frustrated rotational mode is the presence of the low-lying $2\pi^*$ orbital that is sufficiently lifetime broadened so that its low-energy tail dips below the Fermi level. Adsorbate motions cause this orbital to shift slightly in energy, resulting in slight electron transfer into and out of this orbital. To the extent that this electronic motion slightly lags the atomic motion, e-h pair transitions occur. Thus our calculations support the generally accepted mechanism [16]. This fluctuating population of the negative ion state, albeit only fractional, and the induced lag of the electronic motion is closely analogous to the DIMET negative ion formation and destruction. Thus it is not at all clear that a Surface-Hopping description is required for this case. One possible experimentally observable distinction between these mechanisms is that our thermal mechanism would be expected to exhibit an apparent activation energy roughly equal to the ground state binding energy of the adsorbate. In contrast, the DIMET mechanism would be expected to reveal a larger apparent activation energy associated with the energy of the center of the $2\pi^*$ resonance.

4. Conclusions

We have presented a progress report on our current efforts to extend molecular dynamics simulations to processes involving electronic transitions, with particular emphasis on DIET. Promising dynamical methods are now in hand. The bottleneck to widespread application of these methods to real systems is accurate calculation of the required electronic input information: ground and excited potential energy hypersurfaces, excited state lifetimes and nonadiabatic couplings. Significant progress has been made in developing *a priori* methods to compute this input, but more work is clearly needed.

References:

1. See, for example, N. H. Tolk, M. M. Traum, J. C. Tully and T. E. Madey, Eds., *Desorption Induced by Electronic Transitions, DIET I*, (Springer-Verlag, Berlin, 1983). See especially the chapters by J. W. Gadzuk, D. R. Jennison, J. C. Tully, R. Gomer, D. Menzel, P. J. Feibelman, D. E. Ramaker and W. Brenig. See also, J. W. Gadzuk, in *Desorption Induced by Electronic Transitions, DIET IV*, G. Betz and P. Varga, Eds., (Springer-Verlag, Berlin, 1990), p. 2; J. W. Gadzuk, *Phys. Rev.* **B44**, 13466(1991).
2. D. Menzel and R. Gomer, *J. Chem. Phys.* **41**, 3311(1964); P. E. Redhead, *Can. J. Phys.* **42**, 886(1964).
3. J. C. Tully, *J. Chem. Phys.* **93**, 1061(1990); J. C. Tully, *Int. J. Quantum Chem.* **25**, 299(1991).
4. J. C. Tully and R. K. Preston, *J. Chem. Phys.* **55**, 562(1971); J. C. Tully, in *Dynamics of Molecular Collisions, Part B*, edited by W. H. Miller (Plenum, New York, 1976), p. 217; W. H. Miller and T. F. George, *J. Chem. Phys.* **56**, 5637(1972); F. J. Webster, J. Schnitker, M. S. Friedrichs, R. A. Friesner and P. J. Rossky, *Phys. Rev. Lett.* **66**, 3172(1991).

5. P. Nordlander, in *Desorption Induced by Electronic Transitions, DIET V,* edited by A. R. Burns, E. B. Stechel and D. R. Jennison, this volume; See also, P. Nordlander and J. C. Tully, *Phys. Rev. Lett.* **61,** 990(1988); P. Nordlander and J. C. Tully, *Phys. Rev.* **B42,** 5564(1990).
6. J. A. Prybyla, T. F. Heinz, J. A. Misewich, M. M. Loy and J. H. Glownia, *Phys. Rev. Lett.* **64,** 1537(1990); F. Budde, T. F. Heinz, M. M. Loy, J. A. Misewich, F. de Rougemont and H. Zacharias, *Phys. Rev. Lett.* **66,** 3024(1991).
7. J. A. Prybyla, H. W. K. Tom and G. D. Aumiller, *Phys. Rev. Lett.* **68,** 503(1992).
8. W. Schaich, *Solid State Commun.* **15,** 357(1974); E. G. d'Agliano, P. Kumar, W. Schaich and H. Suhl, *Phys. Rev.* **B11,** 2122(1975); A. Nourtier, *J. Phys.* (Paris) **38,** 479(1977); K. L. Sebastian, *Phys. Rev.* **B31,** 6976(1981); B. N. J. Persson and W. L. Schaich, *J. Phys. C: Solid State Phys.* **14,** 5583(1981); K. M. Leung, G. Schon, P. Rudolph and H. Metiu, *J. Chem. Phys.* **81,** 3307(1984); A. Okiji and H. Kasai, *Phys. Rev.* **B38,** 8102(1988).
9. M. Head-Gordon and J. C. Tully, unpublished.
10. W. H. Miller and C. W. McCurdy, *J. Chem. Phys.* **69,** 5163(1978); D. A. Micha, *J. Chem. Phys.* **78,** 7139(1983); Z. Kirson, R. B. Gerber, A. Nitzan and M. A. Ratner, *Surf. Sci.* **137,** 527(1984); S.-I. Sawada, A. Nitzan and H. Metiu, *Phys. Rev.* **B32,** 851(1985).
11. M. Head-Gordon and J. C. Tully, *J. Chem. Phys.* **96,** 3939(1992); M. Head-Gordon and J. C. Tully, *Phys. Rev. B* in press.
12. A similar expression has been derived by B. Hellsing and M. Persson, *Physica Scripta* **29,** 360(1984).
13. C. Lim and J. C. Tully, *J. Chem. Phys.* **85,** 7423(1986).
14. A. R. Burns, E. B. Stechel and D. R. Jennison, *Phys. Rev. Lett.* **58,** 250(1987); E. B. Stechel, D. R. Jennison and A. R. Burns, in *Desorption Induced by Electronic Transitions, DIET III,* R. H. Stulen and M. L. Knotek, Eds., (Springer-Verlag, Berlin, 1987), p. 136.
15. J. A. Misewich, T. F. Heinz and D. M. Newns, unpublished.
16. B. N. J. Persson and M. Persson, *Solid State Commun.* **36,** 175(1980); P. Avouris and B. N. J. Persson, *J. Phys. Chem.* **88,** 837(1984); T. T. Rantala and A. Rosen, *Phys. Rev.* **34,** 838(1986).

Some Current Theoretical Insights on DIET

J.W. Gadzuk

National Institute of Standards and Technology, Gaithersburg, MD 20899, USA

Abstract. An overview of selected post-DIET IV advances in the theoretical treatment and/or understanding of photon and electron induced molecular processes at surfaces will be given.

1. Introduction

It is always interesting to take periodic glances backwards (and forward when possible) in an actively evolving field like Desorption Induced by Electronic Transitions (DIET) not only to see from where one has come but, more importantly, to note what new lines of thinking have emerged beyond a simple adiabatic expansion of knowledge, and to assess what factors have influenced the qualitative new directions in research. The occasion of the DIET Workshops offers a perfect opportunity, as well as a timing device, for such introspection. Since DIET IV [1], a major review article by Ramsier and Yates [2] has appeared which provides an excellent entry into the field of electron-stimulated desorption (ESD). A complimentary review of surface photochemistry at adsorbate/metal interfaces by Zhou, Zhu, and White [3] has also been published which deals with progress in ultra-violet and visible photon initiated DIET. The important connections between laser induced desorption [4,5] photostimulated chemistry at surfaces [3,6-8], standard ESD [2,9], and inelastic hot electron resonance scattering, tunneling and/or desorption [10] will be the unifying theme here, with special emphasis placed on post DIET IV theoretical work. After a brief update, attention will be focused on hot electron processes involving temporary negative-ion shape resonances associated with atoms or molecules adsorbed on metals [11,12]. These processes are similar to those found in gas-phase resonant electron-molecule scattering [13] and inelastic resonant electron tunneling in solid-state junctions [14].

2. DIET Update

As is apparent from many contributions at DIET V, there have been major advances in laser-stimulated surface processes (including desorption) beyond "laser-as-a-Bunsen-burner" thermal effects [4]. Almost without exception, the basic mechanism follows a script starting with photon absorption via electronic excitation involving either a localized electronic state in the surface complex, or a coherent superposition of extended conduction band states of the substrate (i.e., a wavepacket). The first case generally gives rise to desorption by a standard Menzel-Gomer-Redhead (MGR) or Knotek-Feibelman (KF) mechanism described in terms of a Franck-Condon-like projection of the bound desorbate center-of-mass wavefunction onto the continuum desorptive states of a repulsive potential curve [2], or alternatively, as the time-evolution of the nuclear motion wave packet on the excited-state repulsive curve [11,15,16], if the lifetime of the excited repulsive state is long enough for large-yield desorption to occur [17].

An intriguing alternative is based on the idea of using an excited electronic state to "assist" chemistry on the ground-state potential-energy surface [18], as envisioned in femto-second laser chemistry [19] and in

Springer Series in Surface Sciences, Vol. 31
Desorption Induced by Electron Transitions DIET V
Editors: A.R. Burns · E.B. Stechel · D.R. Jennison   © Springer-Verlag Berlin Heidelberg 1993

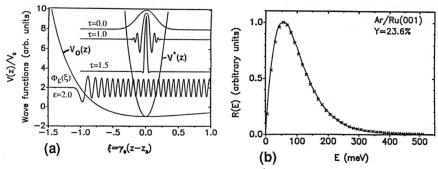

Figure 1 - (a) Surface potentials $V_0(z)$ and $V^*(z)$, continuum wave functions $\Phi_\epsilon(\zeta)$ for $\epsilon/V_0 = 2$, and real parts of $\psi(t)$ for $\tau \equiv \omega^* t = 0.0$, 1.0, and 1.5 where ω^* is the harmonic frequency of $V^*(z)$. (b) Theoretical (solid curve) and experimental (crosses) for kinetic energy distribution of Ar desorbing from Ru(001). [Ref. 21]

charge-transfer molecular beam scattering from surfaces [20]. Of particular utility are transient ionic states of the adsorbate, both positive [15,21] and negative [5,22], since the bonding to the surface is considerably different for charged species compared to neutrals [23].

Gortel and Wierzbicki (GW), building upon the theory of Stechel, Jennison, and Burns [24], have produced theoretical numbers consistent with experimental desorption data for Ar and N_2O physisorbed on Ru(001) [21]. In this model, it is argued that initial creation of a temporary positive "desorbate" ion is followed by a substrate electron hop into an unfilled "screening orbital" of the desorbate [25]. With this orbital, a transient chemical bond can then be formed between the screened ion and the substrate which is characterized by $V^*(z)$, a potential curve much deeper and narrower than $V_0(z)$, the original neutral curve, but having its minimum almost at the same position, as depicted in Fig. 1a. The initial wave packet, $\psi(t=0)$ which is the ground state wavefunction on $V_0(z)$, can be considered as a coherent superposition of excited states on V^* whose subsequent dephasing time evolution is also illustrated in Fig. 1a. Kinetic energy is acquired with little or no net center-of-mass displacement, a purely quantum mechanical effect. After a time t, the screened ion decays to the neutral ground state, in which case the rapidly moving oscillatory (in $V^*(z)$) wave packet is returned to $V_0(z)$ where it projects onto desorptive continuum wave functions such as $\Phi_\epsilon(\zeta)$. Kinetic energy distributions are of the form

$$P(\epsilon) = \tau_R^{-1} \int_0^\infty dt \, e^{-t/\tau_R} |<\Phi_\epsilon|\psi(t)>|^2 \, , \qquad (1)$$

where τ_R is the mean lifetime of the excited, screened-ion state. An example of a kinetic energy distribution obtained by GW, from Eq. 1, for the system of Ar/Ru(001) is shown in Fig. 1b where the consistency between theory and experiment is apparent. Good accord with absolute yields is also obtained.

Temporary negative-ion adsorbate states, regarded as shape resonances, have been the topic of much theoretical activity [12,26]. One reason for this attention is their role in laser-induced desorption in which a postulated flux of laser-excited hot electrons is incident upon the surface from within the substrate [5]. The most dramatic reactive events triggered by the hot electrons are those involving temporary capture of the hot electrons into "bond-relevant" resonance states of the surface complex. The hot electrons can be inelastically scattered back into the substrate, leaving energy behind within the surface molecular complex. This localized energy can then be used to effect an elementary hot-electron-mediated photochemical process such as internal excitation, surface dissociation, or desorption. This will be illustrated in the next section in terms of a widely used "universal" model.

3. Inelastic resonance processes

The abstract problem of localized-oscillator excitation due to a transient driving force associated with electronic transitions between continuum and quasi-localized states or resonances has been of great interest in many different areas of physics [27], including intramolecular vibrational excitation in resonance electron-molecule scattering [13], inelastic resonant tunneling in quantum-well heterostructures [14], and hot-electron-induced resonant desorption [10]. These processes have been formulated within the context of the model Hamiltonian [27]:

$$H = H_{el} + H_{ph} + H_{int},$$ (2a)

with

$$H_{el} = \epsilon_a c^\dagger c + \sum_{\kappa,\alpha} \epsilon_{\kappa\alpha} c^\dagger_{\kappa\alpha} c_{\kappa\alpha} + \sum_{\kappa,\alpha} V_{\kappa\alpha,a}(c^\dagger_{\kappa\alpha} c + c^\dagger c_{\kappa\alpha}),$$ (2b)

$$H = \hbar\omega_o b^\dagger b,$$ (2c)

$$H_{int} = \lambda_o c^\dagger c(b + b^\dagger).$$ (2d)

Equation (2) describes a system in which a discrete localized electronic state [eigenvalue equal to ϵ_a, Fermion operators $(c^\dagger, c)$] is coupled via a set of matrix elements $V_{\kappa\alpha,a}$ to a number of electronic continua specified by index α, with eigenvalues $\epsilon_{\kappa\alpha}$ and operators $c^\dagger_{\kappa\alpha}, c_{\kappa\alpha}$. This electronic system interacts with a dispersionless boson field described by the harmonic oscillator Hamiltonian, Eq. (2c), linearly displaced according to the perturbation, Eq. (2d), when ϵ_a, the localized electronic state, is occupied.

3.1 Scattering

Domcke and Cederbaum (DC) produced a theory of gas-phase resonance-electron scattering from small molecules, originally treated by Herzenberg in terms of the boomerang model [28], based on the model specified by Eqs. 2 [13]. DC have shown that the $0 \to n$ vibrational excitation cross section for a hot electron with incident kinetic energy equal to ϵ_i can be expressed in the form

$$\sigma_n(\epsilon_i) \sim \epsilon_i^{-1} \Gamma(\epsilon_i)\Gamma(\epsilon_f) \left| \sum_{m=0}^{\infty} \frac{\langle n|\tilde{m}\rangle \langle \tilde{m}|0\rangle}{\epsilon_i - \tilde{\epsilon}_m - \Lambda(\epsilon_i) + (i/2)\Gamma(\epsilon_i)} \right|^2$$ (3)

with $\tilde{\epsilon}_m = \epsilon_a - \Delta\epsilon_r + m\hbar\omega_o$, $\Delta\epsilon_r = \lambda_o^2/\hbar\omega_o = \beta\hbar\omega_o$, and with the self-energy operator accounting for the discrete state-continuum coupling simplified to $\Lambda - i\Gamma/2 = \Sigma|V_{\kappa,a}|^2/(\epsilon_i-\epsilon_\kappa)$. In Eq. (3), $\Gamma(\epsilon_{i(f)})/\hbar$ is the electron transition rate at the initial (final) energy into (out of) the quasidiscrete resonance state with unperturbed energy ϵ_a, and $\langle n|\tilde{m}\rangle$, etc., are vibrational overlap integrals where $|\tilde{m}\rangle$ denotes vibrational states of the temporary negative ion and $|n\rangle$ those of the neutral molecule. Using standard formulas for the displaced oscillator overlap integrals or Franck-Condon factors [29] and taking both Λ and Γ as energy-independent constants, Eq. 3 can be worked into an equivalent form for the probability per event that an incident electron inelastically scatters with energy transfer $\Delta\epsilon = \epsilon_i - \epsilon_f$ remaining in the oscillator. The result is that

$$P(\epsilon_f,\epsilon_i) = \Gamma(\epsilon_f)\Gamma(\epsilon_i)e^{-2\beta} \sum_{n=0}^{\infty} \frac{\beta^n}{n!} \delta(\epsilon_i - \epsilon_f - n\hbar\omega_o) \left| \sum_{m=0}^{\infty} \frac{A_{n,m}}{\epsilon_i - (\epsilon_a'-\Delta\epsilon_r)-m\hbar\omega_o+i\Gamma/2} \right|^2$$

$$= \sum_{n=0}^{\infty} P_n(\epsilon_f,\epsilon_i),$$ (4)

$$A_{n,m} = \begin{cases} (-1)^{n-m}L_m^{n-m}(\beta) & \text{for } m \le n \\ \beta^{m-n}(n!/m!)L_m^{m-n}(\beta) & \text{for } m \ge n, \end{cases}$$

where
$$L_m^\alpha(\beta) = \sum_{j=0}^{m} (-1)^j \begin{pmatrix} m+\alpha \\ m-j \end{pmatrix} \beta^j / j!$$

are generalized Laguerre polynomials and $\epsilon' = \epsilon_a + \Lambda$. The inelastic scattering probability, Eq. 4, is the product of incident electron penetration into the molecule $[=\Gamma(\epsilon_i)]$ multiplied by a measure of its resonance with the vibrationally renormalized quasibound state (the energy denominator or propagator), weighted by Poisson factors (or equivalently by vibrational overlap integrals) and interference terms, ultimately multiplied by $\Gamma(\epsilon_f)/\hbar$, the transition rate out of the molecule for an electron with final-state energy $\epsilon_f = \epsilon_i - n\hbar\omega_o$.

3.2 Desorption

A venerable model for a bound molecular system such as an atom or molecule adsorbed on a surface is a harmonic oscillator truncated on one side, as illustrated by V_{ph} in Fig. 2a. Also shown as $V_{ph} + V_{int}$ is the displaced oscillator potential associated with Eqs. 2c and 2d, when the localized electronic state is occupied. The process of vibrational excitation and/or desorption on such a system of curves is identical with that encountered in scattering, the incident electron beam now being the laser-excited hot-electrons from within the substrate. Those excited states of V_{ph} for which $n\hbar\omega_o \geq D$, the desorption energy, will lead to desorption, in analogy with predissociation in isolated molecular systems. The total desorption probability per resonance event for a hot electron with energy ϵ_i, which follows from Eq. 4, is obtained by integrating over final-state energy, that is:

$$P_{des}(\epsilon_i;\tau_R) = \sum_{n=n_d}^{\infty} \int d\epsilon_f \, P_n(\epsilon_f,\epsilon_i;\tau_R) \tag{5}$$

where n_j is the smallest integer greater than $D/\hbar\omega_o$ and the parametric dependence on resonance lifetime $\tau_R = \Gamma/\hbar$ has been made explicit. In the case of the laser-excited hot-electron experiments, the desorption probability is represented by $P_{des}(\epsilon_i;\tau_R)$ multiplied with a (normalized) hot-electron supply function $= j_{in}(\epsilon_i)/j_o$ and integrated over all incident electron energies. With the not-unreasonable approximation of constant $j_{in}(\epsilon_i)$ over the width of

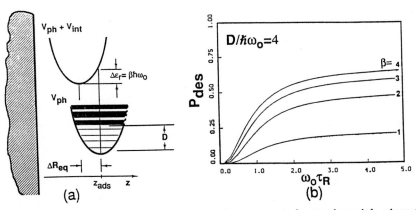

(a)   (b)

Figure 2 - (a) Potential-energy curves for resonant desorption with adsorption bond represented by V_{ph}, a half-truncated harmonic oscillator potential. The predissociative/predesorptive states of the oscillator, those with energy greater than D, are here symbolized by the broadened levels. In this figure, $n_d = 4$. (b) Desorption probability as a function of $\omega_o\tau_R$, from Eq. 6, with $\beta = 1,2,3,$ and 4 and $n_d = 4$. [Ref. 10]

the resonance, the lifetime-dependent total desorption probability which follows from Eqs. 4 and 5 is compactly given by

$$P_{des}(\tau_R) = \int d\epsilon_i (j_{in}(\epsilon_i)/j_o) \, P_{des}(\epsilon_i;\tau_R)$$

$$= e^{-2\beta} \sum_{n=n_d}^{\infty} (\beta^n/n!) \sum_{m,m'} \frac{A_{n,m} \, A_{n,m'}}{1+(\omega_o\tau_R)^2(m-m')^2}. \qquad (6)$$

Some numerical consequencies from Eq. 6 are shown in Fig. 2b where $P_{des}(\tau_R)$ vs. $\omega_o\tau_R$ is displayed for a set value of $D = 4\hbar\omega_o$ and a family of β values. From simple oscillator mechanics it is easily demonstrated that $\Delta R_{eq} = (2\beta\hbar/m\omega_o)^{1/2}$. Succinctly summarized, Fig. 2b shows that longer-lived resonances show larger desorption yields, saturating at a value that is a monotonically increasing function of the oscillator displacement of the intermediate state, as embodied in the value of β. These results [10] are in harmony with previous ideas developed in terms of semi-classical wave packet dynamics [5] and hopefully, with our expanding intuition on hot-electron-induced desorption.

References

1.   "Desorption Induced by Electronic Transitions DIET IV", ed. by G. Betz and P. Varga (Springer, Berlin, 1990).
2.   R. D. Ramsier and J.T. Yates, Jr., Surf. Sci. Repts. 12, 243 (1991).
3.   X.-L. Zhou, X.-Y. Zhu, and J.M. White, Surf. Sci. Repts. 13, 73 (1991).
4.   R.R. Cavanagh, S.A. Buntin, L.J. Richter, and D.S. King, Comments At. Mol. Phys. 24, 365 (1990).
5.   J.W. Gadzuk, L.J. Richter, S.A. Buntin, D.S. King, and R.R. Cavanagh, Surf. Sci. 235, 317 (1990).
6.   W. Ho, in Ref. 1, p. 48.
7.   E. Hasselbrink, Appl. Phys. A 53, 403 (1991).
8.   J.C. Polanyi and H. Reilly, in "Dynamics of Gas-Surface Interactions", ed. by C.T. Rettner and M.N.R. Ashford, (Royal Soc. of Chem., London, 1991) p. 329.
9.   Ph. Avouris and R. Walkup, Ann. Rev. Phys. Chem. 40, 173 (1989).
10.   J.W. Gadzuk, Phys. Rev. B 44, 13466 (1991).
11.   J.W. Gadzuk, Ann. Rev. Phys. Chem. 39, 395 (1988).
12.   R.E. Palmer and P.J. Rous, Rev. Mod. Phys. 64, 383 (1992).
13.   W. Domcke and L.S. Cederbaum, J. Phys. B 13, 2829 (1980); W. Domcke, Phys. Reports 208, 97 (1991).
14.   N.S. Wingreen, K.W. Jacobsen, and J.W. Wilkins, Phys. Rev. B40, 11 834 (1989).
15.   A.R. Burns, E.B. Stechel, D.R. Jennison, and T.M. Orlando, Phys. Rev. B 45, 1373 (1992).
16.   M. Messina and R.D. Coalson, J. Chem. Phys. 92, 5712 (1990); ibid. 95, 5364 (1991).
17.   W. Hübner, W. Brenig, and H. Kasai, Surf. Sci. 226, 286 (1990).
18.   D.J. Tannor and S.A. Rice, Adv. Chem. Phys. 70, 441 (1988).
19.   E.D. Potter, J. L. Herck, S. Pedersen, Q. Liu, and A.H. Zewail, Nature 355, 66 (1992).
20.   J.W. Gadzuk, Chem. Phys. Letters 136, 402 (1987).
21.   Z.W. Gortel and A. Wierzbicki, Phys. Rev. B 43, 7487 (1991).
22.   J.W. Gadzuk and C.W. Clark, J. Chem. Phys. 91, 3174 (1989).
23.   Ph. Avouris, P.S. Bagus, and C.J. Nelin, J. Elec. Spec. Rel. Phen. 38, 269 (1986).
24.   A.R. Burns, E.B. Stechel, and D.R. Jennison, Phys. Rev. Letters, 58, 250 (1987); and in "DIET III", ed. by R.H. Stulen and M.L. Knotek (Springer, Berlin, 1988) p. 137, p. 167.
25.   N.D. Lang and A.R. Williams, Phys. Rev. B 16, 2408 (1977); J.W. Gadzuk and S. Doniach, Surf. Sci. 77, 427 (1978); B. Gumhalter, Prog. Surf. Sci. 15, 1 (1984).
26.   D. Teillet-Billy and J.P. Gauyacq, Surf. Sci. 239, 343 (1990).

27. G.D. Mahan, "Many-Particle Physics" (Plenum, NY, 1981); M. Cini and A. D'Andrea, J. Phys. C $\underline{21}$, 193 (1988).

28. A. Herzenberg, J. Phys. B $\underline{1}$, 548 (1968); D.T. Birtwistle and A. Herzenberg, ibid. $\underline{4}$, 53 (1971).

29. H.J. Korsch, A. Ernesti, and J.A. Núñez, J. Phys. B $\underline{25}$, 773 (1992).

Part VI

Electron Stimulated Desorption
and Chemistry of Adsorbates
on Metals

Isotope Effect in Electron Stimulated Desorption – The Role of Internal Degrees of Freedom in CO Desorption from Pt(111)

A. Szabó and J.T. Yates, Jr.

Surface Science Center, Department of Chemistry,
University of Pittsburgh, Pittsburgh, PA 15260, USA

Abstract. Electron stimulated desorption of CO^+, O^+, metastable neutral CO^* and ground state neutral CO from CO/Pt(111) was studied, using isotopic substitution of CO. Four isotopic versions of CO were compared in their desorption behavior. Contrary to the prediction of theoretical models that suggest a decrease of the ESD yield with increasing mass of a given desorption product, the CO^+ and CO^* ESD yield were found to be larger from the (heavier) $^{12}C^{18}O$ than from the (lighter) $^{13}C^{16}O$ adsorbate. The O^+ ESD yields followed the expected trend $Y(^{16}O^+) > Y(^{18}O^+)$. No isotope effect was observed for ground state neutral CO desorption. Qualitative arguments explaining the anomalous isotope effect, and emphasizing the importance of internal dynamics, in particular the rotation of diatomic desorption products in the ESD process, are presented. Here it is shown that the velocity of departure of the carbon end of the rotating CO molecule may control neutralization or quenching effects for CO^+ or CO^* produced by electron stimulated desorption.

1. Introduction

We have studied the detailed effects of isotopic substitution in the CO molecule on the yields of various products produced by electron stimulated desorption(ESD) from this chemisorbed molecule on Pt(111). The ESD yield is determined by two factors: (1) the probability of excitation from the bound to the excited state, and (2) the probability of desorption of the excited species. The problem has been quantitatively formulated previously, assuming that the observed isotope effect is exclusively due to the mass-related change in the dynamics of the desorbing particle in the excited state [1,2]. The formula describing this isotope effect is:

$$Y = A \bullet \exp(-cM^{1/2}) . \tag{1}$$

Here Y is the yield, A and c are constant for one adsorbate/substrate system, and M is the mass of the desorbing particle. Desorption effects in accordance with this formula have often been observed for desorbing atomic species [1,3-13] and for some species desorbing from chemisorbed CO [3,11-14]. Equation [1] is obeyed because the lifetime of the excited species in the neutralization or quenching region near the surface is proportional to $(mass)^{1/2}$ of the particle.

We have studied ESD from the CO/Pt(111) system previously using the digital ESDIAD method [15]. ESD of O^+, CO^+, and an electronically excited ($a^3\pi$) metastable CO (CO^*) were detected as desorption products [11,16]. By varying the retarding potential we could separate all three of these desorbing particles, and measure their yield and angular distributions.

In the following, we will present ESD measurements of CO^+, CO^*, O^+, and ground state neutral CO. Four isotopic varieties of CO are employed in this work. Electron energies in the range 150 eV - 760 eV are used. For the diatomic excited species, CO^+ and CO^*, an opposite (anomalous) behavior to that predicted from equation [1] is observed, and this observation leads to new insights about the dynamics of neutralization or quenching of the excited diatomic species as they leave the surface.

2. Experimental

A digital ESDIAD analyzer was employed for studies of ionic ESD products and of the metastable CO. A quadrupole mass spectrometer was used (without using the ionization source) for studies of ESD-produced CO^+ and O^+. It was also used for detection of neutral CO by using a pulsed electron beam source and making differential CO pressure measurements above background, keeping the mass spectrometer's thermionic ionizer working.

3. Results

The relative yield of CO^+, CO^*, O^+, and ground state neutral CO has been measured for all four CO isotopic species. As shown in Figure 1, by using mixtures of two isotopic CO molecules

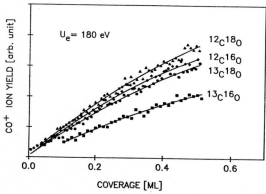

CO$^+$ ESD YIELD FROM CO/Pt(111)
(CONTINUOUS EXPOSURE)

Figure 1. CO^+ ion yields from CO/Pt(111) for the four CO isotopes studied. The data are corrected for gas phase ionization.

incident on the crystal, the relative ion yields were measured in separate cross calibrated adsorption experiments as a function of CO surface coverage. It can be seen that for the entire range of CO coverage, $Y(^{12}C^{18}O) > Y(^{12}C^{16}O) \cong Y(^{13}C^{18}O) > Y(^{13}C^{16}O)$. These results differ from the expectations of equation [1]. Relative O^+ yields were also measured. Here the ^{16}O- containing CO isotopic species exhibit higher relative O^+ yields than the ^{18}O-containing species, in accordance with expectations based on equation [1].

The cross section for CO^* production as a function of coverage was also measured. The desorption cross section, in the 10^{-22} cm^2 range, was measured assuming the collection and counting efficiency for CO^* species was unity for the microchannelplate detector in the ESDIAD analyzer. It was found that the CO^* yield of $^{12}C^{18}O$ exceeds that of $^{13}C^{16}O$. The other two isotopes have more comparable cross sections.

No differences in the CO neutral desorption cross sections were measured for the four isotopic CO species. For three electron energies, the cross section decreases as CO coverage increases. In addition, the cross section for neutral CO production increases slightly with increasing electron energy, remaining in the 10^{-18} cm^2 range, in good agreement with other results [17].

4. Discussion

Figure 2 summarizes the average CO^+ and CO^* yields on a relative basis from this work as a function of the location of the center of mass(CM) of the various CO isotopic species, measured from the carbon atom. A clear correlation exists between the yields of both CO^+ and CO^* species and the position of the center of mass of the diatomic. As the center of mass moves toward the oxygen end of the molecule, the

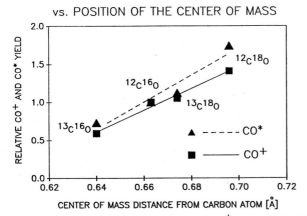

Figure 2. Averaged CO^+ and CO^* ESD relative yield versus the center of mass position in the CO molecule.

relative yield of both CO^+ and CO^* increases in monotonic fashion, changing by almost a factor of three. In contrast, ground state CO neutrals do not exhibit this trend, and O^+ ions follow the expected trend in which the lighter O^+ ions exhibit a higher yield than the heavier O^+ ions, in accordance with the MGR theory. The anomalous isotope effect for CO^+ and CO^* is independent of both CO coverage and primary electron energy. The independence on coverage suggests that structural changes in CO (atop vs. bridged) do not strongly influence the observed effect.

These results also indicate that the observed anomalous effect for both CO^+ and CO^* is independent of the final electronic state of the excited CO species. Since no anomalous effect was seen for neutral CO desorption, where the quenching rate from the excited state is probably less important dynamically, it is believed that the anomaly for CO^+ and CO^* relates to the neutralization or quenching effect on these species by the surface. Since the anomaly is not observed for the monatomic O^+ yield, the effect must be linked with the diatomic nature of the CO^+ or CO^* species. Key differences between monatomic and diatomic excited species are the various internal degrees of freedom possessed by the diatomic particles. The role of internal dynamical motion in contributing to the anomaly is strongly supported by the dependence of yield on the center of mass position in the diatomic species, as shown in Figure 2.

The internal motions that could be responsible for the anomaly of the ESD yield are the frustrated translation, the frustrated rotation, and the vibration normal to the surface. The frustrated translation is unlikely to be responsible for the anomalous isotope effect, because there is considerable thermal excitation of this mode ($\upsilon = 48 cm^{-1}$) in the temperature range 100-200 K, and previous measurements have shown that there is no temperature dependent change in the CO^* ESD yield in this temperature range[11]. Partitioning the zero point vibrational energy normal to the surface is also unlikely to be involved in causing the anomalous isotope effect, since differences in the thermal energy partitioned normally into the various isotopic CO molecules are very small for the different isotopic species [18,19]. The observation of the anomalous isotope effect in ESD for CO on Pt(111) can be qualitatively explained using classical considerations of the rotational behavior of the escaping diatomic particle. This rotation critically governs the motion of the carbon atom in the vicinity of the Pt surface during the onset of the ESD process, and hence the efficiency of neutralization or quenching.

Figure 3 shows a classical picture of a CO molecule departing from the surface with rotational momentum about the CM. A CO/Pt cluster model was employed [19]. It may be seen that as the CM is moved further from the carbon end of the molecule, the magnitude of Δz and Δx increase for a given short rotational time period under the potential free conditions assumed. The more rapid departure of the carbon atom in $^{12}C^{18}O$, compared to $^{13}C^{16}O$, is observed, and may be responsible for the greater ESD yield of the $^{12}C^{18}O$ as observed experimentally.

ROTATION OF DESORBING CO

ABOUT THE CENTER OF MASS

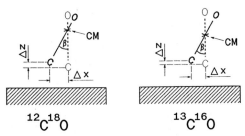

$^{12}C^{18}O$ $^{13}C^{16}O$

Figure 3. Classical dynamics of the CO molecule following ESD. Removal of the carbon end of CO from close proximity to the surface by rotation is more efficient for $^{12}C^{18}O$ than for $^{13}C^{16}O$ because of the position of the center of mass.

On the basis of these results, it seems clear that the neutralization or quenching of the CO diatomic species is governed partially by the rotational dynamics of the departing molecule. The motion of the originally-bound carbon end of the molecule is critical to the electronic deexcitation processes, and classical ideas are consistent with the trends observed for the various isotopic modifications of CO.

These results are consistent with other studies in which the internal dynamics of desorbing particles are considered. For example, Snowdon [20] showed that partitioning of energy in the desorbing molecule can result in excitation of both vibration and rotation. Hasselbrink [21] presented classical trajectory calculations, showing a concomitant increase in rotational and translational energy for a desorbing molecule. Burns, et al. [22] presented a model in which energy stored in the frustrated rotation of adsorbed NO is essential to ESD of neutral NO from Pt(111).

Acknowledgements

This work was supported by AFOSR under contract 82-0311. One of us (AS) acknowledges an Andrew Mellon Predoctoral Fellowship.

References

[1] T.E. Madey, J.T. Yates, Jr., D.A. King, and C.J. Uhlaner, J. Chem. Phys. **52**, 5215, (1970).

[2] R. Gomer in "Desorption Induced by Electronic Transitions, DIET I", Springer Series in Chemical Physics, Vol. 24, Eds. N.H. Tolk, M.M. Traum, J.C. Tully and T.E. Madey, (Springer-Verlag, Berlin 1983) p. 40.

[3] C. Leung, C. Steinbrüchel and R. Gomer, Appl. Phys. **14**, 79 (1977).

171

[4] M.L. Yu, Phys. Rev. **B19**, 5995 (1979).
[5] T.E. Madey, Surface Sci. **36**, 281 (1973).
[6] D.A. King and D. Menzel, Surface Sci. **40**, 399 (1973).
[7] W. Jelend and D. Menzel, Chem. Phys. Lett. **21**, 178 (1973).
[8] C. Klauber, M.D. Alvey and J.T. Yates, Jr., Surface Sci. **154**, 139 (1985).
[9] R.A. Rosenberg, P.R. LaRoe, V. Rehn, J. Stöhr, R. Jaeger, and C.C. Parks, Phys. Rev. **B28**, 3026 (1983).
[10] Q.-J. Zhang, R. Gomer and D.R. Bowman, Surface Sci. **129**, 535 (1983).
[11] M. Kiskinova, A. Szabó, A.-M. Lanzillotto and J.T. Yates, Jr., Surface Sci. **202**, L559 (1988).
[12] W. Riedl and D. Menzel, Surface Sci. **207**, 494 (1989).
[13] W. Riedl and D. Menzel in "Desorption Induced by Electronic Transitions DIET II", Springer Series in Surface Sciences, Vol. 4, Eds.. W. Brenig and D. Menzel (Springer-Verlag, Berlin 1985) p. 136.
[14] J.E. Houston and T.E. Madey, Phys. Rev. **B26**, 554 (1982).
[15] R.D. Ramsier and J.T. Yates, Jr., Surface Sci. Reports **12**, 243 (1991).
[16] M.Kiskinova, A. Szabó, and J.T. Yates, Jr., Surface Sci. **205**, 215 (1988).
[17] R.M. Lambert and C.M. Comrie, Surface Sci. **38**, 197 (1973).
[18] A. Szabó and J.T. Yates, Jr., accepted, J. Chem. Phys.
[19] A. Szabó and J.T. Yates, Jr., submitted to J. Chem. Phys.
[20] K. Snowdon in "Desorption Induced by Electronic Transitions DIET II", Springer Series in Surface Sciences, Vol. 4, Eds., W. Brenig and D. Menzel (Springer- Verlag, Berlin 1985) p. 116.
[21] E. Hasselbrink, Chem. Phys. Lett. **170**, 329 (1990).
[22] A.R. Burns. E.B. Stechel and D.R. Jennison, Phys. Rev. Letters **58**, 250 (1987).

Quantum-Resolved Angular Distributions of Neutral Products in Electron-Stimulated Surface Processes*

A.R. Burns, E.B. Stechel, and D.R. Jennison

Sandia National Laboratories, Department 1114,
Albuquerque, NM 87185, USA

Abstract Two-dimensional imaging of laser resonance-enhanced multiphoton ioniza-tion (REMPI) is used to obtain angular distributions of *neutral* molecules which are pro-ducts of electron stimulated desorption (ESD) and dissociation processes on adsorbate-covered surfaces. Quantum resolution allows the angular distributions to be state-specific. We briefly discuss angular distributions for ESD of metastable $CO^*(a^3\Pi_r, \nu=0)$ and ground-state NO from Pt(111); in addition, we discuss distributions for the NO pro-duct of NO_2 dissociation on clean and O-covered Pt(111). The results are discussed in terms of desorption mechanisms, dissociative forces, site geometries, and disordered coadsorbate layers.

1. Introduction

Quantum resolution of neutral ESD products via laser REMPI has resulted in detailed electronic and dynamical models of stimulated surface processes[1-5]. In much of this work, however, the product translational and internal energy distributions were in-tegrated over all angles. Since stimulated processes are *very* sensitive probes of local electronic and geometric structure of the adsorbate-substrate complex, potentially im-portant dynamical and structural information is lost in angle averaging. As discussed here, neutral desorbate angle resolution can be achieved by two-dimensional (2D) imag-ing of the REMPI signal[6].

The success of ESD *ion* angular distributions, or ESDIAD[7,8], as a adsorbate-substrate probe can be attributed to the correlation between bond directions and ion trajectories, suggesting strong repulsive forces along bond directions. This in turn may be understood to be caused by the localization of multihole excitations in surface-adsorbate and internal adsorbate bonds[9] (e.g., the $5\sigma^{-2}$ excitation which leads to CO^+ and CO^* desorption). However, the products of stimulated surface processes are pre-dominately neutral, since the lifetimes of electronic excitations giving rise to ion desorp-tion are generally very short relative to time scales of atomic motion. By examining the relatively higher probability neutral desorption angular distributions, one should be able to infer more general conclusions about important questions such as the effects of inter-nal rotational and vibrational motion, dissociation forces, and coadsorbates.

2. Experimental

The 2D imaging of ESD neutral desorbates is based on the technique pioneered by Winograd[6] for ion-sputtered neutrals. As shown in Fig. 1, a ribbon-shaped pulsed laser beam parallel to the surface resonantly ionizes desorbates which are in specific quantum states; the ions are then accelerated into a position-sensitive MCP detector (multi-channel plate array and phosphor screen, not shown). The electron gun is pulsed and the detector is gated to avoid desorbed ions. The image is captured by a CCD camera and digitally stored in 240×240 pixel arrays as shown in the intensity map of $CO^*(a^3\Pi,$

---

* Supported by the U.S. Department of Energy under Contract # DE-AC04-76DP00789.

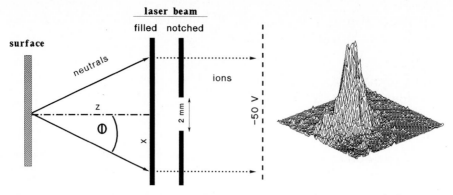

FIG. 1. Left: Schematic of REMPI 2D imaging. Right: Intensity profile of $CO^*(a^3\Pi_r, \nu=0)$

$\nu=0$) ESD in Fig. 1. Determination of the polar angle $\phi = \arctan(x/z)$, where z is known, is accomplished by calibrating the distance x with a temporary, known, "notch," or blank spot, in the laser beam[10]. (In Fig. 1, the 2-mm-notched beam is displaced from the filled beam for display). Each angular distribution presented here is an average of separate cross-sections taken at evenly-spaced azimuths through the center ($\phi=0°$) of the 2D image. It is not symmetrized, thus "noise" will be asymmetrical about the surface normal. All Gaussian fits, however, assume that the distributions are symmetrical about the surface normal.

3. Angular Distributions of CO^* and NO ESD from Pt(111)

Extensive work by Yates and coworkers[8] clearly shows that ESDIAD of electronically-excited species such as CO^* reveals significant information about ground-state adsorbate bonding geometries, adsorbate-adsorbate interactions, and low-energy adsorbate hindered translational motion. In those studies, the long-lived $a^3\Pi$ CO^* is detected by de-excitation at an MCP. Here, we use the laser to resonantly ionize $CO^*(\nu=0)$ via the transition $b^3\Sigma(\nu=0) \leftarrow a^3\Pi(\nu=0)$. The laser-generated ions are then imaged on the MCP as shown in Fig. 1. Although the rotational REMPI spectrum is currently unassigned, the images are completely quantum-resolved with respect to rotational and vibrational level.

There is excellent agreement in most cases between our CO^* REMPI data and the ESDIAD data for CO/Pt(111). As shown at the top of Fig. 2, the CO^* angular distribution from 0.5 ML CO on Pt(111) at 90 K is sharply peaked at the surface normal with a half-width at half-maximum (HWHM) of 6±0.2°. The dashed curve is a Gaussian fit. Yates, et al.[8], were able to show that the normal CO^* angular distribution from 0.5 ML CO/Pt(111) broadened from a HWHM of 5 ±0.5° at 90 K to 8.2±0.5° at 300 K; this is consistent with thermal excitation of the 48-60 cm^{-1} hindered translational motion[11]. A similar temperature broadening was observed for the CO^* REMPI angular distribution from 90 K (HWHM=6±0.2°) to 290 K (HWHM=9.0±0.3°)[10].

At much higher CO doses (>10 L, 90 K), a distinct *narrowing* of the REMPI angular distribution (dashed Gaussian HWHM=4.5±0.2°) was observed, as shown at the bottom of Fig. 2. This result contrasts with off-normal CO^* ESDIAD distributions that were observed for $\theta_{CO} > 0.6$ ML, and attributed to tilted CO species[8] along dense domain "walls." Possibly in our REMPI experiments, the >0.5 ML CO coverage was not dense enough to cause tilting, but did increase lateral CO-CO repulsion. The latter could reduce the amplitude of hindered translational motion and thereby reduce the range of angular-dependent desorption forces.

174

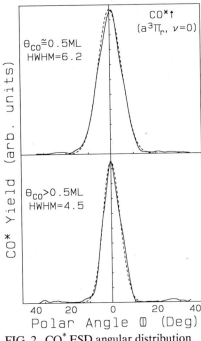

FIG. 2. CO* ESD angular distribution from CO/Pt(111)

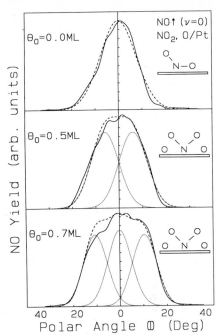

FIG. 3. NO product distribution from NO₂ dissociation on Pt(111)

The angular distribution of *ground-state* NO from a Pt(111) surface pre-covered with 0.25 ML atomic O is Gaussian with a 7±0.5° HWHM about the surface normal. Remarkably, this is comparable to the CO* distributions from both clean and θ_O =0.25 ML pre-covered Pt(111)[8]. For low θ_O, NO and CO (those yielding CO*) occupy atop sites that are surrounded symmetrically by an ordered 2×2 array of O atoms in three-fold hollows[8,12]; the coadsorbed oxygen enhances the NO ESD yield[2-4] relative to the clean surface, but evidently does not alter the desorption trajectories. Furthermore, the sharp NO angular distribution is also similar to that of CO⁺ *ion* distributions from several transition metal surfaces[7-9]. This is very significant because these desorption events involve very different excited states: a one-hole excitation ($5\sigma^{-1}$) in neutral NO desorption and a two-hole excitation ($5\sigma^{-2}$) in metastable CO* and ionic CO⁺ desorption, which result in very different translational energies.

So far no changes in the angular distributions with internal energy have been observed for NO ESD. This is completely expected for vibrational excitation[1], but less so for rotation because the adsorbate hindered rotor rotates the bond angle relative to the surface normal. If there is a net force lateral to the surface due to this motion, there would also be a torque imposed on the desorbate, coupling these two motions. However, *the results suggest that the parallel degree of freedom (hindered translation) dominates the angular distribution for these systems.*

4. NO Product Angular Distributions in the Stimulated Dissociation of NO₂ on Pt(111)

The molecular adsorption of NO₂ on clean Pt(111) has been characterized[12] as a "side-bonded" complex where the nitrogen atom and one oxygen atom are each coordinated to Pt atoms at a bridge site and the other oxygen atom points away from the surface, as shown at the top of Fig. 3. The O-N-O bond angle is estimated to be 115°[12]

175

(vs. 134° in the gas-phase), which implies that the uncoordinated N-O bond is ~25° from the surface normal. Stimulated dissociation of this NO_2 adsorption complex has been examined in detail with REMPI of the gas-phase $N\overset{\circ}{O}$ product[3]; the O fragment remains on the surface, falling into the strong potential well of the three-fold hollow site. Due to dissociative intra-molecular excited-state forces, the vibrational and rotational energies (>850 K) of the NO product are appreciable[3,4,13].

It is immediately apparent from the top of Fig. 3 that the NO product angular distribution is well described by a single Gaussian (dashed curve with a HWHM of 9±0.5°) peaked along the *surface normal*. Furthermore, no change in angular distributions accompanies changes in rotational energy (15-822 cm^{-1}, J=2.5-21.5), and a similar angular distribution was observed for vibrationally-excited NO (ν=1, J=10.5). *Evidently, there is no correlation of the product angle with either the initial bond direction or the degree of rotational or vibrational excitation.* Finally, the trajectories do not reflect the dissociative forces parallel to the surface, but instead, indicate that the dominant force on the NO fragment is normal to the surface.

When Pt(111) is pre-covered with atomic oxygen, the NO_2 adsorption geometry changes to "N-bonded"[12] with both oxygens pointing away from the surface, as shown in the middle (θ_O=0.50 ML) and (θ_O=0.75 ML) bottom of Fig. 3. Electron stimulated dissociation of the N-bonded configuration results in the formation of both NO *and* atomic O in the gas-phase[5]. Unfortunately, the atomic O signal is difficult to image due to the inefficiency of the (2+1) REMPI scheme which generally requires tight laser focussing[5]. The NO angular distributions for the N-bonded adsorbates are characterized by steeper wings and flatter tops which cannot be fit by a single Gaussian as in the case of the side-on data, but can be fit by sums (dashed curves) of Gaussians. The data for θ_O=0.5 ML pre-coverage in the middle of Fig. 3 is most simply fit with 6° off-normal components (HWHM=6±0.5°). At θ_O=0.25 ML (not shown), there is also evidence for off-normal NO trajectories, but the relative amounts of N-bonded and side-bonded NO_2 are not known. At the bottom of Fig. 3, for θ_O=0.75 ML, the NO product data is characterized by different Gaussian components: 9-10° off-normal trajectories together with a distribution normal to the surface (all HWHM=6 ±0.5°).

The data can be understood as follows. The θ_O=0.25 and 0.50 ML overlayers are believed to be 2×2 *ordered* arrays which have *symmetric* atop sites. Since the latter precludes asymmetric forces from the substrate environment (as seen in the NO ESD from θ_O=0.25 ML/Pt), the data indicate that there is an off-normal force which may be due to the intramolecular dissociative force of the N-bonded NO_2. These results thus contrast with the side-bonded NO_2 dissociation on the clean surface, where the substrate-molecule normal force apparently dominates the angular distributions. For the θ_O=0.75 ML covered surface, the O overlayer is substantially *disordered*[14], resulting in NO_2 adsorption at *asymmetric* atop sites, which may cause adsorbate tilting and possibly asymmetric substrate forces. The ESD angular distributions of NO[15] from the disordered θ_O=0.75 ML surface also appear to have off-normal components.

For NO_2 dissociation on both clean and O-covered Pt(111), the absence of any changes in the NO product angular distributions with rotational energy indicates the absence of coupling of the NO_2 bending mode (795 cm^{-1} [14]) to the NO center-of-mass (COM) motion. This result is similar to the NO ESD data discussed above, where the hindered rotor does not appear to be strongly coupled to the COM motion. In addition, there are no apparent angle dependent interactions with neighboring O atoms that might have resulted in rotational excitation.

5. Summary

We have seen that state-resolved detection of neutral angular distributions from stimulated surface processes can yield very important and revealing information:

(1) The narrow (7±0.5° HWHM) angular distribution for stimulated desorption of ground-state NO from θ_O=0.25 ML pre-covered Pt(111) closely resembles that of CO* and CO$^+$ from several transition metal surfaces[7-9]. This is significant because these

desorption events involve very different excited states and translational energies, but similar ground-state hindered translational energies.

(2) The stimulated dissociation of side-on NO_2 on the clean surface is characterized by a normal angular distribution for the NO fragment, fit by a single Gaussian. This does *not* reflect the initial bonding geometry of the adsorbate, nor the dissociative forces which are most likely parallel to the surface.

(3) For N-bonded NO_2 dissociation on ordered O-precovered surfaces ($\theta_O \leq 0.5$ ML), off-normal components appear that may be correlated with dissociative forces. However, at the highest O coverages ($\theta_O = 0.75$ ML), the data show both off-normal and normal components which are likely due to the disordered, close-packed O overlayer. Similar distributions are seen for NO ESD from the same surface. Our results are consistent with asymmetric forces in the excited state and/or asymmetric ground-state geometries (tilting or bending) due to a disordered oxygen overlayer.

(4) In none of these studies was there observable variation in the angular distribution with internal energy of the NO desorbate or product. For ESD this suggests that the parallel degree of freedom (hindered translation) dominates the angular distribution for these systems. The absence of rotational coupling to the angular distribution is also observed in molecular NO_2 dissociation, where dissociative forces result in bond stretching and bond-bending. This appears to be the case even when the trajectories are influenced by disordered O overlayers.

References

1. E. B. Stechel, D. R. Jennison, and A. R. Burns, in <u>Desorption Induced by Electronic Transitions, DIET III</u>, edited by R. H. Stulen and M. L. Knotek (Springer-Verlag, New York, 1988), pp. 67-72, 136-143, 167-172.
2. D. R. Jennison, A. R. Burns, and E. B. Stechel, in <u>Desorption Induced by Electronic Transitions, DIET IV</u>, ed. by G. Betz and P. Varga, (Springer-Verlag, New York, 1990), pp. 41-45, 182-186.
3. A. R. Burns, D. R. Jennison, and E. B. Stechel, Phys. Rev. B <u>40</u> (1989) 9485.
4. A. R. Burns, E. B. Stechel, D. R. Jennison, and T. M. Orlando, Phys. Rev. B <u>45</u> (1992) 1373.
5. T. M. Orlando, A. R. Burns, D. R. Jennison, and E. B. Stechel, Phys. Rev. B <u>45</u> (1992) 8679.
6. Paul H. Kobrin, G. Alan Schick, James P. Baxter, and Nicholas Winograd, Rev. Sci. Intrum. <u>57</u> (1986) 1354.
7. R. H. Stulen, Prog. in Surf. Sci. <u>32</u> (1989) 1.
8. R. D. Ramsier and J. T. Yates, Jr., Surf. Sci. Reports <u>12</u> (1991) 243.
9. W. Riedl and D. Menzel, Surf. Sci. <u>207</u> (1989) 494.
10. A. R. Burns, Surf. Sci. (in press).
11. A. M. Lahee, J. P. Toennies, and Ch. Wöll, Surf. Sci. <u>177</u> (1986) 371.
12. M. E. Bartram, R. G. Windham, and B. E. Koel, Langmuir <u>4</u> (1988) 240.
13. A. R. Burns, E. B. Stechel, and D. R. Jennison, Phys. Rev. Lett. <u>58</u> (1987) 250.
14. D. H. Parker, M. E. Bartram, and B. E. Koel, Surf. Sci. <u>217</u>, 489 (1989).
15. A. R. Burns, E. B. Stechel, and D. R. Jennison, Surf. Sci. (in press).

Electron Stimulated Desorption of the Metallic Substrate at Monolayer Coverage: Sensitive Detection via 193 nm Laser Photoionization of Neutral Aluminum Desorbed from CH₃O/Al(111)

C.E. Young[1], J.E. Whitten[1], M.J. Pellin[1], D.M. Gruen[1], and P.L. Jones[2]

[1]Materials Science/Chemistry Divisions, Argonne National Laboratory,
   Argonne, IL 60439, USA
[2]STI Optronics, Inc., 2755 Northup Way, Bellevue, WA 98004, USA

Abstract. A fortuitous overlap between the gain profile of the 193 nm ArF excimer laser and the Al autoionizing transition $^2S_{1/2}$ (51753 cm^{-1}) ← $^2P^0_J$ (J=1/2,3/2; 0,112 cm^{-1}) has been exploited in the <u>direct</u> observation of substrate metal atoms in an electron stimulated desorption (ESD) process from the monolayer adsorbate system CH₃O/Al(111). The identity of the mass 27 photoion was established as Al$^+$ by (1) isotopic substitution of ^{13}C in the methanol employed for methoxy formation, and (2) tunable laser scans utilizing the ^2D_J (J=3/2, 5/2) intermediate levels at ~32436 cm^{-1} and a 248 nm ionization step. An ESD yield of 1.7 x 10^{-6} Al atoms/(electron at 1 keV) was established by comparison with a sputtering experiment in the same apparatus (Ar$^+$, 3.6 keV). Velocity distributions measured for the desorbed Al species showed some differences in comparison with methoxy velocity data: a slightly lower peak velocity [800 m/s (Al); 1100 m/s (methoxy)] and a significantly less prominent high-velocity component.

1. Introduction

This paper is a summary of recent experiments by our group on the detection of clearly measurable Al$^+$ photoion signal resulting from the laser-ionization of neutral species desorbed in an ESD process from the methoxy/aluminum [CH₃O/Al(111)] system. Our previous measurements on CH₃O/Al(111) [1,2] established the utility of 193 nm nonresonant photoionization for the measurements of yields and velocity distributions of CH₃O resulting from ESD with 3 keV primary electrons. The present work demonstrates the desorption of a substantial yield of substrate atoms from a single-crystal metallic system having only monolayer methoxy coverage. Fuller accounts of these experiments have been prepared for submission elsewhere.

2. Apparatus and Procedures

For the present measurements, a high-sensitivity laser-ionization mass spectrometer built in our laboratory [3,4] was equipped with an electron gun and focusing-lens stack, making either electron or ion bombardment of the sample possible while maintaining the photoion extraction conditions invariant. For ESD experiments, primary electron energies in the hundred volt to kilovolt range were conveniently available; 1 keV was employed in the present measurements. The extraction optics (and other elements) were verified to be operating at potentials low enough for any extraneous electron current arriving at the sample to be negligible. The electron gun could be operated in a pulsed mode (≥ 100 ns) for obtaining time-of-flight (TOF) velocity distribution data or with long pulses (averaging ~1 µA in a 1 mm spot) in order to establish steady-state conditions for yield determinations. Sample preparation involved cleaning via standard heating and sputtering cycles with verification by AES, XPS, and LEED. Dosing with excess methanol (≥50 L) produces a monolayer methoxy coverage; supporting evidence and references have been summarized

Springer Series in Surface Sciences, Vol. 31
Desorption Induced by Electron Transitions DIET V
Editors: A.R. Burns · E.B. Stechel · D.R. Jennison   © Springer-Verlag Berlin Heidelberg 1993

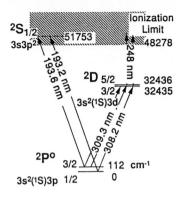

Fig. 1. Energy level diagram for the neutral aluminum atom. Autoionizing transitions are shown on the left and the resonance ionization pathways on the right.

previously [1,2]. The laser excitation scheme is illustrated in Fig. 1. For velocity distribution and yield measurements, 193 nm light from an ArF excimer laser was used under "soft focus" conditions (0.8 x 2.2 mm spot, 20 ns pulse). In Al resonance-ionization experiments, crystal second-harmonic doubling of dye laser output was synchronized with 248 nm radiation from a second excimer for the ionization step.

3. Results and Discussion

Most of the data on neutral Al ESD was obtained in the laser intensity region ≤ 5 MW/cm^2 where background signal from other species is small, in contrast to the regime ≥ 100 MW/cm^2 where the methoxy species is easily ionized and its photofragment ions appear. Two definitive tests established the identity of the mass 27 neutral desorbate: (1) expected mass shifts when CH$_3$O was prepared by dosing with ^{13}C methanol, and (2) the resonance ionization experiment depicted in Fig. 2, where ~309 nm tunable light populated excited levels which were efficiently photoionized by 248 nm photons (~14 MW/cm^2). Despite some nonresonant signal from the intense 248 nm light, characteristic neutral Al resonances are clearly observed. Additional experiments established the following facts. (1) No mass 27 signal was observable when the ESD experiment was performed on clean Al(111). (2) Subsequently, the Al signal appeared clearly upon deposition of a saturation methoxy coverage. (3) Pyrolysis at 600K, a process known to leave only surface oxide and carbide species [5] caused a drastic reduction in the mass 27 ESD.

Velocity distribution data, shown in Fig. 3, was obtained via TOF from sample to photoionization volume (2.2 and 4.4 mm paths) with 193 nm radiation and electron

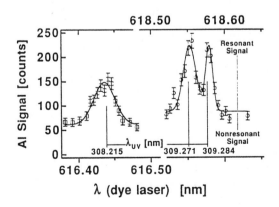

Fig. 2. Resonance ionization spectrum of Al atoms detected in the ionizing sequence of Fig. 1. Resonances in the Al photoionization spectrum occur at the expected ultra-violet wavelengths, generated by crystal frequency doubling.

179

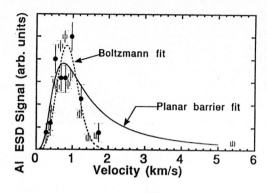

Fig. 3. Velocity distributions obtained for the desorbed neutral Al species by 193 nm photoionization. Data are for two flight distances - open points: 2.2 mm, solid points: 4.4 mm.

pulse lengths of ~1 µs. The peak velocity of 800 m/s was somewhat less than the corresponding value of 1100 m/s observed for methoxy desorption from Al(111). The Al signal is better represented by a narrow parametric model (Boltzmann) rather than the "planar barrier" formulation containing some high-energy component, needed to fit typical CH_3O desorption distributions [2].

An absolute yield for Al neutral desorption was reliably determined through a calibration experiment in which aluminum atoms were sputtered into the gas phase with a known yield, for Al(111), of $Y_s = 2.2$ atoms/(Ar^+ ion at 3.6 keV) [6]. The velocity distribution needed to convert the density data proportional to the detected signal to flux data proportional to yields was measured (Fig. 3) for ESD, and modelled by an accurate and established formulation [7] for sputtering. In the ESD experiment, the neutral Al density, n, in the ionization volume is given as n = $I_e Y G/(2v_m \sqrt{\pi})$, where I_e is the current of bombarding electrons (e^-/sec), $v_m = \sqrt{(2kT/mass)}$, i.e., the Boltzmann peak velocity (a fitting parameter here), and G = $<\cos\theta/R^2>$ a geometric average over the takeoff angle θ and the sample-to-volume distance R. A corresponding formula applies to the sputtering data, with the factor $2\sqrt{\pi}$ replaced by 4, I_e replaced by I_s, the ion current, and Y by Y_s, given above. Also, v_m is replaced by $v_b = \sqrt{(2E_b/mass)}$, E_b being represented by the heat of sublimation of aluminum, 3.36 eV. Assuming the factor G to be the same in both experiments, we calculate Y = 1.7 x 10^{-6} Al atoms/(1 keV electron) for the ESD process. The latter is significantly smaller than the methoxy yield (~10^{-3} at 3 keV [1, 2]), but larger than the corresponding measured ESD yield of ions (~10^{-6} protons/(3 keV electron)).

Our attempts to observe the parent neutral desorbate leading to the observed Al^+ photoions have thus far been unsuccessful, despite the ability to utilize a large range of laser intensities from ~100 MW/cm^2 to 0.1 kW/cm^2. However, some indication that the initial desorbate may be molecular is contained in laser saturation curves, i.e., Al^+ photoion signal versus laser intensity. The ESD process is compared with sputtering, where the released species is overwhelmingly atomic. The ESD signal was observed to decrease more rapidly than that from sputtering as the laser intensity is reduced. Thus, a photodissociation scheme AlX ->Al ->Al^+ is suggested, in which the parent molecule, AlX, is laser-dissociated to Al atoms which subsequently ionize efficiently due to the 193 nm overlap with a 1-photon autoionizing transition (see Fig.1). Quantitative modelling is in progress. It can be noted that the AlOCH$_3^+$ species is known in SIMS monitoring of the CH$_3$O/Al(111) system [8]. The Al-O bond in the neutral species $AlOCH_3$ has recently been determined to be 5.44 ± 0.1 eV by G2 ab initio theory [9]. Calculations at a lower accuracy level with the methoxy species bound on various clusters $(Al)_n$ confirm the picture of a strong Al-O bond [9].

Finally, we note that ESD followed by 193 nm laser mass spectrometry on the analogous system - water-dosed Al(111) - exhibits numerous easily detectable neutral aluminum molecular species as indicated in Fig. 4. The AlOH parent is seen, as well as various AlH_n fragments, the latter being expected in view of the recent observation of electron-bombardment fragments from AlH_3 and Al_2H_6 thermally desorbed from hydrogen-dosed aluminum [10]. Consequently, the $AlOCH_3$ neutral

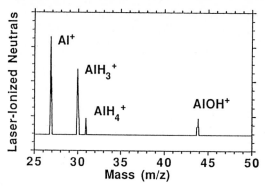

Fig. 4. Photoionization mass spectrometry at 193 nm on neutral species desorbed by 1 k eV electrons from slightly oxidized, water-dosed Al(111). Laser intensity was ~1 MW/cm^2.

desorbate (or other parent precursor) may be detectable in another wavelength range where the photodissociation process may be less dominant.

Acknowledgement

This work was supported by the U.S. Department of Energy, BES-Materials Sciences, under Contract No. W-31-109-ENG-38.

References

1. C. E. Young, J. E. Whitten, M. J. Pellin, D. M. Gruen and P. L. Jones. Desorption Induced by Electronic Transitions (DIET IV). G. Betz and P. Varga eds. (Springer-Verlag, Berlin, 1990), p.187.
2. J. E. Whitten, C. E. Young, M. J. Pellin, D. M. Gruen and P. L. Jones. Surf. Sci. **241**, 73 (1991).
3. C. E. Young, M. J. Pellin, W. F. Calaway, B. Jørgensen, E. L. Schweitzer and D. M. Gruen. Inst. Phys. Conf. Ser. **84**, 163 (1987).
4. C. E. Young, M. J. Pellin, W. F. Calaway, B. Jørgensen, E. L. Schweitzer and D. M. Gruen. Nucl. Instrum. Methods Phys. Res. **B27**, 119 (1987).
5. J. G. Chen, P. Basu, L. Ng and J. T. Yates Jr. Surf. Sci. **194**, 397 (1988).
6. M. T. Robinson and A. L. Southern. Journal of Applied Physics. **38**, 2969 (1967).
7. P. Sigmund. Phys. Rev. **184**, 383 (1969).
8. I. F. Tindall and J. C. Vickerman. Surf. Sci. **149**, 577 (1985).
9. L. A. Curtiss and D. Kock. Private communication.
10. M. Hara, K. Domen, T. Onishi, H. Nozoye, C. Nishihara, Y. Kaise and H. Shindo. Surf. Sci. **242**, 459 (1991).

Structure and Kinetics of Electron Beam Damage in a Chemisorbed Monolayer: PF₃ on Ru(0001)

T.E. Madey[1], H.-S. Tao[1], L. Nair[1], U. Diebold[1], S.M. Shivaprasad[1], A.L. Johnson[1], A. Poradzisz[1], N.D. Shinn[2], J.A. Yarmoff[3], V. Chakarian[3], and D. Shuh[3]

[1]Department of Physics and Laboratory for Surface Modification, Rutgers, The State University of New Jersey, Piscataway, NJ 08855, USA

[2]Sandia National Laboratories, Albuquerque, NM 87185, USA

[3]Department of Physics, University of California-Riverside, Riverside, CA 92521, USA

Abstract. We have used a combination of methods to probe the structure and kinetics of electron beam induced damage in a monolayer of PF₃ on Ru(0001). This is a particularly rich system, in which molecularly adsorbed PF₃ is reduced to PF₂, PF and P by electron bombardment. The concentrations and kinetics of damage by 550 eV electrons are measured as a function of surface temperature (100 to 300K) and PF₃ coverage using soft x-ray photoemission spectroscopy (SXPS) excited by synchrotron radiation. Structures of fragments and ion desorption kinetics are measured using electron stimulated desorption ion angular distribution (ESDIAD). Evidence is seen for quenching of DIET processes via intermolecular interactions at high coverages. Damage rates and product distributions vary with temperature, due to a competition between DIET and thermal kinetic processes.

1. Introduction

Electron and photon induced decomposition of PF₃ adsorbed on metal surfaces has been the subject of several recent studies [1-7]. PF₃ bonds to metals via the P atom, with the F atoms pointing away from the surface [1-10]. DIET (Desorption Induced by Electronic Transitions) of PF₃ is particularly rich and interesting. Upon irradiation with electrons or soft x-ray photons, PF₃ decomposes sequentially to form adsorbed PF₂, PF and P [2, 4-7]. Evidence for the formation of these surface decomposition products was first obtained in electron stimulated desorption ion angular distribution [ESDIAD] studies of PF₃ on Ni(111) [1,2] and Ru(0001) [3-7], where distinct electron beam-induced changes in ESDIAD patterns and intensities signaled their appearance. The surface decomposition products were then identified unambiguously by soft x-ray photoemission (SXPS) of the P2p level using synchrotron radiation [5]: there is a nearly 6 eV chemical shift in P2p emission between adsorbed PF₃ and adsorbed P.

The present study focuses on a number of issues that were not addressed in the previous work. We make a direct quantitative comparison of electron beam damage in PF₃ adsorbed on Ru(0001) using SXPS and ESDIAD. Our goals are to determine the kinetics and mechanisms for beam damage in this complex system, and to determine the role of intermolecular interactions in the DIET processes. To accomplish this, we

Springer Series in Surface Sciences, Vol. 31
Desorption Induced by Electron Transitions DIET V
Editors: A.R. Burns · E.B. Stechel · D.R. Jennison   © Springer-Verlag Berlin Heidelberg 1993

- Identify and measure concentrations of beam-induced damage products quantitatively using SXPS,
- Measure kinetics of disappearance and formation of surface species as a function of initial PF$_3$ coverage, and
- Correlate F$^+$ ESDIAD (structures, intensities, damage rates) with SXPS over a wide range of coverage and temperature.

The preliminary results reported here indicate that DIET of adsorbed PF$_3$ is far more complex than had been recognized previously. The rates of electron stimulated appearance and disappearance of the PF$_x$ species depend strongly on both coverage and temperature. The key observations are: (a) At low coverages, the cross section for disappearance of PF$_3$ is larger than at high coverages, indicating a competition between substrate mediated deexcitation and intermolecular deexcitation processes, and (b) The temperature effects indicate a competition between temperature independent DIET processes and slow temperature dependent thermal reactions.

2. Experimental Procedures

The experiments described here were performed in two different stainless steel ultrahigh vacuum (uhv) systems. The SXPS measurements were made at the National Synchrotron Light Source, Brookhaven National Laboratory, Beamline U16, in a uhv chamber containing a sample manipulator (coolable to ~100K), a double pass cylindrical mirror analyzer, an electron gun for damaging the PF$_3$ layer, and a gas doser [11]. The combined resolution of the monochromator and the CMA is ~0.3 eV. The ESDIAD measurements were made at Rutgers in a uhv system containing a concentric hemispherical analyzer for AES, gas doser, electron gun, and LEED/ESDIAD optics that contain planar grids and a resistive anode detector [12,13]. The two-dimensional ESDIAD signal is digitized and processed using a PC-based system. Typical electron beam currents for ESDIAD measurements are ~1 nA; currents used to damage the surface are in the 100 nA range. The beam damage results reported herein are for 300 to 550 eV electrons.

3. Results and Discussion

3.1 SXPS of Electron Beam Damage

SXPS measurements of electron beam damage to PF$_3$ adsorbed on Ru(0001) are carried out using procedures similar to those described in Ref. [6], although in a different vacuum chamber [11]. Briefly, the clean Ru(0001) crystal is exposed to PF$_3$ at a sample temperature of either 100K or 300K. The PF$_3$ coverage is determined from the relative intensity of PF$_3$-derived valence band features, and checked by the relative intensity of P2p photoemission, as compared to the saturation coverage. The sample is irradiated with 550 eV electrons for increasing doses (expressed as electrons/cm^2). P2p photoemission spectra are measured using 160 eV photons for the undamaged surface and for the damaged area after each electron dose (the photon spot size of ~2 mm^2 is smaller than the electron irradiated area).

Figure 1 shows a series of P2p photoemission spectra for 0.33 ML (6 Langmuirs) of PF$_3$ adsorbed at 300K. Initially, only the P2$p3/2, 1/2$ photoemission from

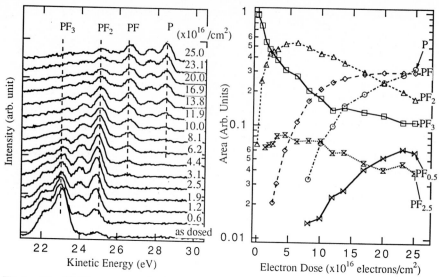

Fig 1   P2p SXPS spectra of electron
stimulated decomposition of PF₃ on
Ru(0001). Electron energy = 550 eV.

Fig 2   The concentration for each of
the surface products in Fig 1, plotted
as a function of electron dose.

molecular PF₃ is observed.   As the electron dose increases, the sequential
appearance of P2p-emission due to PF₂, PF and P occurs. The data are analyzed
using a curve fitting routine in which Gaussian line shapes are assumed. The
areas of each P2p₃/₂, ₁/₂ pair are plotted as a function of beam dose on a semilog
plot in Figure 2. The drop in PF₃ intensity is accompanied by a rapid increase of
PF₂ intensity, which passes through a maximum. After 2.5 x 10¹⁷ electrons/cm²,
the dominant surface products are P and PF. Traces of chemically shifted species
labeled as PF₂.₅ and PF₀.₅ are believed to be PF₂ and P, respectively, interacting
with dissociation fragments.

     The initial decomposition process is PF₃ + e → PF₂ + F, with sequential loss of
F leading to the other products. The initial decomposition cross section is
obtained from the slope of the PF₃ intensity vs. time for data such as those in
Figure 2. For 0.09 ML of PF₃, the cross section is 6 ± 1 x 10⁻¹⁷cm²; for 0.33 ML, the
cross section is 3 ± 0.5 x 10⁻¹⁷cm². To within experimental error, there is no
temperature dependence.   The coverage dependence is striking, however,
suggesting that intermolecular interactions at high coverages quench the DIET
processes that cause beam-induced decomposition of PF₃.

3.2 ESDIAD of Beam Damage

A sequence of ESDIAD patterns accompanying the electron beam induced
decomposition of PF₃ on Ru(0001) is given in Figure 3. These data are for ~0.25
ML at 300K, and are similar to observations of Alvey and Yates for PF₃/Ni(111)
[1,2]. The halo of F⁺ emission on the undamaged surface is identified with freely
rotating PF₃. (Note that rotation is hindered at 0.33 ML, the saturation coverage
of PF₃ [3,4]). As the electron dose increases, successive changes in the patterns are

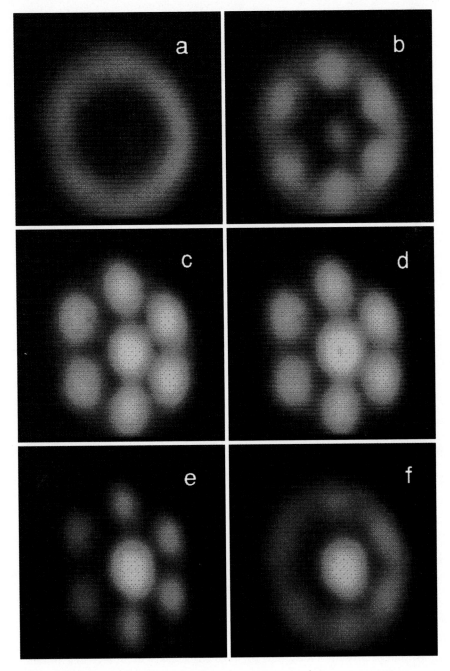

Fig 3   Beam damage sequence of ESDIAD patterns of F^+ from 0.25ML of PF$_3$/Ru(0001) at 300K. a. As dosed, b. 1.6×10^{15} electrons/cm^2, c. 2.2×10^{16} electrons/cm^2, d. 4.2×10^{16} electrons/cm^2, e. 1.2×10^{17} electrons/cm^2, f. annealed.

185

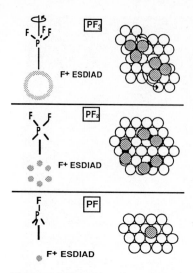

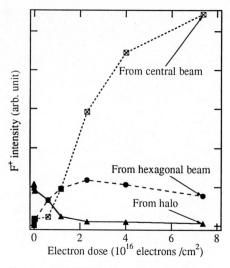

Fig 4   Model of PF_x (x=1, 2, 3) on Ru(0001) consistent with ESDIAD ans SXPS data[6].

Fig 5   ESDIAD F^+ intensity v.s. electron intensity. In each case, the F^+ signal is measured over the same solid angle.

observed. A hexagonal array of beams grows in as the halo fades, and a normal (central) beam grows. Based on the SXPS data and by analogy with ESDIAD data in [1-5], we identify the hexagonal beams with 3 domains of PF_2, and the central beam with PF. A model of these structures, consistent with the calculations of Chan and Hoffman for PF_x on Ni(111) [14], is given in Figure 4. Figure 5 is a plot of ESDIAD intensity as a function of electron dose. The qualitative similarity to Figure 2 is clear. However, there are quantitative differences due to coverage dependent ion desorption cross sections.

3.3 Temperature Dependence of Damage

Although the initial damage cross section for loss of PF_3 exhibits little temperature dependence, the rates of appearance (and disappearance) of the PF_x fragments (x = 0, 1, 2) depend strongly on temperature. A particularly sensitive indicator of temperature effects is the rate of appearance of PF as a damage product: at 100K, the PF product appears almost immediately in both SXPS and ESDIAD signals upon damaging with electrons, whereas there is a substantial delay in the appearance of PF at 300K [13,15]. This is believed to be a consequence of slow thermal annealing effects that change the effects of DIET processes at ~300K and higher temperatures [6].

Thermal effects are manifest directly in the present work by observing the time evolution of a damaged surface at 300K in the <u>absence</u> of electron bombardment. Figure 6 illustrates the time dependence of surface PF_x product concentrations for a saturation PF_3 layer exposed to a dose of ~ 2.5 x 10^{17} electrons/cm^2, and then held at 300K without further electron damage. The PF_3 signal grows as the PF_2 signal drops; PF increases very slightly. Some possible thermal kinetic processes include:

186

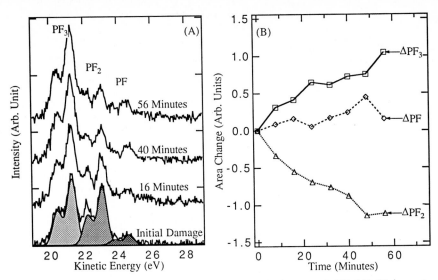

Fig 6 Time evolution of PF_x (x=1, 2, 3) on Ru(0001) using SXPS; (A) P2p spectra, (B) Concentration change v.s. time.

$$PF_2(a) + F(a) \rightarrow PF_3(a) \quad ,$$
$$PF_2(a) + PF_2(a) \rightarrow PF_3(a) + PF(a) \quad ,$$
$$PF_2(a) + PF(a) \rightarrow PF_3(a) + P(a) \quad \bullet$$

Evidence for irreversible thermal formation of PF_3 is seen also using ESDIAD in Fig. 3e, f, where heating the damaged surface causes the PF_3 halo to reappear as the PF_2 hexagon weakens in intensity; a similar effect occurs for PF_3/Ni(111) [2].

4. Summary and Conclusions

The rates of electron stimulated appearance and disappearance of the PF_x species depend strongly on both coverage and temperature. At low coverages, the cross section for disappearance of PF_3 is larger than at high coverages, indicating a competition between substrate mediated deexcitation and intermolecular deexcitation processes. The temperature effects indicate a competition between temperature independent DIET processes and slow thermal reactions.

5. Acknowledgment

S. M. Shivaprasad acknowledges partial support from a BOYSCAST Fellowship (India) and A. Poradzisz acknowledges support from the Maria Sklodowska-Curie Foundation (US-Polish Joint Committee). That part of this work carried out at NSLS was supported in part by US Department of Energy, BES.

187

References

[1] M. D. Alvey, J. T. Yates, Jr., K. J. Uram, J. Chem. Phys. <u>87</u>, 7221 (1987).
[2] M. D. Alvey, J. T. Yates, Jr., J. Amer. Chem. Soc. <u>110,</u> 1782 (1988).
[3] A. L. Johnson, S. A. Joyce, T. E. Madey, Phys. Rev. Lett. <u>61</u>, 2578 (1988).
[4] S. A. Joyce, A. L. Johnson,T. E. Madey, J. Vac. Sci. Technol. <u>A7</u>, 2221 (1989).
[5] T. E. Madey, S. A. Joyce, J. A. Yarmoff, Springer Series in Surface Sciences <u>22</u>, 55 (1990).
[6] S. A. Joyce, J. A. Yarmoff ,T. E. Madey, Surf. Sci. <u>254</u>, 144 (1991).
[7] S. A. Joyce, A. L. Johnson, J. A. Yarmoff, L. Nair, T. E. Madey, in press.
[8] F. Nitschke, G. Ertl, J. Kuppers, J. Chem. Phys. <u>74</u>, 5911 (1981).
[9] K. L. Shanahan, E. L. Muetterties, J. Phys. Chem. <u>88</u>, 1996 (1984).
[10] Y. Zhou, G. E. Mitchell, M. A Henderson, J. M. White, Surf. Sci. <u>214</u>, 209 (1989).
[11] N. D. Shinn and K. L. Tsang, J. Vac. Sci. Technol. <u>A9</u>, 1558 (1991).
[12] R. A. Baragiola, L. Nair, T. E. Madey, Nucl. Instr. Meth. <u>B58</u>, 322 (1990).
[13] L. Nair, A. Johnson, S. Shivaprasad, A. Poradzisz, T. Madey, in preparation.
[14] A. W. E. Chan, R. Hoffmann, J. Chem. Phys. <u>92</u>, 699 (1990)
[15] H. S. Tao, U. Diebold, V. Chakarian, D. Shuh, J. A. Yarmoff and T. E. Madey, in preparation.

Electron-Induced Decomposition
of Metal Carbonyls on Ag(111)

R.D. Ramsier, M.A. Henderson, and J.T. Yates, Jr.

Surface Science Center, Department of Chemistry,
University of Pittsburgh, Pittsburgh, PA 15260, USA

Abstract. The adsorption of $Fe(CO)_5$ and $Ni(CO)_4$ on Ag(111) at 90 K proceeds molecularly with minimal decomposition, with both molecules desorbing in the temperature range 150-190 K. Bombardment of the adsorbed molecules with low energy electrons induces fragmentation and the formation of adsorbed $M_x(CO)_y$ clusters. The cluster species thermally decompose (240 - 400 K) liberating CO(g) and leaving pure metal particles on the Ag substrate. The total cross section for electron induced decomposition of the parent molecules is in the range 2-14 x 10^{-16} cm^2.

1. Introduction

Organometallic complexes have become increasingly important in the chemical vapor deposition of metallic films [1] and thus the use of electron bombardment to activate the decomposition of metal carbonyls has received particular attention [2-6]. This paper will summarize recent results on the thermal and electron-induced chemistry of $Fe(CO)_5$ and $Ni(CO)_4$ adsorbed on Ag(111) at 90 K [5,6]. These results suggest that, in both cases, low-energy electrons efficiently produce surface-bound $M_x(CO)_y$ clusters, which thermally decompose leaving metal deposits on the Ag surface and liberating gas phase CO. These cluster species are more stable with respect to thermal and electron-induced decomposition than the parent molecules, which is interpreted in terms of an increased coupling of the Fe and Ni cluster atoms to the substrate.

2. Experimental

The specifics of the measurements to be reported here have been described in detail previously [5,6]. Ag(111) was chosen as a substrate since it will not bind CO at 90 K [7,8], which was important due to the fact that the carbonyl molecules decomposed slightly in the gas handling/dosing system [5,6,9]. Ag(111) was obtained by thermal evaporation and epitaxy of Ag multilayers on a clean, well-characterized Pt(111) crystal [10]. All molecular dosing, electron bombardment, Auger electron spectroscopy (AES) and electron stimulated desorption ion angular distribution (ESDIAD) measurements were performed with the Ag(111) surface held at 90 K. Temperature programmed desorption (TPD) data were collected with the crystal biased (-150 V) to prevent stray electrons from reaching the adsorbed molecules.

Springer Series in Surface Sciences, Vol. 31
Desorption Induced by Electron Transitions DIET V
Editors: A.R. Burns · E.B. Stechel · D.R. Jennison   © Springer-Verlag Berlin Heidelberg 1993

3. Results

3.1 Electron Bombardment Effects Measured by TPD

Figure 1 shows TPD spectra (monitoring CO^+ from $Fe(CO)_5$ or CO) from one monolayer of $Fe(CO)_5$ on Ag(111) following various exposures to 132 eV electrons (drawn from the mass spectrometer by biasing the crystal positively) [5]. As the electron fluence increases, the molecular $Fe(CO)_5$ desorption state (181 K) diminishes, and a 330 K CO desorption feature grows. This high temperature state yields no Fe species in TPD and is assigned to the thermal decomposition of $Fe_x(CO)_y$ clusters formed by electron bombardment. Electron fluences up to 10^{17} cm^{-2} did not measurably change the integrated area in the 330 K feature, implying that the clusters have a much smaller cross section for electron-induced dissociation than the parent $Fe(CO)_5$ molecules (the $Fe(CO)_5$ ESD total cross section was 14×10^{-16} cm^2, as determined from the data of fig. 1).

Figure 2 illustrates the effects of 115 eV electron bombardment on the TPD spectra of $Ni(CO)_4$ on Ag(111) [6]. The molecular $Ni(CO)_4$ desorption feature (150–170 K) diminishes with increasing electron fluence, concomitant with an increase in CO desorption from $Ni_x(CO)_y$ species near

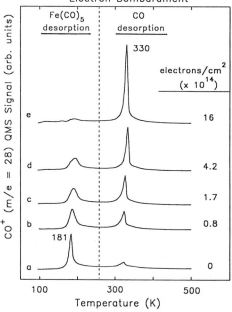

Fig. 1. TPD spectra after 132 eV electron bombardment of 1 monolayer of $Fe(CO)_5$ on Ag(111) at 90 K. A linear temperature ramp of 4 K/s was employed. CO^+ (m/e = 28) was monitored [5b].

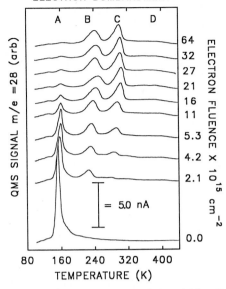

TEMPERATURE PROGRAMMED DESORPTION
FROM Ni(CO)$_4$ ON Ag(111) FOLLOWING
ELECTRON BOMBARDMENT

Fig. 2. TPD spectra after 115 eV electron bombardment of
Ni(CO)$_4$ on Ag(111) at 90 K (gas exposure =
13 x 10^{14} cm^{-2}). A linear temperature ramp of
3.3 K/s was employed. CO$^+$ (m/e = 28) was monitored
[6].

240, 300 and 400 K. No Ni-containing TPD products were
observed above 200 K. Note that the upper traces in fig. 2
are essentially identical, i.e. the clusters are insensitive
to electron bombardment as compared to the Ni(CO)$_4$
molecules. The total electron-induced decomposition cross
section for Ni(CO)$_4$ on Ag(111) is 2 x 10^{-16} cm^2 (determined
from the data in fig. 2).

3.2 ESDIAD

Electron bombardment of the parent molecules induces the
desorption of positive ions (CO$^+$ and O$^+$) and neutral
fragments (unidentified), predominantly in the direction of
the surface normal. The Fe(CO)$_5$ ESDIAD patterns (not shown)
are rather broad from the unbiased crystal, with a half-
width at half maximum (HWHM) of about 12^0 +/- 3^0, whereas
the Ni(CO)$_4$ patterns are more narrow (HWHM = 7.6^0 +/- 1^0).
By collecting data before and after prolonged exposure to
electrons from the ESDIAD gun, it was determined that the
centrally located ESDIAD features originate from the parent
molecules. In both cases, the M$_x$(CO)$_y$ clusters formed via
ESD gave broad, featureless patterns with much lower signal-
to-noise ratios than those from the parent molecules. These
observations are consistent with the formation of clusters

having lower ESD cross sections and an increased orientational randomness of the CO ligands.

3.3 AES

Following adsorption on Ag(111) at 90 K and annealing to 400 K, AES analysis shows that Fe(CO)$_5$ undergoes minimal thermal decomposition (<1/8 monolayer), with essentially no C or O and only small amounts of Fe being detected [5]. Under the same conditions, Ni(CO)$_4$ leaves no detectable impurities on the Ag surface [6]. After bombardment by low energy electrons and heating to 400 K, only Fe(Ni) was detected by AES, with very little increase in C or O signals. Annealing to 750 K drove the Fe(Ni) into the Ag film, as indicated by a sharp decrease in the Fe(Ni) AES signals.

4. Discussion

The TPD data of figs. 1 and 2 show that both Fe(CO)$_5$ and Ni(CO)$_4$ adsorb molecularly on Ag(111) at 90 K. They reversibly desorb upon heating with the desorption peak temperatures and profiles suggesting that both molecules adsorb weakly [5,6]. The large ESD cross sections determined from figs. 1 and 2 for the parent molecules are comparable to gas phase values [11-13].This is due to a low quenching rate indicative of weak adsorbate-substrate coupling.

The weak interaction between the adsorbed carbonyls and the Ag surface may result in a low probability for the de-excitation of excited-state carbonyls via electron hopping, electron-hole pair creation or surface plasmon excitation. Although the excitation mechanisms and transient excited-states involved in the ESD of these adsorbates are unknown, it is possible that excitation is similar to that for the gas phase carbonyls. If the CO ligands closest to the surface were to act as a spacer between the metal atom and the substrate, M-CO bond scission in the ligands which project furthest from the surface might occur as efficiently as in the gas phase. The decarbonylated species can then react with neighboring molecules following ESD to form $M_x(CO)_y$ clusters.

The ESD cross sections measured here for the $M_x(CO)_y$ clusters are at least 1000 times smaller than for the parent molecules. This may be due to to an increased M-Ag coupling in the cluster species, leading to more efficient quenching and energy dissipation of electronic excitations.

5. Conclusions

This paper has shown that low energy electrons efficiently decompose adsorbed metal carbonyls, which then react to form metal clusters containing CO ligands. These species exhibit greater thermal and ESD stability on the Ag(111) substrate than the parent molecules. These differences can be attributed to an increase in the metal-substrate interaction upon cluster formation.

Acknowlegements

This work was supported by AFOSR under contract 82-0311.

References

[1].   R.J. Bourcier, G.C. Nelson, A.K. Hays and A.D. Romig, Jr., J. Vac. Sci. Tech. **A4**, 2943 (1986).

[2].   J.S. Foord and R.B. Jackman, Chem. Phys. Lett. **112**, 190 (1984); Surf. Sci. **171**, 197 (1986).

[3].   R.R. Kunz and T.M. Mayer, Appl. Phys. Lett. **50**, 962 (1987); J. Vac. Sci. Tech. **B6**, 1557 (1988).

[4].   J.R. Swanson, C.M. Friend and Y.J. Chabal, Mater. Res. Soc. Symp. Proc. **75**, 559 (1987).

[5].   M.A. Henderson, R.D. Ramsier and J.T. Yates, Jr., (a) J. Vac. Sci. Tech. **A9**, 1563 (1991); (b) Surf. Sci. **259**, 173 (1991).

[6].   R.D. Ramsier, M.A. Henderson and J.T. Yates, Jr., Surf. Sci. **257**, 9 (1991).

[7].   G. McElhiner, H. Papp and J. Pitchard, Surf. Sci. **54**, 617 (1976).

[8].   B. Roop, P.M. Blass, X.-L. Zhou and J.M White, J. Chem. Phys. **90**, 608 (1989).

[9].   M.A. Henderson, R.D. Ramsier and J.T. Yates, Jr., J. Vac. Sci. Tech. **A9**, 2785 (1991).

[10].   P.W. Davies, M.A. Quinlan and G.A. Somorjai, Surf. Sci., **121**, 290 (1982).

[11].   B.C. Hale and J.S. Winn, J. Chem. Phys. **811**, 1050 (1984).

[12].   P.M. George and J.L. Beauchamp, J. Chem. Phys. **76**, 2959 (1982).

[13].   R.N. Compton and J.A.D. Stockdale, Int. J. Mass Spectrom. Ion Phys. **22**, 47 (1976).

Quantum-Resolved Stimulated Surface Reactions

T.M. Orlando*, A.R. Burns, E.B. Stechel, and D.R. Jennison

Sandia National Laboratories, Div. 1151, Albuquerque, NM 87185, USA
*Permanent address: Molecular Science Research Center K2-14,
 Pacific Northwest Laboratory, Richland, WA 99352, USA

Stimulated reactions of coadsorbates on Pt(111) surfaces have been probed using laser resonance-enhanced multiphoton ionization (REMPI) spectroscopy on the gas-phase neutral products. In particular, electron stimulated *dissociation* of NO_2(a) coadsorbed with up to 0.75 ML of atomic O on Pt(111) has been studied. Adding coadsorbed O to the surface enhances the specific dissociation yield, narrows the NO translational energy distribution, reduces the NO internal energy, and produces gas-phase O dissociation fragments. Reactive scattering between coadsorbates has also been studied. Specifically, NO_2(d) *production* has been observed during electron-beam irradiation of NO coadsorbed with O_2 on Pt(111). The NO_2(d) was indirectly observed as NO(v=5) and O(^3P_J) gas-phase photodissociation fragments. We assign NO_2 production to a collision between NO(a) and a hot O atom which was produced by electron-stimulated dissociation of adsorbed O_2.

1. Introduction

The chemical reactivity of atoms and molecules adsorbed on solid surfaces is a topic of general importance in many areas of physics and chemistry. Recent work has shown that chemical reactions on adsorbate-covered transition metal, semiconductor, and insulator surfaces can be enhanced or stimulated using either photon, electron, or ion-beam impact[1-3]. Information concerning the nature of the electronic excitations involved in stimulated reactions can be inferred from product state distributions, measured using quantum-resolved laser techniques. We have completed a detailed study of the stimulated dissociation of NO_2 in the presence of coadsorbed O which focuses on the final state energy partitioning in the gas-phase fragments. In this study, both the NO and O ground state fragments were detected above the surface. Stimulated dissociation can also produce energetic fragments which may initiate reactions with coadsorbed species. We have examined this possibility by studying the production of NO_2(d) via a stimulated reaction between coadsorbed NO and O_2 on Pt(111) at 90 K. This particular study demonstrates that reactions involving energetic dissociation fragments (O atoms produced via dissociation of adsorbed O_2) and coadsorbed molecules (NO) can occur. A similar reaction has been demonstrated previously by Ho and coworkers[2] in their study of photostimulated production of CO_2(d) from O_2 coadsorbed with CO on Pt(111). These types of reactions are particularly interesting since they occur under a restricted range of geometries, impact parameters, and collision energies. The possibility of selectively controlling reactions on surfaces therefore exists.

The experimental arrangement for REMPI detection of neutrals produced in electron -stimulated processes has been described in detail elsewhere[1,6]. The dosing procedures for the NO_2 + O/Pt(111) and the NO + O_2/Pt(111) experiments have also been previously published[4,5].

Springer Series in Surface Sciences, Vol. 31
Desorption Induced by Electron Transitions DIET V
Editors: A.R. Burns · E.B. Stechel · D.R. Jennison   © Springer-Verlag Berlin Heidelberg 1993

2. Results and Discussion

2.1. Electron-stimulated dissociation of $NO_2(a)$ on $O/Pt(111)$

At temperatures below 150 K, NO_2 adsorbs molecularly on clean Pt(111) in a μ-N,O-nitrito side-bonded complex at bridge sites[7]. Previous work has demonstrated that electron-stimulated dissociation of this species primarily results in surface retention of the O atom fragment[6]. However, with increasing O-coverage, the NO_2 is constrained to occupy atop (terminal) sites in a nitro (N-end down) complex[7]. Fig. 1 shows the $O(^3P_2)$ yield of $NO_2(a)$ dissociation as a function of these bonding configurations (see the inset in Fig. 1). It is clear that $O(^3P_J)$ *does* leave the surface due to the N-end down bonding geometry and the extensive site blocking by preadsorbed O atoms. These "free" O atoms could react with coadsorbates *if* they had the proper trajectories, however, the majority of the O atoms simply leave the surface with translational energies ranging from 0.05 to 0.5 eV. The $O(^3P_J)$ spin-orbit state distribution is (5.0):(2.5):(1.0) for J=2,1,0 respectively; this is within experimental error of the 2J+1 statistical limit of (5.0):(3.0):(1.0). ESD of chemisorbed O has a cross section below our present detection limits ($\sim 10^{-19}$ cm^{-2}).

In previous work[6], the cross section for electron-stimulated dissociation on clean Pt(111) was estimated to be $\sim 5 \times 10^{-18}$ cm^2 for 350 eV excitation. We have observed a dramatic increase in the specific $NO_2(a)$ dissociation yield as a function of preadsorbed atomic oxygen coverage. At $\theta_O = 0.75$ ML, this yield increases by a factor of 26 implying an overall cross section of $\sim 10^{-16}$ cm^2. We interpret the large increase in the dissociation cross section mainly in terms of increased excited state lifetimes which result from the O-atom induced reduction of substrate screening[1]. Since the longest-lived excitations are nondegenerate with the substrate valence bands and predominantly decay via intramolecular Auger processes[1], reduced screening results in a slower decay rate and an enhanced lifetime. We assigned the highest probability excitations leading to dissociation of NO_2 on clean Pt(111) to the $3b_2^{-1}$ one-hole and $1a_2^{-2}$ two-hole states[6]. The threshold energy is similar for dissociation on the O-covered surface and we tentatively assign it to the $1a_2^{-2}$ excitation.

Rotational and vibrational excitation of the NO fragment is expected since screening charge occupies antibonding molecular levels. Average rotational energies of 560 ± 84 K and 703 ± 105 K for the $\upsilon=0,1$ levels respectively have been measured. Spectra for the $\upsilon=2$ and $\upsilon=3$ were too weak for an accurate rotational analysis. These energies are significantly lower than those for dissociation on the clean surface: 862 ± 30 K and 821 ± 35 K for $\upsilon=0,1$ respectively. The vibrational distribution which has been obtained by integrating the rotational spectra for each vibrational level is (1.0):(.86):(~.16):(~.15) for $\upsilon=0,1,2,3$, respectively. This is lower than the distribution

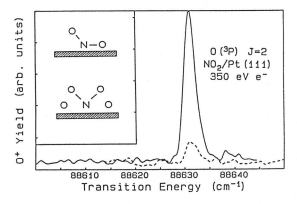

Fig. 1. The $O(^3P_2)$ yield of NO_2 dissociation on Pt(111) as a function of the bonding geometries which are displayed in the inset. The dotted line is for the bridge-bonded species (top); the solid line is for the nitro (N-end down) complex which exists at $\theta_O = 0.75$ ML (bottom).

195

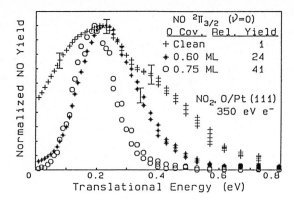

Fig. 2. Comparison of the normalized TOF E_{trans} distributions for the NO $^2\Pi_{3/2}$ ($\upsilon=0$) fragment of NO_2 dissociation as a function of O atom coverage on Pt(111).

of (1.0):(.96):(.52) ($\upsilon=3$ is below detection limits) obtained for dissociation on the clean surface[6]. We believe this decrease in fragment internal energy is due to reduced excited state forces resulting from diminished charge transfer screening.

TOF E_{trans} distributions of the NO $^2\Pi_{\Omega=3/2}$ ($\upsilon=0$) dissociation fragments for several preadsorbed atomic oxygen coverages are shown in Fig. 2. These distributions, normalized by their maxima, have pronounced variations in their widths with θ_O. The broad NO E_{trans} distribution found from NO_2 dissociation on clean Pt(111)[6] collapses to a relatively narrow distribution peaked at 0.2 eV as θ_O approaches saturation coverage. The E_{trans} distributions at $\theta_O=0.75$ ML are the same within experimental error for both spin-orbit levels ($^2\Pi_{\Omega=3/2}$ and $^2\Pi_{\Omega=1/2}$).

Changes in E_{trans} distributions reflect changes in excited state forces that result in part from variations in screening charge density. The reduction of the screening charge by the O atoms and possibly the change to an N-bonded coordination can "flatten" the excited state potential energy surface. The definite collapse of the high energy features in the E_{trans} distributions of the NO fragment supports the idea that reduced excited state forces result from diminished screening. The disappearance of the low energy tail is most likely due to an increased excited state lifetime[1]. Our results clearly show that in the case of NO_2 dissociation on O-covered Pt(111), *both* the dissociation forces and excited state lifetimes change. However, we see that small changes in lifetimes have larger consequences on the yield than reduction of excited state forces.

2..2. Electron-stimulated production of NO_2(d)

Fig. 3 shows the NO_2 TOF E_{trans} distribution following pulsed 350 eV electron-beam irradiation of a Pt(111) surface with 0.44 ML of O_2 and 0.30 ML of NO at 90 K. The NO_2(d) reaction product was detected indirectly using two REMPI schemes that are based on photodissociation of NO_2 *above* the surface; direct REMPI detection of NO_2 is inefficient. The first detects the NO($\upsilon=5$) product of NO_2(d) photodissociation and the other detects the O(^3P_J) product[5]. The absolute NO($\upsilon=5$) and O($^3P_2$) signals have been normalized to the peak maximums to emphasize comparison of the relative shapes of the fragment TOF distributions plotted on an NO_2 energy scale. The identical nature of the distributions verify that both fragments originate from the same desorbing parent molecule which promptly leaves the surface upon formation and is dissociated in the laser beam. Stimulated production of NO_2 followed by desorption is unlikely in view of the efficient stimulated dissociation of NO_2(a).

The NO_2 yield varies monotonically with the NO coverage at atop sites. Since NO bonds normal to the surface (at atop sites) and O_2(a) bonds parallel (side-on), the NO_2 production is most likely initiated by the dissociation of O_2 (a) followed by subsequent

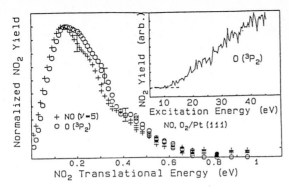

Fig. 3. TOF E_{trans} distribution of the $NO_2(d)$ stimulated reaction product measured by monitoring the $NO(\upsilon=5)$ and $O(^3P_2)$ photofragments. The inset shows the excitation energy threshold for NO_2 formation measured via the $O(^3P_2)$ yield. Identical threshold behavior is seen with the $NO(\upsilon=5)$ signal.

reaction of a "hot" O atom with a neighboring NO: $O^* + NO(a) \rightarrow NO_2(d)$. The ~10 eV threshold for $NO_2(d)$ production (see inset of Fig. 3), is consistent with the dissociative ionization of the O_2 $3\sigma_g$ bonding molecular orbital. The $NO_2(d)$ yield is considerably higher than that observed for photoexcitation (< 5 eV) because the $3\sigma_g^{-1}$ hole can not resonantly decay[8].

In summary, the outcome of a stimulated surface process is dependent upon several factors such as excited state lifetimes and forces, fragment trajectories and energies, and adsorbate bonding geometries. Long-lived, deep valence excitations of molecular bonding orbitals can produce fragments which may react with coadsorbates if the energy is sufficient to overcome reaction barriers and the trajectories are parallel to or towards the surface.

Acknowledgments

This work was performed at Sandia National Laboratories, supported by the U.S. Department of Energy under Contract DC-AC04-76DP00789.

References

1. A. R. Burns, E. B. Stechel, D. R. Jennison, and T. M. Orlando, Phys. Rev. B. **45**, 1373 (1992).
2. W. Ho in <u>Desorption Induced by Electronic Transitions</u>, DIET IV ed. by G. Betz and P. Varga, (Springer-Verlag, New York, 1990) pp. 48-64.
3. L. Sanche and L. Parenteau, Phys. Rev. Lett. **59**, 136 (1987).
4. T. M. Orlando, A. R. Burns, D. R. Jennison, and E. B. Stechel, Phys. Rev. B. **45**, 8679 (1992).
5. T. M. Orlando, A. R. Burns, E. B. Stechel, and D. R. Jennison, J. Chem. Phys. **93**, 9197 (1991).
6. A. R. Burns, E. B. Stechel, and D. R. Jennison, Phys. Rev. B. **40**, 9485 (1989).
7. M. E. Bartram, R. G. Windham, and B. E. Koel, Langmuir **4**, 240 (1988).
8. W. Mieher and W. Ho, J. Chem. Phys. (in press).

Part VII

Spatially Resolved DIET

Atomic Scale Desorption and Fragmentation with the STM

Ph. Avouris

IBM Research Division, T.J. Watson Research Center,
P.O. Box 218, Yorktown Heights, NY 10598, USA

Abstract. The STM is not only a powerful probe of the atomic structure of surfaces, but is also a unique tool for the atomic scale modification and manipulation of materials. I will illustrate this new application of the STM with three examples: First, I will discuss a general approach for the selective breaking of strong chemical bonds, which we use to manipulate Si atoms on Si(111) at room temperature. I will then use the manipulation of a Au(111) surface to show that "local" modifications can couple with long-range elastic surface interactions to lead to large scale atomic rearrangements. Finally, I will discuss the use of electrons from the STM tip to induce the dissociation of individual molecular adsorbates.

1. Introduction

In the scanning tunneling microscope (STM) the tip and sample interact in a variety of ways. These interactions can be chemical, leading to attractive or repulsive forces acting on a set of sample atoms. There may be electrostatic interactions due to the electric field that exists between sample and tip because of an applied bias voltage, or magnetic interactions, if magnetic materials are involved. The tunneling or field-emitted electrons from the STM tip can lead to the elecronic or vibrational excitation of the sample, dissociative electron attachment processes or even heating under certain conditions. The strength of the above interactions can, in most cases, be tuned and controlled by the experimenter. Coupling the capability to apply local pertubations of the types mentioned above with the ability of the STM tip to address individual atomic sites leads to a unique tool for the atomic scale modification and manipulation of materials. Already, there is a growing list of applications involving depositing, moving, desorbing or dissociating surface atoms or adsorbates /1,2/. Here I will illustrate this unique use of the STM using three examples from our recent work: (a) the selective breaking of strong Si-Si chemical bonds at the Si(111)-7x7 surface, and the room-temperature manipulation of Si atoms and clusters /3,4/, (b) the use of local surface modification coupled with the long-range elastic stress of the surface layer to manipulate the reconstruction of Au(111)-22x$\sqrt{3}$ /5/, and (c) the dissociation of individual decaborane molecules chemiscrbed on Si(111) using electrons emitted from the STM tip /6/.

Springer Series in Surface Sciences, Vol. 31
Desorption Induced by Electron Transitions DIET V
Editors: A.R. Burns · E.B. Stechel · D.R. Jennison   © Springer-Verlag Berlin Heidelberg 1993

2. Moving Strongly Bonded Atoms by a Chemically-Assisted Field-Evaporation Process: Manipulation of Si Atoms

A fundamental capability in the area of the atomic and nanometer scale modification of materials with the STM involves the ability to selectively break particular surface chemical bonds in order to remove individual atoms or clusters of atoms and place them on the tip. The tip can then be moved to another, preselected area of the surface where the material can be redeposited. Recently we have shown that these goals can be accomplished with the help of a process which we termed "chemically-assisted field-evaporation" (CAFE). To explain the principles on which chemically-assisted field-evaporation is based, we refer to the potential energy (P.E.) diagram shown in Fig. 1.

In panel A, the solid line shows the P.E. curve of an atom bound to the surface of the sample (left well) or to the tip (right well). In order to transfer this atom from the surface to the tip one must overcome the potential energy barrier present. The barrier can be reduced by bringing the STM tip closer to the surface so as to establish a degree of chemical bonding between tip and

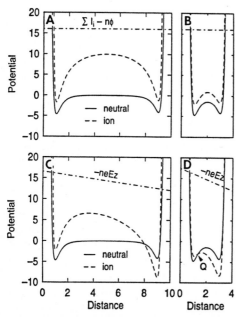

Fig. 1. Schematic potential energy (P.E.) curves illustrating a chemically-assisted field evaporation process. The solid line P.E. curve represents the neutral ground state where an atom (A) is bonded either to the substrate (S, left well) or the the tip (T, right well). The dashed line corresponds to an ionic state, $S^{n-}A^{n+}$. Panels A and B pertain to field free conditions, while in C and D an electric field E is present.

sample (Fig. 1B). However, the preference of a particular atom for the sample surface or the tip would still be determined entirely by their respective chemical compositions. To gain control over the direction of atom transfer we introduce electrostatic forces whose magnitude and direction can varied. As shown in Fig. 1A, at energies higher than that of the neutral ground state curve there will be ionic excited state curves such as the one denoted with the dashed line which depicts the situation where a surface atom, A, has given the substrate, S, n electrons leading to a configuration $S^{n-}A^{n+}$. When an electric field E is applied between tip and sample, the neutral P.E. curve is only weakly perturbed through polarization effects, while the energy of the ionic curve is reduced by -neEz. The ionic curve may cross the neutral curve but, as Fig. 1C suggests, this may happen too far from the surface to have an effect on the P.E. barrier. If, however, the tip is first brought close to the surface and then the electric field is generated, curve-crossing takes place close to the surface so that only a small effective barrier Q needs to be surmounted. If the field direction is reversed, the direction of the electrostatic force is also reversed. The close proximity of tip and sample leads to a significant reduction of the field strength required to initiate material transfer in the STM configuration as compared to the field required for field evaporation in the field ion microscope (FIM). The essential features of the proposed CAFE atomic manipulation mechanism described above were also confirmed by a recent first-principles electronic structure calculation /7/.

In our experiment, we first bring the tip close to the surface over the site we want to modify. The distance scale is established by determining the electronic contact point where the apparent tunneling barrier, ϕ_{eff}, becomes zero /3/. Strong tip-sample chemical interactions develop in the distance range where ϕ_{eff} is drastically reduced /8,9/. The electric field which provides the directional electrostatic forces is then generated by applying a voltage pulse to the sample. Fig. 2 shows an example of a nanometer scale surface modification. Panel A shows the Si(111)-7x7 surface before modification. The tip is then moved to ~3Å from electronic contact above the surface and a +3V pulse is applied to the sample. This leads to the formation of a characteristic structure composed of a central hill surrounded by a depression (panel B). Next, the tip is placed over the hill and another +3V voltage pulse is applied. As a result of this second pulse, the material forming the hill is transferred to the tip. This Si cluster on the tip can now be moved and redeposited anywhere on the surface by applying an opposite polarity (negative) voltage pulse to the sample. In the case shown in panel C, we have deposited the cluster to the left of the hole. Because of the interaction of the desorbing atoms with both sample and tip, the threshold field required for the removal of Si is under the conditions of Fig. 2 only ~1V/Å; significantly lower than the ~4V/Å field measured with FIM /10/. The close proximity of tip and sample is also responsible for the generation of the characteristic hill plus depression structure (panel B). When a voltage pulse is applied, an electric field gradient is generated. Atoms tend to move towards the apex of the tip where the field is strongest, piling up there to form a bridge connecting the tip with the surface

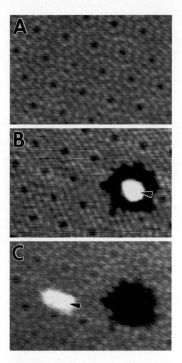

Fig. 2. (A) Section of a Si(111)-7x7 surface. (B) A + 3V pulse applied to the sample at ~3Å from electronic contact has led to the formation of a structure involving a central hill surrounded by a depression. (C) Another + 3V pulse has been applied and the hill has been transferred to the tip. The tip was then moved to the left of the hole and the cluster was redeposited by applying a -3V pulse.

/3,4/. As the tip is retracted the bridge breaks, leaving a cluster of atoms in the center of the hole.

The size of the affected area (i.e. number of removed atoms) can be controlled to great extent by varying the tip-sample distance and the magnitude and duration of the voltage pulse. For single atom manipulation, a very sharp tip is essential. The tip is brought very close (~1Å) to the atom to be removed. The chemical interaction of this atom with the tip weakens its bonding to the substrate. Then a low voltage pulse (~1V) is applied so that this particular atom is removed to the tip. Fig. 3 shows a series of atomic scale manipulations /4/. Panel A shows a section of a Si(111)-7x7 surface containing a defect in the lower right (dark site) which is used as a marker. In B the voltage pulse was applied while the tip was centered over the site indicated by the arrow and three Si atoms were removed, leaving a fourth under the apex of the tip. This fourth Si atom was left in an unstable configuration, and it migrated to the left to occupy a center-adatom site of the 7x7 lattice (panel C).

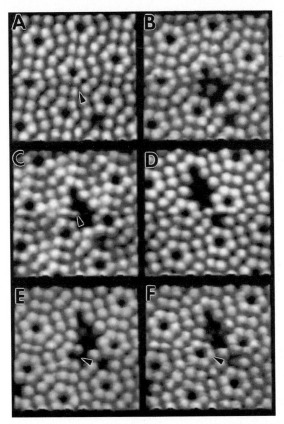

Fig. 3. A series of atomic scale manipulations: (A) The tip is placed at ~1Å from electronic contact over the site indicated by the arrow. (B) A 1V pulse removes 3 atoms leaving the fourth under the tip. (C) The first attempt to remove this atom leads to its migrating to the left (see arrow) and bonding as a center adatom. (D) A second pulse removes this fourth atom. (E) A new corner-adatom is removed and in (F) it is placed back to its original position.

This atom was then removed with a second pulse (panel D). Finally, with another pulse a corner-adatom was removed (panel E). One question that arises is: can the STM undo a modification, e.g. can it put back a removed atom or cluster of atoms and reform the original structure? We find that, for Si, in general, and especially for clusters, the answer is no. Although we can place the cluster over the original site, it does not incorporate itself. This is the result of the rebonding that takes place at both the substrate and the cluster after the removal of a number of atoms. The redeposited atoms must overcome a sizeable activation barrier to occupy their original sites. However, as panel F shows, in the case of single atom removal, incorporation occasionally can be achieved by bringing the tip over the vacancy site and applying a negative pulse to the sample. From the above examples, it is clear that CAFE provides a powerful and general approach to the manipulation of strongly bonded atoms.

Finally, we note that CAFE processes may occur while taking constant current topographs of partially reacted semiconductor surfaces. During scanning, the tip follows contours of constant local density of states (LDOS). On semiconductor surfaces the LDOS near E_F is dominated by contributions from dangling-bond surface states. Upon chemical reaction, the dangling-bonds are eliminated and are replaced by new bonding and anti-bonding levels whose DOS peaks are far removed from E_F. Thus, over reacted sites, the tip approaches closely the adsorbates, to maintain a constant current, and if the bias is high enough, conditions for CAFE may arise. This was demonstrated in the case of the reaction products of H_2O with Si(111) /11/.

3. Manipulation of Metal Surfaces: Long Range Effects of "Local" Modifications

The same CAFE technique can be used to manipulate metal surfaces. Here I will use local modifications of a Au(111) surface to induce long range surface restructuring /5/.

The surface structure of the Au(111) is the result of two opposing tendencies. First, the surface layer would like to contract in order to compensate for its reduced coordination, i.e. it is under tensile stress. Opposing this contraction is the underlying substrate potential which favors a commensurate layer. The result of these opposing forces is a reconstruction with an $nx\sqrt{3}$ surface unit cell (n = 22) containing $2n + 2$ Au atoms /12,13/. This reconstruction allows an ~4.5% contraction along the $\langle 1\bar{1}0 \rangle$ direction. The surface Au atoms can occupy two different hollow sites of the bulk (111) lattice leading to either an fcc (ABC ABC...) stacking or an hcp (AB AB...) stacking. In the $22x\sqrt{3}$ reconstruction both types of sites are occupied, giving rise to alternating domains having fcc or hcp stacking. This arrangement provides the optimal decrease in the surface bond length. The energy of the fcc stacking is somewhat lower than that of the hcp stacking, and as a result, the width of the fcc domains is larger /12,13/. Separating the fcc and hcp domains are transition regions where the surface Au atoms occupy bridge-like sites. These transition regions appear in STM images as bright bands (~0.1-0.2Å high) running along the $\langle 11\bar{2} \rangle$ direction /12,13/. These transition zones are partial dislocations which are usually described as Frenkel-Kontorova solitons /14,15/.

Returning to the issue of surface stress, we note that the $22x\sqrt{3}$ reconstruction has an anisotropic stress tensor since the incorporation of the extra two Au atoms per unit cell relieves the tensile stress along the $\langle 1\bar{1}0 \rangle$ direction but not in the transverse $\langle 11\bar{2} \rangle$ direction /14/. On large terraces a herringbone-like superstructure composed of rotational domains is formed which partially relieves stress along $\langle 11\bar{2} \rangle$. Near steps, the stress is partially relieved by orienting the F-K solitons and therefore the stress perpendicular to the steps.

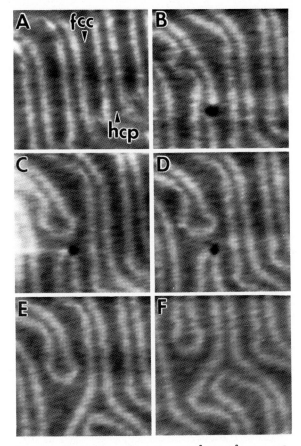

Fig. 4. (A) Topograph of a 200Åx200Å area of a Au(111)-22x $\sqrt{3}$ surface. (B) After a surface hole was generated at the position of the arrow by bringing the tip at 2.5Å over this site and applying a 10ms, +1V pulse to the Au sample (0-6 min). (C) After 6-12 min. (D) After 12-18 min. (E) After 18-24 min. (F) After 60 min. All images were obtained with the sample bias of -0.5V and a current of 1nA.

We have reasoned that by using the CAFE approach we can remove surface atoms and that the resulting surface hole will act locally as a step to relieve the surface stress. We can then image in a time-resolved manner the resulting changes in surface structure. In Fig. 4 we show some typical results. Panel 1A shows the Au surface before modification. In 4B a hole (diameter ~ 1.5nm) was made on top of a soliton line. Atomic relaxation at the hole relieves the stress along the soliton line but the resulting stress-imbalance leads to the generation of a dislocation in the adjacent soliton. The long range nature of the surface stress is clearly demonstrated by the fact that the dislocation is ~80Å away from the STM generated hole. In 4C obtained 6-12 min later we see that the original two soliton lines have split. The solitons in the upper half have been joined to form a U-shape loop, while the lower portion of the solitons has been pinned to the hole. The higher stability of this configuration is probably due to the fact that relaxation around the hole is more effective in relieving stress in one direction than in two opposing directions. As time progresses the size of the hole decreases (4D, 12-18 min) as diffusing Au atoms enter the hole. Then at panel E (18-24 min) we observe that the hole has suddenly disappeared and a new arrangement of the solitons has emerged.

This arrangement involves the two lower branches of the solitons, previously stabilized by the hole, fusing with the neighboring soliton pair to their right to give the characteristic forked structure of panel E. This forked structure is quite stable as can be seen in the topograph shown in panel F taken one hour later.

The above sequence of topographs illustrates a number of important conclusions reached by these studies: 1) It provides a direct verification of the existence of long range elastic interactions at metal surfaces. 2) The long range atomic rearrangements involved occur over long time scales of the order of minutes. 3) Surface diffusion occurs not only by single atom motion, but also by the concerted motion of large numbers of atoms (see transition from D to E). 4) Through "local" modification the structure of large areas can be manipulated. Since it has been found that the reactivity of Au(111) involves the soliton lines, by rearranging them we can change the spatial distribution of surface reactions.

4. Electron-Stimulated Dissociation of Molecules by the STM

In the previous examples we have used the static electric field between tip and sample as well as chemical interactions to induce surface modifications. One can also take advantage of the fields generated by the tunneling (or field-emitted) electrons in the STM to induce local electronic excitation of adsorbates to dissociative states. Such a process involves inelastic tunneling in which the tunneling electron gives its energy to an adsorbate excitation while dropping into an empty level of the conduction band of the sample. The range of available energies which is given by the energy difference of the Fermi level of the tip and of the bottom of the conduction band of the sample, is tuned by varying the bias voltage. Such processes can be used to desorb an adsorbate /16/ or to dissociate an adsorbed molecule /6/. Here we will discuss molecular dissociation using the case of $B_{10}H_{14}$ (decaborane, DB) adsorbed on Si(111)-7x7 as an example /6/. In earlier work /17/ we found that DB molecules adsorb on Si(111)-7x7 at room temperature and can be directly imaged with the STM as roundish protrusions with a diameter of ~7Å and can be used as a source of B-doping of Si. Fig. 5A shows a number of adsorbed DB molecules. Upon thermal annealing the surface mobility of the adsorbed molecules is increased with most of the molecules diffusing to steps or domain boundaries where they dissociate, while others dissociate locally.

Fig. 5. (A) Topograph of a Si(111)-7x7 surface with adsorbed decaborane molecules which appear as roundish protrusions ~7Å in diameter. Sample bias: +2V. (B) Topograph of a decaborane-exposed (0.2 L) Si(111)-7x7 surface after brief thermal annealing to 600°C. Sample bias: +2V.

Fig. 5B shows that the result of the thermal dissociation of DB molecules is the generation of dark patches on the surface. The darkness (low LDOS) of the patches results from the elimination of the Si surface dangling-bonds by the fragmentation products /17/. At sufficiently high annealing temperature the boron atoms occupy special sub-surface sites and become electrically active /18/. For nanoelectronics applications it would be desirable to avoid the diffusion process and be able to induce local doping.

Thus, it will be desirable to be able to dissociate DB locally via an electronic excitation. Dissociation adsorbed DB as a result of high energy synchrotron or electron irradiation has been reported /19/. Tunneling spectroscopy studies of the chemisorbed DB molecules indicate /6/ that they have occupied (unoccupied) states 1.5eV below (above) the Fermi energy, i.e. within the STM excitation energy range. To induce dissociation, the selected molecules(s) is scanned by the STM tip as in a topographic scan. Scans are repeated as a function of the bias voltage. Large current densities are delivered to the sample in this way. In a single constant current (100pA) scan each DB molecule receives a dose of ~10^5 electrons. The experiment indeed showed that single molecule dissociation can be achieved. Fig. 6 shows the dependence of the dissociation probability on the applied bias, at +8V it is ~80%.

In Fig. 7A we show two adsorbed DB molecules imaged at +2V sample bias, while Fig. 7B shows the same area after it has been scanned once at +8V and then imaged again at +2V. The two DB molecules are gone and in their place there are several dark sites, in complete analogy with the behavior observed upon thermal dissociation of adsorbed DB (Fig. 5B). It is clear that molecular dissociation can be achieved with the STM. The available evidence indicates that the electron energy is the main factor controlling the dissociation. However, the static electric field may also be involved in the dissociation process. Further experiments are under way to delineate the role of these two factors.

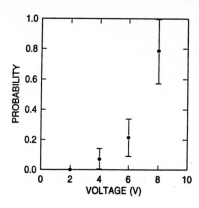

Fig. 6. The dissociation probability of adsorbed decaborane molecules on Si(111)-7x7 as a function of bias voltage.

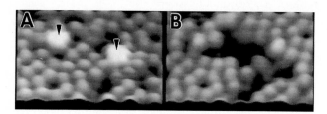

Fig. 7. (A) Topograph showing two decaborane molecules on Si(111)-7x7. (B) The same area of the surface imaged after a scan at a bias voltage of + 8V. Both images obtained with a sample bias of + 2V.

Acknowledgement. I gratefully acknowledge my co-workers in the work described in this talk: In-Whan Lyo, Yukio Hasegawa, Gerald Dujardin and Robert E. Walkup.

References

1. C.F. Quate, in Highlights and Prospects in Condensed Matter Physics, Ed. L. Esaki (Plenum, New York, 1992), and references therein.
2. J.A. Stroscio and D.M. Eigler, Science, 254, 1319 (1991), and references therein.
3. I.-W. Lyo and Ph. Avouris, Science 253, 173 (1991).
4. Ph. Avouris, I.-W. Lyo, Appl. Surf. Sci., in press.
5. Ph. Avouris and Y. Hasegawa, to be published.
6. G. Dujardin, R.E. Walkup and Ph. Avouris, Science, 255, 1232 (1992).
7. N.D. Lang, Phys. Rev. B, in press.
8. C.J. Chen, J. Phys. (Condens. Matter), 3, 1227 (1991).
9. S. Ciraci, E. Tekman, M. Gokcedag and I.P. Batra, Ultramicroscopy, in press.

10. T.T. Tsong, Atom-Probe Field Ion Microscopy (Cambridge University Press, Cambridge, 1990).

11. Ph. Avouris and I.-W. Lyo, Surf. Sci. **242**, 1 (1991); I.-W. Lyo and Ph. Avouris, J. Chem. Phys. **93**, 4479 (1990).

12. Ch. Woell, S. Chiang, R.J. Wilson, and P.H. Lippel, Phys. Rev. B, <u>39</u>, 7988 (1989).

13. J.V. Barth, H. Brune, G. Ertl and R.J. Behm, Phys. Rev. B, <u>42</u>, 9307 (1990).

14. Y.I. Frenkel and T. Kontorova, Zh. Eksp. Teor. Fiz., <u>8</u>, 1340 (1938).

15. S. Narasimhan and D. Vanderbilt, to be published.

16. R.S. Becker, G.S. Higashi, Y.J. Chabal and A.J. Becker, Phys. Rev. Lett., <u>65</u>, 1917 (1990).

17. Ph. Avouris, J. Phys. Chem., <u>94</u>, 2247 (1990), Ph. Avouris et al. J. Vac. Sci. Technol. B, <u>8</u>, 3405 (1990).

18. I.-W. Lyo, E. Kaxiras and Ph. Avouris, Phys. Rev. Lett., <u>63</u>, 1261 (1989).

19. Y.G. Kim, P.A. Dowben, J.T. Spencer and G.O. Ramseyer, J. Vac. Sci. Technol. A, <u>7</u>, 2796 (1989).

Atomic Force Microscopy of Surface Defects on NaCl(100)

R.T. Williams, R.M. Wilson, and G.P. Williams, Jr.

Department of Physics, Wake Forest University,
Winston-Salem, NC 27109, USA

Abstract Scanning force microscopy reveals features interpreted as chloride ion vacancies or substitutional surface impurities on NaCl (100) at room temperature. Monatomic step edges have been imaged as well. Using an evacuable chamber backfilled with helium at controlled humidity, studies of image resolution versus ambient moisture and tip force are described. The goal of this work is an atom-resolving probe of surface modifications resulting from radiation-induced desorption and damage phenomena.

1. Introduction

Scanning tunneling microscopy (STM) and atomic force microscopy (AFM) have opened intriguing possibilities for direct imaging of atomically resolved surface modifications caused by DIET phenomena, in addition to more general studies of surface defects.   The ability of STM to resolve surface atomic structure including point defects in ultrahigh vacuum (UHV) is already well established. Results presented at this workshop represent some of the first applications of STM to spatially resolved desorption phenomena. [1]   However, STM can only be used with electrically conducting samples. Atomic force microscopy appears to offer the complementary tool for spatially resolved study of desorption and surface defect phenomena on insulating substrates. However, it must be pointed out that atomically resolved AFM is still in its infancy, compared even to STM. Despite atomic periodicities that are readily observed, there is still debate on whether the AFM can image individual atomic features such as point defects. Whereas STM is very amenable to UHV, there are real physical factors limiting the performance of atom-resolving AFM in vacuum, at present. At the force levels presently required to achieve atomic resolution, the AFM tip is very intrusive on the sample being studied. Furthermore, a high degree of surface ion mobility is found experimentally on alkali halides, at least. The surface features generally do not sit still long enough for protracted imaging studies at room temperature. [2]

The goal of our work in this area is to develop the foundation for atom-resolving studies of defects produced on the surfaces of alkali halides and other insulators by exposure to electron, ion, and photon beams. In addition, atom-resolving characterization of the surfaces before irradiation may help to identify defect sites which are important in some of the desorption processes, e.g. laser-induced phosphorous emisssion from GaP. [3]

Springer Series in Surface Sciences, Vol. 31
Desorption Induced by Electron Transitions DIET V
Editors: A.R. Burns · E.B. Stechel · D.R. Jennison   © Springer-Verlag Berlin Heidelberg 1993

2. Point Defects and Steps on NaCl (100)

Until 1990, atomic resolution AFM images had been reported only for layered materials, such as highly oriented pyrolytic graphite [4], and transition metal dichalcogenides. [5]   In 1990, E. Meyer et al reported atomic-scale periodic images on LiF (100) [6], and later the same year, G. Meyer and N. Amer reported periodic atomic resolution on NaCl (100) in ultrahigh vacuum. [7]   We also have found predominantly well-ordered lattice arrays on NaCl [2], as well as LiF, NaF, KCl, RbCl, KBr, CaF_2, SrF_2, and BaF_2 imaged in air. Typically, it is the halogen sublattice that is seen.   For example, the ionic radii of Na^+ and Cl^- are 0.98 and 1.81 Å, respectively, and the (100) surface of NaCl is known to reconstruct by outward relaxation of the anions and inward relaxation of the cations. [8]   It should be noted from the outset that ion diffusion on the surfaces of these halide crystals is significant at room temperature.   Using approximate diffusion data, we estimated in Ref. [2] that the rate of halogen ad-ion jumps on NaCl (100) is of the order of 260 s^{-1} at 293 K.   This partly accounts for the typical appearance of perfection on freshly cleaved alkali halide surfaces observed by AFM.   What is imaged is the average of halide ion positions over the scan time of ~8 s.   Under room-temperature scanning conditions, only those surface defects which are stabilized in some way should be observable.   In Ref. [2], we described one such stabilized point imperfection. It appeared as a vacancy on the halogen sublattice.   Since there was little relaxation of the halide ions surrounding the vacancy, we tentatively surmise that the charge state was that of an F center or of a vacancy occupied by a negatively charged impurity.   In fact, the deepest depression was not measured at the center of the halide vacancy site, but was distributed asymmetrically in two corners of the "box" formed by the neighbors.   The hill which could be resolved in the center of the vacancy may in fact be a substitutional surface impurity. The stabilized vacancy was recognizable for 4 consecutive image frames, or about 30 s.

Figure 1(a) shows an atomically resolved image of a different NaCl (100) sample presented as a 10 nm x 10 nm constant-force topograph, where lighter color represents higher elevation. These images were acquired in air on cleaved NaCl single crystals, using a commercial AFM. [9] The cantilever tip was Si_3N_4, and a tip force of about 60 nN was used in repulsive mode. There are several vacancy-like features in the field of Fig. 1(a). It probably is significant that such a density of defects is not typically found on freshly cleaved NaCl, including samples cleaved, mounted, and imaged under dry helium. The image of Fig. 1 was acquired 3 weeks after cleaving in air. A magnified 2.7 nm x 2.7 nm topview of the area framed in the upper left of (a) is displayed in Fig. 1(b). Slightly to the upper right of center in the magnified view, a vacancy-like feature can be seen. In the false color display, black is the lowest elevation, while blue and white are successively higher.   Whereas the vacancy described in Ref. [2] was surrounded by eight chloride neighbors in a symmetric square, this one has an interesting variation: Figure 1(b) bears an astonishing resemblance to a vacancy-interstitial pair. We hasten to acknowledge that a vacancy-interstitial pair of this separation in the interior of a pure crystal would have a lifetime of only

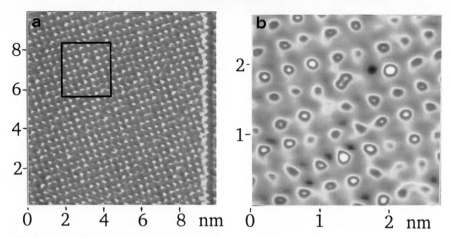

Fig. 1. Topview images of NaCl (100) in air at room temperature. (b) is an enlarged view of the marked area in (a), showing a surface defect.

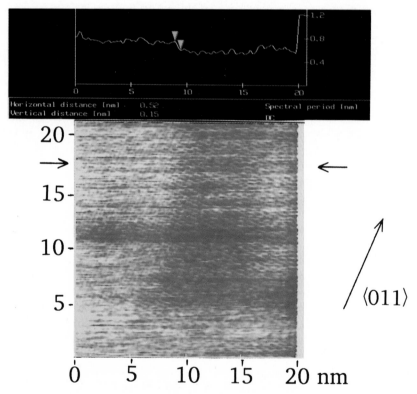

Fig. 2. Topview image of NaCl (100) showing a step edge bisecting the picture vertically, as well as resolved atomic structure. A height profile taken along the indicated line perpendicular to the step is shown above.

microsceconds even at cryogenic temperatures. [10]   It is essential to the proposed interpretation of this image that the vacancy in fact be occupied by a stabilizing impurity with sufficient charge or size to keep the interstitial species at bay, yet one which appears weak in the AFM image compared to the rare-gas-like chloride ions. Substitutional OH⁻ comes to mind as a common surface impurity that might fit this description.

Such images are fascinating objects of speculation, but we hope to move beyond that as soon as possible. There are several approaches. One is to continue accumulating and cataloging such images from hundreds of AFM scans, with the expectation that various defect configurations, if real, must begin to repeat in characteristic structure and form. By studying surfaces under different environments and with different pretreatments, it should be possible to manipulate the hypothesized stabilizing impurities and see corresponding manipulations of the observed surface defects. In the mean time, we are looking at another type of surface defect to test the ability of the AFM to image isolated structures of atomic dimension: monatomic step edges. In this case the defect has sufficient size in one dimension to allow identification and relocation in successive images, but has atomic size in the other two dimensions to test instrument resolution. Figure 2 shows the topview image of NaCl (100) on a 20 nm x 20 nm frame size, with a monatomic step bisecting the picture along a vertical line, and a deeper feature bisecting the picture along a horizontal line. Resolved atomic periodicity can be seen as rows of dots along the <011> directions. A line scan was selected in the upper part of the image at the position indicated by arrows. The height profile along this line is shown above the image in Fig. 2. The difference between the average heights extending 4 nm to either side of the step is about 2 Å, to be compared to the monatomic step height of $a_0/2 = 2.82$ Å on NaCl. The two cursors placed at the top and bottom edges of the step have a lateral separation of 5 Å and vertical separation of 1.5 Å.

3. Effects of Adsorbed Water and Tip Pressure

The images presented so far were all obtained in air. For normal laboratory air of about 50 % relative humidity, one expects at least a monolayer film of water adsorbed on the sample and the tip, preventing the formation of strong adhesive bonds between the tip and surface. [11]   The water film may help support the load of the tip over a fairly large area, while the highest atomic-scale protrusion of the tip makes hard contact with the surface under the film and yields the image as a small modulation of the average repulsive force on the whole tip. With so much discussion of water's role in AFM imaging, some concern about the state of the NaCl surface in air is warranted. Although the high solubility of NaCl in bulk water is well known, it has been shown by SIMS analysis of NaCl cleaved and exposed to air for 30 minutes that only very small amounts of $(NaOH)Na^+$ and other potential water reaction products were formed. [12]   Furthermore, exposure to water vapor in vacuum at room temperature showed very low sticking probability as judged by SIMS. More relevant to the present case, Tepper et al have shown that adsorbed water on NaCl, as detected by anisotropic surface second harmonic generation (SHG), does not reach ordered monolayer coverage until after 2 days exposure in air at 50 % relative humidity. [13] Thus there are grounds for expecting that a reasonable representation of the NaCl surface can be obtained by scanning in air, but undoubtedly the water will influence the defects being probed, such as the hypothesized stabilization of the vacancies discussed earlier. It clearly is advantageous to work without the water film if atomic resolution can be maintained.

A few UHV atomic force microscopes have been built [7, 14]. Giessibl et al have constructed a low-temperature UHV AFM. They could resolve atomic periodicity on graphite, but it was not possible to achieve atomic resolution on KBr as of the first report. [14] Our approach early in this study has been to use a commercial AFM in dry atmosphere. Requirements for cooling the laser diode and incompatibility of some materials of construction with vacuum bakeouts prevent UHV operation of our Digital Instruments AFM in its standard form. We have provided an evacuable chamber which can be backfilled with helium and desiccated by means of a cryosorption pump operating at liquid nitrogen temperature. A dew point monitor operating down to 78 K is used to measure the humidity. Using this apparatus, images were recorded as the humidity was increased from 63 ppm (water/helium) up to 5100 ppm. There is an unmistakeable trend toward better atomic resolution at higher humidity. This has been confirmed repeatedly. The best images have been obtained around 40 % relative humidity ($\sim$10,000 ppm)

At the tip forces we have had to employ so far to achieve good atomic resolution, the AFM can break bonds, especially in absence of a lubricating film. On the one hand, this suggests that the tip may be used as an agent of defect formation in some future studies. More immediately, it means that the tip is a major perturbation of the scene being imaged, and may destroy the evidence of desorption or radiation damage before the complete image can be acquired. By scanning a small area and then quickly enlarging the scan size to look at the borders of the eroded area, we have seen that under conditions so far used for atomic resolution on NaCl, the sample is eroded at the rate of a monolayer in several minutes. [2] Our preliminary studies of image quality versus tip force suggest that the force required for a good image increases in dry helium. In one such study at 1240 ppm water in helium, the tip force was varied from 37 nN up to 286 nN. Only at 286 nN was good atomic resolution obtained in the dry atmosphere. As noted earlier, imaging in room air can be done successfully with considerably less tip force, but we have not yet identified a tip pressure which gives atomic resolution on NaCl without eroding the surface.

Acknowledgments

Research supported by NSF grant # DMR-8901103. We wish to thank W. E. Pendleton and G. E. Matthews for discussions and experimental assistance, and L. M. Slifkin for helpful comments.

References

1. Ph. Avouris, in these proceedings.
2. R. T. Williams, R. M. Wilson, and H. Liu, Nucl. Inst. & Meth. B 65, 473 (1992).
3. K. Hattori, A. Okano, Y. Nakai, N. Itoh, and R. F. Haglund, Jr., J. Phys.: Cond. Matter 3, 7001 (1991).
4. G. Binnig, Ch. Gerber, E. Stoll, T. R. Albrecht, and C. F. Quate, Europhys. Lett. 3, 1281 (1987).
5. E. Meyer, D. Anselmetti, R. Wiesendanger, H. J. Guntherodt, F. Levy, and H. Berger, Europhys. Lett. 9, 695 (1989).
6. E. Meyer, H. Heinzelmann, H. Rudin, and H. J. Guntherodt, Z. Phys. B 79, 3 (1990).
7. G. Meyer and N. M. Amer, Appl. Phys. Lett. 56, 2100 (1990).
8. A. M. Stoneham, Cryst. Latt. Def. and Amorph. Mat. 14, 173 (1987).
9. Nanoscope II, Digital Instruments, Inc.
10. R. T. Williams and K. S. Song, J. Phys. Chem. Sol. 51, 679 (1990).
11. N. A. Burnham, R. J. Colton, and H. M. Pollock, J. Vac. Sci. Technol. A 9, 2548, (1991).
12. J. Estel, H. Hoinkes, H. Kaarman, H. Nahr, and H. Wilsch, Surf. Sci. 54, 393 (1976).
13. P. Tepper, J. C. Zink, H. Schmelz, B. Wasserman, J. Reif, and E. Matthias, J. Vac. Sci. Technol. B 7, 1212 (1989).
14. F. J. Giessibl, Ch. Gerber, and G. Binnig, J. Vac. Sci. Technol. B 9, 984 (1991).

Electron Microscopy of Electron-Stimulated Processes at Oxide Surfaces

D.J. Smith, M.R. McCartney, and M. Gajdardziska-Josifovska

Center for Solid State Science, Arizona State University,
Tempe, AZ 85287, USA

Abstract. This paper provides a brief overview of recent electron microscopy studies of electron-stimulated desorption(ESD) processes occurring at maximally valent oxide surfaces, with emphasis upon the profile imaging mode of observation. The products that accumulate on the oxide surfaces as a result of ESD depend upon characteristics of the electron beam (energy, current density), the residual vacuum level, and the specimen temperature, and include reduced oxides, metallic particles and facetted exit-surface sputter pits. The diversity of results is an indication of competing electronic, diffusional and ballistic mechanisms: determination of the dominant process must be inferred from the final state of the surface under a variety of experimental conditions.

1. Introduction

The high-energy electron beam used for observation in the electron microscope can induce permanent physical and chemical changes to the specimen by either a ballistic knock-on process involving electron-nuclear interactions or by way of radiolytic processes involving electron-electron interactions[1]. Electron energy thresholds for ballistic processes in bulk materials are functions of binding energy and incident electron energy, and are typically in the range of 100-1000keV. Above the threshold energy, cross sections are proportional to electron energy but are typically less than $3 \times 10^{-23} cm^2$ at 400keV for ionic oxides. Cross-sections for radiolytic processes fall off rapidly with energy, so that at 100keV they are 10^3 less likely to occur than at 100eV. Both mechanisms are more prevalent at surfaces because of lower binding energies.

Real-time observations of beam-induced changes can be made on the atomic scale using the surface profile imaging mode of the high-resolution electron microscope (HREM)[2]. Particular attention has recently been given to electron-stimulated surface processes for a range of maximally valent transition metal oxides(TMO), prompted by discoveries of reduced phases on TMO surfaces after HREM study[3,4]. Recent studies show, for example, that the end-products are strongly influenced by the current density of the incident illumination [5] and the sample temperature during observation [6]. These results can not be simply described only in terms of electron-stimulated desorption (ESD) processes [7,8]. In this paper, we provide an overview of electron-beam-induced processes at maximally-valent oxide surfaces as observed by electron microscopy, as well as describing some recent results for MgO at elevated temperatures using HREM and reflection electron microscopy (REM).

Springer Series in Surface Sciences, Vol. 31
Desorption Induced by Electron Transitions DIET V
Editors: A.R. Burns · E.B. Stechel · D.R. Jennison   © Springer-Verlag Berlin Heidelberg 1993

2. Observations

When the maximally-valent transition-metal oxides TiO_2, V_2O_5, and Nb_2O_5 were irradiated at current densities of 5-50A/cm^2 at 400keV in conventional microscope vacuum (10^{-6} - 10^{-7}Torr), the final reaction products were the respective metallic monoxides, invariably found to have well-defined three-dimensional epitaxial relationships with the original oxides [9-11]. In the case of WO_3, we observed an epitaxial rocksalt phase tentatively identified as WO [9]; for TiO_2, an intermediate epitaxial phase of TiO_2-II was formed as a precursor before formation of the final TiO product [10]; for V_2O_5, two separate intermediary phases have been reported [11]. Figure 1 illustrates the direct reduction of Nb_2O_5 to NbO [10].

Structural drawings of these TMO and their respective reduced oxides revealed that the nature of the three-dimensional epitaxy was determined by the spacing and orientation of the cation sub-lattice of the original TMO [10]. The reduced oxides TiO, NbO and VO have the defective rocksalt structure, and it is significant that they are all known to be metallic in nature: the ready availability of their conduction electrons is believed to stifle the inter-atomic Auger process thought to give rise to oxygen desorption [12]. These observations and their explanations are in concurrence with earlier Auger studies in which it was found that surface regions of TMO were reduced by ESD processes, but not completely to the respective metal [13].

An interesting contrast is provided by SnO_2 which is isostructural with TiO_2 and also has a stable monoxide phase. Surface layers developed marked facetting during microscopy observations, but no reduced oxide, as shown in figure 2. The absence of a reduced phase was predicted for SnO_2 on the basis of the smaller electronegativity difference of its constituents: the size of this difference was originally suggested as a criterion for effective inter-atomic Auger processes [12].

Reduction beyond the monoxide phase occurs when the same TMO are irradiated in the electron microscope with a small focussed probe of extremely high current density ($\sim10^3$-10^4A/cm^2). Typically, reduction proceeded directly to the base metal and there was no epitaxy with the bulk oxide. For example,

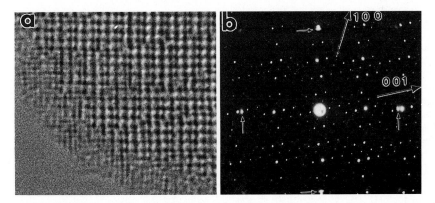

Fig.1.(a) Profile image of Nb_2O_5 crystal in [010] projection recorded at 400kV after 30min irradiation at 15A/cm^2. Note well-established epitaxial layer of NbO; (b) Selected-area electron diffraction pattern from edge of Nb_2O_5 crystal showing extra spots (arrowed) due to NbO surface phase.

Fig.2. Profile image of [001] SnO_2 crystal at 400kV after 25min irradiation at $35A/cm^2$, showing electron-beam-induced facetting of the (010) surface.

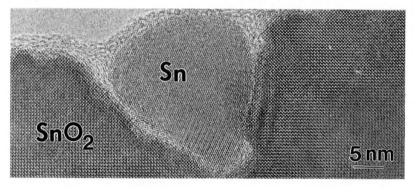

Fig.3. Profile image of [001] SnO_2 crystal and Sn single crystal which resulted from irradiation with focussed probe of $10^4 A/cm^2$ at 100kV.

intense irradiation of WO_3 led to a region, with the approximate size of the probe, of disordered and randomly oriented microcrystals of W [5]. Similar irradiation of TiO_2 caused damage pits to develop: these consisted of regions of lighter contrast which were shown by high-resolution imaging and convergent beam electron diffraction to correspond to substoichiometric oxides of titanium [5]. Small crystalline particles of Sn develop on SnO_2 surfaces under extreme dose rate conditions, as shown in figure 3. Similar irradiation of the mixed TMO $SrTiO_3$ produced amorphization but no obvious crystalline phase[14].

Irradiation under ultrahigh vacuum conditions helps to identify the influence of the local environment on the reaction products. Observations with a conventional UHV HREM confirmed the same reduction sequence of TiO_2/TiO_2-II/TiO as previously observed, indicating that residual oxygen in the microscope vacuum had played no role in this system [6]. It was also significant that no evidence for metallic titanium was observed despite the extremely low oxygen partial pressure. Irradiation of WO_3 under normal HREM imaging conditions is reported to lead to W when the sample is at UHV [15].

Fig.4. REM micrographs from cleaved MgO (100) surface after irradiation with 300keV electrons: (a) no visible damage at room temperature; (b) extensive surface pitting develops at 600°C. Insets show corresponding RHEED patterns.

Observations of TiO_2 at 100keV in a UHV scanning transmission electron microscope equipped with entrance and exit surface secondary electron detectors revealed that the exit surface of the sample was far more susceptible to damage than the entrance surface, thus indicating that both ballistic and radiolytic processes were occurring at the surfaces [16]. Further results reported elsewhere [17], such as the presence of an amorphous carbon overlayer apparently stifling desorption, reinforce the conclusion that both the reaction path and the final product(s) may depend on the local specimen environment during irradiation.

The effects of irradiation of MgO crystals have been studied over a range of sample temperatures, mostly under UHV conditions. MgO is a maximally-valent oxide that is not expected to be particularly susceptible to ESD processes [12]. Intense irradiation of a transmission sample held at room temperature under UHV eventually resulted in the formation of small facetted pits but there was no sign of any reaction products. When MgO was heated to 1000°C, mild irradiation with both 100keV and 300keV electrons produced facetted holes at a rapid rate under both conventional and ultrahigh vacuum. The geometry of the beam-induced pitting depended on the relative orientation of the sample to the incident beam direction indicating that the prevailing damage mechanism was exit-surface sputtering. Reaction products were sometimes visible decorating the edges of some holes. Their composition and crystal structure remain to be determined but they may be related to Ca segregation which is known to occur when MgO is annealed at temperatures above 950°C [18].

Careful studies of cleaved MgO surfaces using REM indicated that the onset of pitting of (100) surfaces under normal current density conditions occurred at about 600°C and 900°C for irradiation with 300 and 100keV electrons respectively. Figure 4 compares REM images obtained with 300keV electrons from a cleaved (100) surface. Radiation damage in the form of surface pitting is observed to be rapid at high temperatures but minimal at room temperature.

Finally, it is interesting to compare these changes in MgO surfaces with similar effects observed when large single crystals of rutile were irradiated at elevated temperatures under UHV conditions [6]. Well-facetted holes developed rapidly at temperatures in excess of about 570°C but there was no evidence for the intermediary phase nor the lower oxide seen at room temperature under conventional vacuum and UHV. Extensive regions of both TiO_2-II and TiO, as well as facetted pits, were visible at lower temperatures.

3. Discussions and Conclusion

Standard studies of electron-stimulated desorption of oxygen from TMO surfaces primarily involve detection of the desorbed neutral atoms or charged ions [13,19]. Electron microscopy provides complementary information by allowing observation of the irradiated surface as the desorption proceeds. Surface modification to a reduced crystalline phase is the prevailing result of electron irradiation for many TMO during observation in the electron microscope. Observations of rutile, the archetypal material for ESD studies, have revealed, however, that damaged crystals may reoxidize from the bulk outwards under low flux irradiation [10]. When TiO_2 is irradiated at 5A/cm^2 to the same total dose as that which produced a well-developed layer of TiO, no surface reduction is observed [10]. Annealing of rutile under UHV in the absence of the electron beam also led to restoration of the surface stoichiometry[6]. Thus, irradiation under varying current density conditions highlights the competition between diffusion-enhanced and displacement processes.

For some materials, such as TiO_2 and MgO, irradiation above certain temperatures that depend upon beam energy leads to rapid development of facetted holes, suggesting that a possible means for enhancing nanolithographic writing speed would be to execute the irradiation step at elevated temperature. The failure of SnO_2 to reduce under normal microscope viewing conditions is not unreasonable given that the reduced electronegativity difference between Sn and O makes core-initiated events less efficient.

Overall, we conclude that a rich variety of interesting events occurs at maximally valent oxide surfaces as a result of electron irradiation within the electron microscope. The high-resolution electron microscope is able to provide atomic-scale information about the structural damage. Despite the high energy, penetrating probe used in the HREM, the results described here provide experimental evidence that the prevailing damage mechanisms for these oxides is ESD rather that ballistic. Moreover, it is clear that both thermal and radiation-enhanced diffusion processes play an important role in influencing surface damage mechanisms [8]. Care must always be taken to define the experimental conditions in such a way that the results are meaningful and useful to the community interested in ESD processes.

Our work utilized the Center for High Resolution Electron Microscopy at Arizona State University supported by NSF Grant DMR-89-13384.

References

[1] L.W. Hobbs, in Quantitative Electron Microscopy, eds. J.N. Chapman and A.J. Craven (SUSSP Publications, Edinburgh, 1984) Chapter 11.

[2] D.J. Smith, R.W. Glaisher, P.Lu and M.R. McCartney, Ultramicroscopy 29,123(1989).

[3] D.J. Smith and L.A. Bursill, Ultramicroscopy 17,387 (1985).

[4] A.K. Petford, L.D. Marks and M. O'Keefe, Surface Sci. 172,496(1986).

[5] M.R. McCartney, P.A. Crozier, J.K. Weiss and D.J. Smith,Vacuum 42,301(1991).

[6] M.R. McCartney, and D.J. Smith, Surf. Sci. 250,169(1991).

[7] M.I. Buckett, J. Strane, D.E. Luzzi, J.P. Zhang, B.W. Wessels and L.D. Marks, Ultramicroscopy 29,217(1989).

[8] M.I. Buckett and L.D. Marks, MRS Symp. Proc. 235,405(1992).

[9] D.J. Smith, M.R. McCartney and L.A. Bursill, Ultramicroscopy 23,299(1987).

[10] M.R. McCartney and D.J. Smith, Surface Sci. 221,214(1989).

[11] H.J. Fan and L.D. Marks, Ultramicroscopy 31, 357(1989).

[12] M.L. Knotek and P.J. Feibelman, Surface Sci. 90,78(1979).

[13] D. Lichtman, Ultramicroscopy 23,291(1987).

[14] M.R. McCartney and D.J. Smith, MRS Symp. Proc. 183,311(1990).

[15] P.A. Crozier, M.R. McCartney and D.J. Smith, Surface Sci. 237,232(1990).

[16] M.I. Buckett, S.R. Singh, H. Fan, T. Wagner and L.D. Marks, in Proc. 47th. Ann. Meet. EMSA, edited by G.W. Bailey (San Francisco Press, San Francisco,1989) pp.636-637.

[17] M.I. Buckett and L.D. Marks, Surface Sci. 232,353(1990).

[18] M. Gajdardziska-Josifovska, M.R. McCartney and J.M. Cowley, in Proc. 49th. Ann. Meet. EMSA, edited by G.W. Bailey (San Francisco Press, San Francisco, 1991) pp.624-625.

[19] M.J. Drinkwine and D. Lichtman, Prog. Surface Sci., 8,123(1977).

Part VIII

Semiconductors

State-Resolved Studies of the Laser-Induced Desorption of NO from Si(111): The Importance of Localized Excitations

L.J. Richter, S.A. Buntin, D.S. King, and R.R. Cavanagh

National Institute of Standards and Technology, Gaithersburg, MD 20899, USA

Abstract. Quantum-state-resolved techniques have been applied to the laser-induced desorption (LID) of NO from Si(111). When contrasted with LID from metal substrates, where prevailing theories indicate the crucial role of the rapid screening response of the substrate, the LID results from silicon indicate the overriding importance of the localization of electronic excitations inherent to the covalent bonds of the substrate.

1. Introduction

Recently, there has been significant interest in photoreactions of adsorbed molecules on metal surfaces.[1] In a number of systems it has been determined that the dominant mechanism for transferring the energy of the incident photon into the adsorption complex involves the participation of photogenerated 'hot' carriers in the substrate. The transfer of energy between the carrier system and the adsorbate is often attributed to the formation of short lived ionic species via charge transfer processes. The chemistry resulting from these processes is determined by the rapid screening response of the metallic substrate, which strongly influences both the excited state lifetimes and the excited state potential energy surface (PES). State-resolved studies of the energy partitioning in gas phase products resulting from surface photon induced reactions have been instrumental in establishing both the importance of substrate photogenerated carrier mediated reactions and the influence of substrate screening.[2,3] For the specific case of the LID of NO from Pt(111)[2], it was proposed that desorption was the result of capture of a hot *electron* by the adsorbate. The equilibrium geometry of the resulting NO⁻ will in general not coincide with that of the neutral NO and forces will be exerted on the molecule in the excited state. If these forces are strong enough, and the excited state lifetime long enough, desorption may occur following neutralization back to the ground state PES. For the low photon energies used in the NO/Pt(111) studies, desorption of the ion was energetically forbidden. Wave packet calculations[4] based on the Antoniewicz model[5] for desorption of physisorbed species, in which the ionic PES is viewed as simply the ground state PES augmented by an image potential due to the screening response of the substrate, were qualitatively consistent with the data. In the Antoniewicz model, the image force pulls the NO⁻ towards the surface and desorption follows when neutralization places the NO on the repulsive wall of the Pt-NO potential. A similar model involving formation of a temporary NO⁺ has been proposed for the electron-stimulated desorption of NO from Pt(111)[6]. However, the postulated shape of the excited PES differed significantly, due to the assumption that charge transfer screening was dominant.[6,7]

In the area of photochemistry of semiconductor substrates, chemical reactions due to the interaction of adsorbates with photogenerated carriers have been long established.[8] Models for these processes have generally been presented in the context of electrochemistry. For example, photodesorption from wide-band gap substrates has been described by a model in which the original chemisorption event is due to the formation of an adsorbed ion. Neutralization of the ion by a photogenerated substrate carrier results in a physisorbed neutral, that subsequently thermally desorbs. Variations of this model have been used to correlate photodesorption and photoconductivity measurements.[8]

We highlight here the results of state-resolved studies of the laser-induced desorption of NO from Si(111). The studies indicate that localization of the excitation is key to understanding the photochemistry on the covalently bound substrate.

2. LID of NO from Si(111)

The details of the experiments have been described previously.[9] The LID was driven by photons derived from a Q-switched, Nd^{+3}:YAG laser. In general, the incident beam energy was sufficiently low that pulsed heating of the surface is negligible. The gas phase NO LID product was detected via LIF by exciting the $A^2\Sigma^+-X^2\Pi_\Omega$ transition near 226 nm. State-specific kinetic energy distributions were obtained from optically detected time-of-flight (TOF) spectra resulting from varying the time delay between the firing of the desorption and probe lasers. Both the flux-weighted mean kinetic energy and the shape of the velocity distribution were observed to change as a function of internal state probed; therefore, internal-state population distributions were obtained by time integrating the TOF spectra recorded for various NO quantum states.

Both n- and p-type Si wafers were used as substrates. They were used as received and cleaned via Ne ion sputtering and annealed. The 7x7 reconstruction was confirmed by LEED. NO exposures were performed at substrate temperatures of nominally 95 K. This also was the initial temperature for all LID experiments. The total fluence (J cm^{-2}) incident on a given spot was restricted so that the desorption signal decreased by no more than 10% during the acquisition of a TOF spectra.

The thermal chemistry of the NO/Si(111) system is complex but has been characterized by HREELS and TPD following adsorption of NO at low temperature.[10] Additionally, irradiation of the saturated surface with monochromated light from a cw-arc lamp has been demonstrated to result in photon-induced desorption of N_2O, N_2, and NO, along with photodissociation.[11]

It was observed that virtually all aspects of the LID, i.e., energy partitioning in the desorbed NO, wavelength dependence of LID yield, and relative branching ratios between observed photoreaction channels, varied significantly with the initial NO coverage.[9] This was attributed to the presence of two distinct desorption mechanisms, one that dominates at low coverages, but is quenched by increasing NO exposure, and a second that dominates at high coverages. These two mechanisms

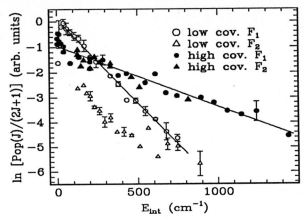

Figure 1. Internal-state-population distributions for LID of the two spin-orbit states of NO (F_1 and F_2) in the ground vibrational state from Si(111) with 355 nm photons. The initial NO coverages were $\approx$ 0.03 ML (low cov.) and saturation (high cov.).

will be referred to as the low coverage channel and the high coverage channel. The differences in the energy partitioning characteristic of the two mechanisms are highlighted in Fig. 1, which shows internal-state-population distributions for NO in the ground vibrational state following desorption with 355 nm pulses.

The results of complete characterization of the energy-partitioning of both desorption channels have been presented.[9] With a desorption laser wavelength of 355 nm, the low coverage channel was characterized by moderate vibrational and translational energy release and substantially less rotational energy release. Additionally, a strong selection for the ground spin-orbit state ($\Omega_{1/2}$, F_1 in Fig. 1) was observed. There was little quantitative change in the energy partitioning for excitation wavelengths spanning 1907-355 nm. Again at 355 nm, the high coverage channel was characterized by greater energy release in all degrees of freedom. Also the selectivity for the ground spin-orbit state was absent. When the desorption wavelength was decreased to 1064 nm, the energy content in all degrees of freedom decreased.

3. Discussion

It is manifestly obvious from the energy partitioning that the desorption process for the two channels is quite different. Based on studies of the wavelength and substrate doping dependence of the LID yield, it was proposed that the low coverage channel involved surface state excitation.[9] A study of the effects of selected coadsorbates on the LID yield confirmed this and established that the desorption was due to the photogeneration of holes in the normally doubly occupied, rest-atom localized, surface state of the 7x7 reconstruction (the S_2 surface state). It should be stressed that the desorption is not due to the creation of bulk electrons from the 'ionization' of the S_2 state, but is directly the result of the surface-localized electronic excitation.

The S_2 surface state is a relatively dispersionless band and thus has a narrow energy width. This accounts for the observed weak dependence of the energy partitioning on the desorption wavelength. The role of the S_2 state is also consistent with the quenching of the low coverage channel with increasing NO coverages; at low coverages the adsorbed NO preferentially interacts with the adatoms of the 7x7 structure (quenching the S_1 surface states at mid gap), while at higher coverages the NO interacts with the S_2 surface state (the rest atoms).

The formation of temporary ionic species, as postulated for desorption from metals, is not likely to be relevant to the low coverage channel. This is because an NO^+, after formation via interaction with the photoexcited S_2 state, would have to return the hole to the Si substrate with about 1 eV less energy than the S_2 state. The only possible place to return the hole is to occupied, mid-gap surface states (S_1). This was inconsistent with the coadsorption experiments which showed no diminution of the LID yield with quenching of the S_1 state. We therefore have proposed that the NO desorbs due to a direct process that formally looks like electron transfer from an adsorbed NO^- to the S_2 hole. STM studies and theoretical calculations have determined that charge transfer from the S_2 state to the adatom-localized dangling bond (presumed to be involved in the Si-NO bonding) is facile.[12]

The energy partitioning observed for the low coverage channel is very similar to the energy partitioning observed for the photolysis of gas phase carbon analogs of NO adsorbed onto Si. Specifically, the photolysis of $(CH_3)_3CNO$ at wavelengths < 670 nm resulted in a rapid (<10 ns) dissociation that was attributed to intersystem crossing from the optically prepared singlet excited state to a triplet excited state, followed by dissociation over a small exit channel barrier. The presence of the barrier was invoked to explain the high translational energy release compared with rotational energy. The gas phase photolysis exhibited a similar selection for the ground spin-orbit state, attributed to the specific subtleties of the electronic potentials.[13] The similarity of the energy partitioning in the LID and carbon analog studies suggests that the PES relevant to the LID may be understood in localized, "molecular", terms.

Our understanding of the microscopic nature of the high coverage channel is much more limited than the low coverage channel. Earlier studies of the wavelength dependence of the state integrated yield over the range 600-250 nm lead to the conclusion that the desorption of NO involved an interaction with 'hot' carriers: those with significant excess energy with respect to the band edges and an associated short ($\approx$20 nm) attenuation length.[11] In our studies, we have observed desorption from saturated surfaces with 1064 nm radiation, which is approximately resonant with the Si indirect gap and thus creates bulk carriers with negligible excess energy. The yield at 1064 nm was linear in fluence over a factor of 16 in variation, indicating that multi-photon generation of hot carriers is not occurring. Free carrier heating leading to desorption is unlikely, as similar yields were observed on both n- and p-type substrates. Since at saturation coverage photoemission studies indicate that essentially all of the surface states have been swept out of the gap by bonding to the adsorbate,[14] it seems that hot carriers are not a necessary condition for the LID of NO in the high coverage channel.

A desorption mechanism consistent with increased yield for hot carriers, but similarly applicable to band-edge carriers, is a carrier recombination mechanism.

Since early work by Lang and Kimerling,[15] it has been recognized that a variety of bulk processes, such as defect annealing and vacancy migration, proceed at greatly enhanced rates upon the introduction of non-equilibrium carrier densities. Weeks, *et al.*[16] qualitatively explained the enhancement in terms of a RRK-type[17] mechanism. The release of the electronic energy of the carriers into vibrations of the defect that mediates recombination results in a localized hot spot, which can promote processes involving the defect. Surface recombination enhanced reactions have recently been invoked in the photoassisted oxidation of GaAs.[18] The RRK-type mechanism would be expected to produce more 'thermal' energy partitioning (although dynamical constraints may still be important in both the recombination itself, and the desorption) in accord with the observed equilibration between the spin-orbit states and rotational states, and the better agreement between T_{rot} and $<KE>$. Additionally, it is interesting to note that the branching ratios between the various photostimulated reactions in the high coverage channel are the same as for thermal excitation. (The branching ratios for the low coverage channel deviate significantly from thermal excitation.)

4. Summary

The fundamental aspects of the LID of NO from Si(111) appear to be dominated by localization of the excitations of the system, consistent with the covalent nature of the substrate (and adsorbate bonding). This localization is manifest in the identification of surface state photogenerated holes as mediating the low coverage channel, and may be reflected in the proposed 'molecular' nature of the resulting excited state. This localization is also implicit in the vibrational localization necessary for the RRK-type coupling proposed for recombination driven processes which may be relevant for the high coverage channel. This localization implies that modern *ab initio* theoretical treatments of the low energy excited states of adsorbates on semiconductors may be both tractable and illustrative.

This work was partially supported by the US Department of Energy, Office of Basic Energy Sciences (D.E.-AI05-84ER13150).

References

1. X.-L. Zhou, X.-Y. Zhu and J.M. White, Surf. Sci. Rept. 13, 74 (1991).
2. S.A. Buntin, L.J. Richter, D.S. King, and R.R. Cavanagh, J. Chem. Phys. 91, 6429 (1989); S.A. Buntin, L.J. Richter, R.R. Cavanagh, and D.S. King, Phys. Rev. Lett. 61, 1321 (1988).
3. K. Mase, S. Mizuno, Y. Achiba, and Y. Murata, Surf. Sci. 242, 444 (1991).
4. J.W. Gadzuk, L.J. Richter, S.A. Buntin, D.S. King, and R.R. Cavanagh, Surf. Sci. 235, 317 (1990).
5. P.R. Antoniewicz, Phys. Rev. B 21, 3811 (1980).
6. A.R. Burns, E.B. Stechel, and D.R. Jennison, Phys. Rev. Lett. 58, 250 (1987).
7. A.R. Burns, E.B. Stechel, D.R. Jennison, and T.M. Orlando, Phys. Rev. B 45, 1373 (1992).

8. L.J. Richter and R.R. Cavanagh, Prog. Surf. Sci. **39** (1992) and references therein.

9. L.J. Richter, S.A. Buntin, D.S. King, R.R. Cavanagh, Phys. Rev. Lett. **65**, 1957 (1990); J. Electron Spectrosc. Relat. Phenom. **54**, 181 (1990); J. Chem. Phys. **96**, 2324 (1992).

10. Z.C. Ying and W. Ho, J. Chem. Phys. **91**, 2689 (1989); *ibid* **91**, 5050 (1989).

11. Z.C. Ying and W. Ho, Phys. Rev. Lett. **60**, 57 (1988); Z.C. Ying and W. Ho, J. Chem. Phys. **93**, 9089 (1990).

12. Ph. Avouris and In-Whan Lyo in *Chemistry and Physics of Solid Surfaces VIII*, ed. by R. Vanselow and R. Howe (Springer, Berlin, 1990) pg 371.

13. M. Noble, C.X.W. Qian, H. Reisler, and C. Wittig, J. Chem. Phys. **85**, 5763 (1986).

14. F. Bozso and Ph. Avouris, Phys. Rev. B **43**, 1847 (1991).

15. D.V. Lang and L.C. Kimerling, Phys. Rev. Lett. **33**, 489 (1974).

16. J.D. Weeks, J.C. Tully, and L.C. Kimerling, Phys. Rev. B **12**, 3286 (1975).

17. P.J. Robinson and K.A. Holbrook, *Unimolecular Reactions* (Wiley, London, 1972).

18. K.A. Bertness, P.H. Mahowald, C.E. McCants, A.K. Wahi, T. Kendelewicz, I. Lindau, and W.E. Spicer, Appl. Phys. A **47**, 219 (1988).

Laser-Induced Electronic Processes on Semiconductor Surfaces

N. Itoh, Y. Nakai, K. Hattori, A. Okano, and J. Kanasaki

Department of Physics, Faculty of Science, Nagoya University, Chikusa-ku, Nagoya 464-01, Japan

ABSTRACT. Recent studies of laser-induced DIET (desorption induced by electronic transitions) processes from surfaces of semiconductors are discussed. It is pointed out that the active sites for particle emission from semiconductors are defects on surfaces, different from alkali halides, in which self-trapping of an exciton can form a Menzel-Gomer-Redhead-type antibonding adiabatic potential energy surface (APES) at any site on the surface. The defect-initiated particle emission from semiconductors is divided into three types: that arising from adatoms, steps and vacancies. For all processes, the yield is a superlinear function of fluence, and the former two processes are induced most efficiently by excitation at the surface states. The superlinear yield-fluence relation indicates that desorption is induced not by a single excitation event but by interaction of more than one excitation event. The mechanisms of formation of an anti-bonding APES by multi-hole localization are discussed.

1. INTRODUCTION

Irradiation of surfaces of semiconductors with laser beams results in emission of particles, modification of surface morphology and even stoichiometry change. Although melting of the surface layers has been considered to be the main cause of these processes [1], recent high sensitivity measurements of laser-induced particle emission from compound semiconductors by Hattori et al. [2-4] have revealed that non-thermal particle emission from the surfaces of compound semiconductors is induced at the initial stage of laser irradiation. According to these studies, the several types of defects on surfaces are the source of the particle emission, depending on laser fluence and wavelength.

Since the primary interaction of laser beams with solids induces electronic transitions, the non-thermal laser-induced particle emission described above should be of electronic origin. It is therefore of interest to compare the laser-induced electronic processes in semiconductors with those on the surfaces of other types of solids and on the surfaces with adsorbates and to elucidate the mechanisms of laser-induced particle emission for each case.

2. A SUMMARY OF EXPERIMENTAL RESULTS OF LASER-INDUCED PARTICLE EMISSIONS FROM THE SURFACES OF COMPOUND SEMICONDUCTORS.

Measurements of Ga^o emission induced by laser pulses at high sensitivities, capable of detecting 10^{-6} monolayer, have been

Springer Series in Surface Sciences, Vol. 31
Desorption Induced by Electron Transitions DIET V
Editors: A.R. Burns · E.B. Stechel · D.R. Jennison   © Springer-Verlag Berlin Heidelberg 1993

carried out for GaP (110) [3-5] and ($\bar{1}\bar{1}\bar{1}$) [2] and GaAs (110) [6] surfaces, and the major experimental results described below have been observed for all surfaces. First we point out that there is a sharp threshold laser fluence above which the surfaces are damaged. This ablation threshold can be detected by measuring macroscopic damage by LEED. Furthermore, the emission yield induced by repeated irradiation with laser pulses above the threshold is found to increase, while it decreases as the irradiation is repeated below the ablation threshold. The change in the yield below the ablation threshold, where no macroscopic damage is detected even after prolonged irradiation, has been ascribed to laser-induced progressive change of the defect morphology. Additional experimental observation of enhancement of the yield in this fluence range by Ar ion bombardment and its reduction by annealing have substantiated the presumption stated above. It is worthwhile mentioning that the emission yield for GaP below the ablation threshold is the highest for electron transitions involving the surface states [5,6].

The defect-initiated emission can be observed at fluences about 1/10 of the ablation threshold, and the yield-fluence dependence exhibits a power function with power indices between 2 and 6. The results that the emission is induced at fluences of 1/10 of the ablation threshold and that the yield is enhanced by surface excitation indicate clearly that the emission is an electronic process. The process, however, cannot be explained in terms of either the Menzel-Gomer Readhead (MGR) mechanism [7,8] or the Feibelman-Knotek (FK) mechanism [9], since both mechanisms predict that the yield is proportional to fluence. Since no core-electron excitation is involved in the laser-induced particle emission, the FK mechanism is not relevant to the defect-initiated, laser-induced particle emission.

3. MECHANISM OF LASER-INDUCED PARTICLE EMISSION FROM SOLID SURFACES

3. 1. Desorption from solid surfaces by the MGR mechanism

According to the MGR mechanism, electronic excitation of a adsorbate-substrate system from a bonding state to an anti-bonding state causes emission. The MGR mechanism [7,8] has been suggested originally to explain desorption of adsorbates on surfaces and hence assumes the presence of a molecular system having localized electronic states on surface. The concept of formation of an anti-bonding state can be used to explain emission of particles induced by band-to-band excitation or by exciton formation in insulating solids, in which excitons can be localized by self-trapping [10]. An anti-bonding path leading to desorption may be constructed in a multi-dimensional adiabatic potential energy surface for a self-trapped exciton. An example of such adiabatic instability for an exciton on surfaces of alkali halides is illustrated in Fig. 1. The figure describes three coordinates, which are not necessarily normal coordinates: Q_1 describes formation of an X_2^--molecular ion by combining two halogen ions and a hole, Q_2 describes off-center displacement of the X_2^--molecular ion and Q_3 describes decomposition of the X_2^--molecular ion into an X ion and an X atom that is emitted. The Q_1 and Q_2 relaxations have been well established from the atomic structures of the self-trapped excitons in the bulk [11]. The adiabatic poten-

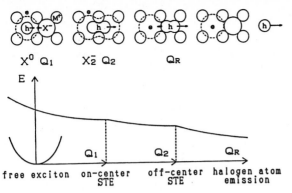

X^0 Q_1 X_2^- Q_2 Q_R

free exciton on-center off-center halogen atom
                 STE   STE   emission

Fig. 1   An adiabatic potential energy surface describing emission of a halogen atom by formation of an exciton in alkali
halides.   Coordinate Q_1 is the distance between two neighboring
halogen ions, Q_2 denotes the translational displacement of a X_2^-
molecular ion, where X denotes a halogen atom and Q_3 describes
the dissociation reaction leading to emission of a halogen
atom.

tial energy surface for the Q_3 relaxation has not yet been
proved experimentally, but a theoretical calculation has been
made to support the anti-bonding nature [12].

3. 2. Mechanism for defect-initiated particle emission from
semiconductor surfaces

For semiconductors, unlike alkali halides, the multi-photon
band-to-band transition as a cause of particle emission or
production of defects on surfaces can be excluded since it is
known that single excitation of semiconductors can induce none
of these processes.*   As is the case for alkali halides, two
crucial processes, localization and adiabatic instability, must
occur for defect-initiated particle emission from the semiconductor surfaces. A defect on the surface can act as a site for
trapping an electron or a hole, and eventually the other carrier will be trapped in an excited state, producing a localized
excited state. We presume that the localized excited state of
defects on surfaces does not lead to particle emission, since
so far no linear response of the yield on laser fluence has
been observed.
     It is well established that localization of an exciton on a
defect in the bulk of semiconductors produces a metastable
state, which is accompanied by a large lattice relaxation [15].
Not only localization of a single charge carrier but of two
carriers of the same sign can be induced if the lattice
relaxation energy exceeds the on-site Coulomb repulsion energy
(negative-U) [16]. We emphasize in particular that the localization of a hole weakens a bond surrounding the defect and
makes the lattice more deformable. Furthermore the defects on
surfaces are placed on low-symmetry sites and hence are highly

----
* In semiconductors, the subthreshold effects refer to defect
formation by photons below the band-gap energies [13].   These
processes, however, are often related to defects and surfaces,
and their origin is not necessarily clear [14].

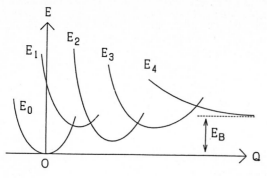

Fig. 2. An adiabatic potential energy surface describing emission of an atom from a defect on semiconductor surfaces. E_0 is that for the ground state and E_i's are those for the states excited i times. In the figure, emission can occur after being excited four times. E_B is the binding energy of the emitted atom.

deformable. Thus formation of a metastable state by excitation or localization of a hole or two holes on a surface defect is very plausible.

The probability that a metastable state thus formed is excited within the same laser pulse is given by $F\sigma\tau$, where σ is the cross section, F is the flux and τ is the lifetime of the metastable state. Under laser irradiation with a fluence ($\sim 10^{-18}$ cm^{-2}) usually used for emission studies, the probability that an allowed transition occurs exceeds unity. Furthermore, a metastable state usually has an occupied or unoccupied level in the forbidden gap and lifetimes often in the range of milliseconds. Therefore we expect that the metastable state formed by localization of a hole or two holes, the latter due to the negative-U interaction, can be multiply re-excited during the duration of a pulse.

A schematic configuration coordinate diagram which leads to particle emission or bond breaking by multiple excitation is shown in Fig. 2. The defect has a configuration at the minimum of curve E_0 before excitation. After trapping an electron-hole pair at the lowest excited state, the defect relaxes to the local configuration at the minimum of curve E_1. After re-excitation, the defect includes two holes and then relaxes to another relaxed configuration at the minimum of curve E_2. The energy of this minimum is drawn lower than that of curve E_1, because of possible negative-U interaction. Direct relaxation to this minimum may occur if the density of electron-hole pairs is sufficiently large [16]. Further excitation from the doubly excited relaxed state will lead to curve E_3 and so forth. Desorption will occur if the excited state has an antibonding adiabatic potential energy surface, as depicted by curve E_4. Although the photon energy is fixed for all transitions, the continuum state overlaps with the excited state, thus the cascade excitation is considered to be induced.

A recent computer simulation of phonon-assisted multi-hole localization on defects on surfaces carried out by Ong et al. [18] shows that the processes described above are highly plausible. They chose a Ga-adatom and an anti-site Ga defect in the second topmost layer as examples, and found that excitation

236

of a positively charged Ga adatom produces an anti-bonding
adiabatic potential energy surface. For the other type of
defect, electronic excitation produces a highly distorted metastable state, which can emit atoms after further excitation.

4. CONCLUSION

In conclusion, we presented a new DIET process induced from
semiconductor surfaces. The MGR process can describe the
emission, the yield of which is proportional to the number of
photons absorbed, as observed in solids in which excitons are
self-trapped. In semiconductors, emission of atoms can occur
by localization of the electron-hole pairs at defect sites and
only after multiple excitation or phonon-assisted multi-hole
localization of holes. We can depict DIET in alkali halides
and semiconductors in the following way: A localized hole can
weaken the bond of atoms near the site, since a bonding electron is missing, and hence the electronic excitation energy can
now deform the lattice. In alkali halides a bond is broken by
single hole localization, while multi-hole localization is
needed to break the bond in semiconductors. The multi-hole
localization in this case requires lattice relaxation after
localization of a hole and is completely different from the
Feibelman-Knotek process, in which localization occurs within a
femtosecond and has an extremely small lifetime. Localization
of the Feibelman-Knotek type can be reached only by the Auger
transitions but not from densely produced electron-hole pairs
because of the Coulomb repulsion, unless the negative-U interaction is effective [14].

REFERENCES

1. M. Von Allmen, Laser-Beam Interactions with Materials
(Springer-Verlag, Berlin, 1987).
2. K. Hattori, Y. Nakai and N. Itoh, Surf. Sci. 227, L115
(1990).
3. K. Hattori, A. Okano, Y. Nakai, N. Itoh and H. F. Haglund,
Jr, J. Phys. Condens. Matter 3, 7001 (1991).
4. K. Hattori, A. Okano, Y. Nakai, and N. Itoh, Phys. Rev.
B45, 8424 (1992).
5. A. Okano, K. Hattori, Y. Nakai and N. Itoh, Surf. Sci. 258,
L671 (1991).
6. K. Kanasaki, A. Okano, K. Ishikawa and N. Itoh, to be
published.
7. D. Menzel and R. Gomer, J. Chem. Phys. 41, 3311 (1964).
8. P. Redhead, Can. J. Phys. 42, 886 (1964).
9. M. L. Knotek and P. J. Feibelman, Phys. Rev. Lett. 40, 964
(1978).
10. N. Itoh, Adv. Phys. 31, 491 (1982) and Cryst. Lattice
Defects Amorph. Mater. 12, 103 (1985).
11. R. T. Williams, K. S. Song, W. L. Faust and C. H. Leung,
Phys. Rev. B33, 7232 (1986).
12. N. Itoh, A. M. Stoneham and A. H. Harker, J. Phys. C.
13. P. M. Mooney and J. C. Bourgoin, Phys. Rev. B29, 1962
(1984).
14. N. Itoh, Nucl. Instrum. Methods B27, 155 (1987).
15. G. D. Watkins, Materials Science Forum 39, 38-41 (1989).
16. P. W. Anderson, Phys. Rev. Lett. 34, 953 (1975).
17. H. Sumi, Surf. Sci. 248, 382 (1991).
18. K. S. Ong, G. C. Khoo and N. Itoh, to be published.

Mechanisms of Metastable Atom Quenching on GaAs(100)

A. Ludviksson and R.M. Martin

Department of Chemistry, University of California,
Santa Barbara, CA 93106, USA

Abstract. Metastable quenching spectroscopy (MQS) with He*(^{1}S), He*(^{3}S), and Ne*(^{3}P) metastable atoms was used to probe the surface electronic structure of gallium-rich c(8x2) surfaces of GaAs(100). The study of n-type versus p-type GaAs and the addition of adsorbates provided further information which was useful in interpreting the MQ spectra. We find that an electronic band with binding energy 2.8-2.9 eV below the valence band maximum (VBM) dominates the quenching, which takes place by both the Auger deactivation (AD) and the resonance ionization (RI) + Auger neutralization (AN) mechanisms. The data suggests that RI of He*(^{3}S) and Ne*(^{3}P) occurs via a previously observed Ga dangling bond state lying in the band gap 0.9 eV above the VBM.

1. Introduction

Photoemission studies have been used quite extensively to perform bandmapping of GaAs surfaces. Separating surface states from bulk electronic states is often accomplished by observing changes in the spectra due to the addition of adsorbates on the surface. Table 1 summarizes the surface and bulk electronic states of GaAs(100) that have been found using such methods. [1-4] Also listed are observed bulk states of a cleaved GaAs(110) sample studied with XPS. [5] The UPS spectra reported by Chiang et al [4] has a broad feature peaking at 1.5 eV below the VBM with a prominent shoulder toward higher binding energy. Judging from the other data in Table 1, it is likely that this feature is due to overlap of surface

Table 1. Observed GaAs(100) bulk and surface electronic states.[a]

Binding Energy (eV)	Assignment	Surface structure	Method	Reference
-0.9	Ga dangling bond	c(8x2)	LELS	1
0.3	Ga dangling bond	(4x2), (2x4)	UPS	2
0.4	Ga dangling bond	(4x1), c(8x2)	LELS	1
0.5	Ga-dimers	(4x1)	UPS	3
0.7	As-dimers	(4x1)	UPS	3
0.8	As dangling bond	(4x2), (2x4)	UPS	2
1.3	As dangling bond	c(2x8)	LELS	1
1.3	dangling bonds (4x1)		UPS	3
3.0, 6.8	bulk states	(4x1)	UPS	3
2.7, 6.5	bulk states	(4x2), (2x4)	UPS	2
1.5, 6.2, 11.2	bulk states	c(8x2)	UPS	4
2.4, 6.6, 11.4	bulk states	cleaved GaAs(110)	XPS	5

[a] The binding energies are given with respect to the valence band maximum.

Springer Series in Surface Sciences, Vol. 31
Desorption Induced by Electron Transitions DIET V
Editors: A.R. Burns · E.B. Stechel · D.R. Jennison   © Springer-Verlag Berlin Heidelberg 1993

states and bulk states. In Table 1, filled surface states have binding energies of 0.3 - 1.3 eV while bulk states have binding energies of 2.4-3.0, 6.2-6.8, and 11.2-11.4 eV.

Most studies of metastable quenching electron spectroscopy (MQS) on clean crystal surfaces have been on metals. The quenching of metastable rare gas atoms Rg* on transition metal surfaces usually takes place by resonant ionization (RI) to form the Rg^+ ion, followed by Auger neutralization (AN) of the ion by the surface: this is known as the RIAN mechanism, and produces the ion neutralization electron spectrum (INS) of the surface. In AN an electron from the surface drops into the Rg^+ vacancy, coupling with another surface electron which may be ejected. As Rg* approaches a metal surface the ionization potential is decreased due to the image-charge potential, giving an effective long range ionization potential $I_{eff} = I - 3.6(Å-eV)/d$, where I is the ionization potential of the unperturbed Rg* and d is the distance between Rg* and the metal surface. RI can occur when I_{eff} of Rg* is less than the work function of the metal, so that the excited electron level is above the Fermi level.

Quenching of Rg* can also occur by the Auger deactivation (AD) or "Penning ionization" mechanism, in which an electron from the surface drops into the Rg* core vacancy, coupling with the excited electron which is ejected. MQ electron spectroscopy of adsorbates in the monolayer range on metal substrates often involves a competition between RI (followed by AN) and AD: this competition is determined by the energetics of RI described above, and the relative overlap of metal orbitals versus adsorbate orbitals with the excited and core Rg* orbitals.

In the present work He* and Ne* metastables were used to study the MQ spectra of clean n-type and p-type GaAs c(8x2)Ga surfaces. With semiconductors the existence of a band-gap, possible surface states in the band gap, and band-bending at the surface add additional parameters which can affect the quenching mechanisms. The addition of adsorbates provided further information which was useful in interpreting the MQ data.

The experimental methods have been described in detail previously [6]. The GaAs(100) n-type (p-type) wafers had final layers grown in the UCSB MBE Lab, with Si (Be) doping of 10^{18} cm^{-3}. A 0.8 cm^2 sample was cut from the wafer and attached to the molybdenum face of a cylindrical Spectra-Mat heater by a thin layer of indium metal, and directly installed in the UHV chamber without chemical cleaning. The sample was cleaned by 250 eV Ar$^+$ ion bombardment at 700 K for 30 min., with the ion beam at a 45° incident angle. After annealing at 800 K for 2 min., LEED showed a c(8x2) pattern characteristic of the Ga-rich c(8x2)Ga reconstructed (100) surface.

2. Results and Discussion

Fig. 1 shows the He*(^{1}S), He*(^{3}S) and Ne*(^{3}P) MQ spectra of the clean p-type GaAs(100) c(8x2)Ga surface at a sample temperature of 100 K. Similar spectra were obtained with the n-type sample. The energy scale is the measured electron kinetic energy as determined from the retarding potential difference (RPD) of the analyzer. The Fermi levels on the RPD scale (i.e., the energies of electrons ejected by AD from the Fermi level) are at 12.8 eV for Ne*(^{3}P), 16.0 eV for He*(^{3}S), and 16.8 eV for He*(^{1}S). The work function of the sample ϕ_s can be determined from $\phi_s = RPD_o + \phi_a$, where RPD_o is the low energy onset of the spectrum and $\phi_a = 3.8$ eV is the work function of the analyzer. This gives work functions of 5.2 eV and 4.6 eV for the p-type and n-type samples, respectively.

Fig. 2 shows electronic energy diagrams for the n-type and p-type GaAs together with the energy levels of the Rg* atoms. Photoemission yield spectroscopy experiments on GaAs(100) [7] have placed the conduction band minimum (CBM) 4.0 eV below the vacuum level. The bulk Fermi level has been shown to lie close to the top of the band gap for n-type GaAs [8,9]. From this and the work function we obtain upward band bending of 0.6 eV at the surface for the n-type (Fig. 2a). With the bulk Fermi level just above the VBM for the p-type, and a band gap of 1.5 eV at 100 K [10], a downward band bending of -0.3 eV is obtained (Fig. 2b). This gives an ionization energy of 5.5 eV from the VBM at the surface, which agrees with the

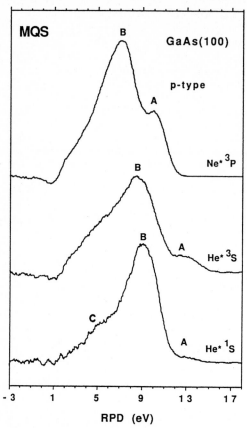

Fig. 1. MQ spectra of p-type GaAs(100) c(8x2)Ga at 100 K with He*(^{1}S), He*(^{3}S), and Ne*(^{3}P) metastable atoms.

measurement of 5.4 eV at room temperature [7], where the band gap is 1.4 eV. As shown in Fig. 2a, for the n-type the Fermi level is positioned 0.9 eV above the VBM at the surface. This suggests that the Fermi level is pinned at the surface by the empty Ga dangling-bond state that has been observed [1] at this level in the band gap (Table 1).

With Fig. 2 as a guide to the energy relationships, we can proceed to the interpretation of the MQ spectra shown in Fig. 1. The spectra show two main features, labeled A and B, and it is seen that A increases in prominence in going from He*(^{1}S) to He*(^{3}S) to Ne*(^{3}P). There is also an indication of a third feature C at low energy, which is most discernible in Fig. 1 with He*(^{1}S) at about 5 eV. This feature was most prominent with He*(^{3}S) on the n-type GaAs (not shown), where it was centered at about 4 eV. The main question to be addressed is whether features A, B, and C are due to RIAN or AD. If they are due to RIAN, then the spectra should be nearly the same as INS spectra from the corresponding low energy ions. No INS spectra of GaAs(100) surfaces are available from the literature, but Pretzer and Hagstrum [12] studied the INS spectra of 10 eV He$^+$ and Ne$^+$ on GaAs(110) and (111). The most important difference between their INS spectra and our MQS spectra is that feature A is absent in the INS spectra, indicating that it is due to the AD mechanism. Except for the absence of feature A, the rise of the INS signal from the high energy onset to the maximum agrees quite closely with feature B of the MQ spectra. The IN and MQ spectra cannot be compared at

240

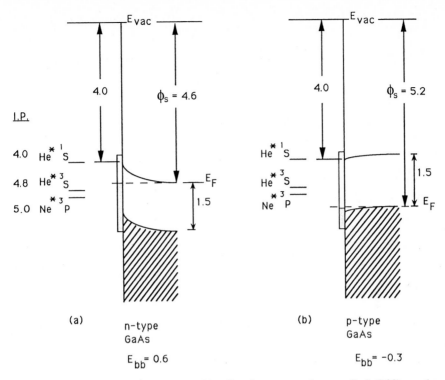

Fig. 2. Electronic energy levels and band bending for n-type and p-type GaAs(100) together with the Rg* energy levels.

energies below peak B, due to differences in the intensity of low energy secondary electrons in the two experiments. The qualitative agreement of the INS results with feature B indicates that it is due to the RIAN mechanism. However this mechanism is expected to give identical MQ spectra with He*(^{1}S) and He*(^{3}S), since both would be produced by AN of He$^+$. As seen in Fig. 1, the shapes of the ^{1}S versus ^{3}S spectra are quite different, and the energy of peak B is 0.4 eV lower for ^{3}S. The differences in feature B with He* ^{1}S versus ^{3}S ionization indicate that AD is also contributing significantly to feature B, or that the RIAN mechanism produces different spectra on n-type versus p-type GaAs(100).

Table 2 shows the calculated binding energies E_b with respect to the VBM of peaks A, B, and C assuming the AD versus RIAN mechanisms. To obtain the binding energies for the RIAN mechanism, we assumed that the two electrons involved in the AN transition originate from the same level. This is a simplifying approximation, but should give a reasonably accurate value for the average binding energies of the bands producing the peaks. The E_b values for the A peak are given only for the He*(^{3}S) and Ne*(^{3}P) cases, since the A peak was too small with He*(^{1}S) to be accurately measured. The A peak E_b values from the four spectra (He*(^{3}S) and Ne* on n-type and p-type) are in good agreement by both mechanisms: for AD the average E_b(A)=2.8 eV, and for AN the average E_b(A)=1.0 eV. The B peak E_b values for the AN mechanism are also in good agreement for the six spectra, giving an average E_b(B)=2.9 eV. However the E_b(B) values calculated for the AD mechanism are not in agreement for He* versus Ne*: the averages are 7.4 eV for He* and 5.5 eV for Ne*. From this we conclude that AN is the dominant mechanism producing the B peak, and that the A and B peaks are produced from the same electronic band at the surface with E_b=2.8-2.9 eV. This corresponds to the

Table 2. Binding energies (eV) of the MQ peaks for the AD and RIAN mechanisms.

	n-type GaAs(100)			p-type GaAs(100)	
	A	B	C	A	B
AD Mechanism					
He*(^{1}S)		7.5			7.5
He*(^{3}S)	2.9	7.5	11	2.9	7.1
Ne*(^{3}P)	2.8	5.7		2.5	5.3
RIAN Mechanism					
He*(^{1}S)		3.0			2.8
He*(^{3}S)	1.0	3.4	5	1.0	3.2
Ne*(^{3}P)	1.2	2.6		1.0	2.5

highest occupied bulk state band at 2.7-3.0 eV found with UPS [2,3] for GaAs(100) (Table 1). We assign the C peak which was observed clearly with He*(^{3}S) on n-type as an AD peak with E_b(C)=11 eV. This corresponds to the bulk state with the same energy (Table 1) seen in UPS of GaAs(100) [4] and XPS of GaAs(110) [5].

With the above assignments for the A and B peaks, differences in the A/B intensity ratio seen for different Rg* are due to differences in the relative AD/RIAN quenching probabilities. To explore this model further, we used NH_3 as an adsorbate to investigate the effect of lowering the work function. TPD spectra showed that monolayer NH_3 coverage which desorbed over a broad temperature range was obtained at about 1 L, and a narrow multilayer peak at 115 K developed at higher exposures. From the shift in the low energy cutoff, it was found that ϕ_s decreased by 1.4 eV with 0.25 L NH_3 and 2.6 eV with 2.5 L NH_3. In spite of very effective shielding of the surface by the NH_3 molecules, the GaAs A peak was strongly enhanced up to a NH_3 coverage of about 0.25, while the B peak was strongly attenuated. This shows that lowering ϕ_s by 1.4 eV changes the Ne* quenching mechanism on GaAs from being predominantly RIAN to predominantly AD. In Fig. 2 it is seen that for the clean p-type surface the Ne* level is above E_F, but when ϕ_s is reduced by 1.4 eV it will be well below E_F. This greatly reduces the probability of RI, and correspondingly increases AD. A similar increase in the A/B ratio was seen when ϕ_s was lowered by the adsorption of K or H_2O on the surface. We also found that the He*(^{3}S)/He*(^{1}S) peak A intensity ratio fell from a value of about 5.1 on the clean p-type surface to about 1.6 for 0.25 L NH_3 exposure. The ratio of 1.6 is the same as the ratio of He*(^{3}S)/He*(^{1}S) metastable atom intensities in the He* beam, which is further evidence that when ϕ_s is reduced by 1.4 eV, most of the quenching on GaAs occurs by AD to give peak A.

We next consider the question of how RI can occur with the energy levels shown in Fig. 2. Since the He*(^{1}S) excited electron energy level coincides with the CBM, there should be little or no energy constraint on RI in this case. However He*(^{3}S) and Ne*(^{3}P) lie in the gap, 0.8-1.0 eV below the CBM, and there are no known surface states at this level. With the n-type this energy range is also below E_F, and any surface states would be filled. RI of He*(^{3}S) or Ne*(^{3}P) by an empty or half-filled Ga dangling bond state located 0.6 eV below the CBM (0.9 eV above the VBM, Table 1) will become possible if there is an attractive coulombic interaction of 0.2 eV or 0.4 eV, respectively, upon formation of Rg$^+$. A potential of this strength may be present from the polarizability of the surface. From Fig. 2 and the above discussion we see that the probability of RI is expected to decrease as one goes from He*(^{1}S) to He*(^{3}S) to Ne*(^{3}P), giving a corresponding increase in peak A. This is the trend that was observed with both n-type and p-type. The model also agrees with the trend that is observed when NH_3 is used to decrease ϕ_s: peak A is enhanced most strongly with Ne*(^{3}P), less with He*(^{3}S), and least with

He*(^{1}S). Comparison of the p-type spectra in Fig. 1 with the corresponding n-type spectra shows that for He*(^{1}S) the A peaks are the same, for He*(^{3}S) the A peak is slightly stronger on p-type, and for Ne*(^{3}P) the A peak is much stronger on p-type. The He*(^{1}S) result agrees with the expectation that facile RI should occur on both n-type and p-type via the conduction band. The He*(^{3}S) and Ne*(^{3}P) results suggest that the coulombic potential which is needed energetically for RI is stronger on n-type than on p-type. This may be due to a greater polarizability of the n-type surface, as a result of the filled surface state(s).

Acknowledgement

This work was supported by The NSF Science and Technology Center for Quantized Electronic Structures, Grant #DMR 91-20007.

References

1. R. Ludeke and L. Esaki, Surface Sci. **47**, 132 (1975); R. Ludeke and A.Koma, CRC Critical Rev. Solid State Sci. **5**, 259 (1975).
2. P.K. Larsen, J.H. Neave and B.A. Joyce, J. Phys. Chem. Solid State Phys. **12**, L868 (1979).
3. Y. Xu, G. Dong, X. Ding, S. Yang and X. Wang, Chinese Physics **4**, 547 (1984).
4. T.-C. Chiang, R. Ludeke, M. Aono, G. Landgren, F.J. Himpsel and D.E. Eastman, Phys. Rev. **B27**, 4770 (1983).
5. L. Ley, R.A. Pollak, F.R. McFeely, S.P. Kowalczyk and D.A. Shirley, Phys. Rev. **B9**, 600 (1974).
6. F. Bozso, J.T. Yates,Jr., J. Arias, H. Metiu and R.M. Martin, Surface Sci. **78**, 4256 (1983); C.P. Dehnbostel, A. Ludviksson, C. Huang, H.J. Jänsch and R.M. Martin, Surface Sci. **265**, 305 (1992).
7. J. Szuber, Thin Solid Films **120**, 133 (1984); J. Szuber, J. Electron Spectrosc. Rel. Phenom. **46**, 419 (1988).
8. Z.J. Gray-Grychowski, R.G. Egdell, B.A. Joyce, R.A. Stradling and K. Woodbridge, Surface Sci. **186**, 482 (1987).
9. A. Zangwill, Physics at Surfaces, Cambridge Univ. Press, 1988.
10. J.S. Blakemore, J. Appl. Phys. **53**, 123 (1982).
11. D.D. Pretzer and H.D. Hagstrum, Surface Sci. **4**, 425 (1966).

The Synchrotron Radiation Induced Chemistry of the Adamantane/Si(111)-7×7 Surface

J.K. Simons[1], S.P. Frigo[2], and R.A. Rosenberg[3]

[1]Department of Chemistry, Synchrotron Radiation Center,
University of Wisconsin-Madison, 3731 Schneider Drive,
Stoughton, WI 53589-3097, USA
[2]Department of Physics, Synchrotron Radiation Center,
University of Wisconsin-Madison, 3731 Schneider Drive,
Stoughton, WI 53589-3097, USA
[3]Advanced Photon Source, Argonne National Laboratory,
9700 South Cass Avenue, Argonne, IL 60439, USA

Abstract. The synchrotron radiation (SR) induced photochemistry of adamantane adsorbed on Si(111)-7x7 surfaces has been studied using photoemission spectroscopy (PES) and photon stimulated desorption (PSD). The clean Si substrates were first dosed with adamantane at a temperature of 85 K, then exposed to broadband SR with the temperature held at 85 K, and finally flash annealed to 525 K. When annealed, the unexposed adlayer dissociated forming predominantly hydrated Si-C species. When the adlayer was exposed to high energy (hv>284 eV) broadband SR, the adamantane layer again dissociated but reacted with the surface to form a layer similar to SiC. However, when the surface was exposed to low energy broadband SR, 1-adamantanyl radicals were generated which in turn reacted with the surface to form adamantyl-Si species. The adamantyl-Si reactants were stable up to temperatures of at least 525 K.

1. Introduction

The successful growth of thin diamond films by chemical vapor deposition (CVD) techniques on nondiamond substrates has generated a great deal of effort in trying to understand and control the nucleation and growth mechanisms which govern diamond film growth.[1,2] While there now exist several techniques for growing diamond films, little is known of the nucleation process itself. To gain insight into the nucleation process of diamond films on Si surfaces, we have performed PES and PSD experiments of adamantane ($C_{10}H_{16}$) adsorbed on Si(111)-7x7 surfaces at 85 K. Adamantane was chosen as a model compound since its structure is essentially that of a very small hydrogen terminated diamond crystal (see inset in Figure 1). Changes in the adsorbate and the surface structure were systematically followed after exposing the surfaces to broadband synchrotron radiation (SR) and annealing the surfaces to higher temperatures.

2. Experimental

Experiments were performed at the Synchrotron Radiation Center (SRC), University of Wisconsin-Madison using the SRC 3-meter and the Vanderbilt 6-meter toroidal grating monochromators (TGM) and the SRC Mark II grasshopper monochromator. All experiments were carried out under ultrahigh vacuum (UHV) conditions ($1x10^{-10}$ torr). Details of the experimental set up have been described elsewhere.[3] The Si(111) samples (p-type, 9.6 ohm-cm) were cleaned by cycles of Ar^+ sputtering, followed by annealing to 1000° C. The cleaned surfaces were then cooled to 85 K and

Springer Series in Surface Sciences, Vol. 31
Desorption Induced by Electron Transitions DIET V
Editors: A.R. Burns · E.B. Stechel · D.R. Jennison   © Springer-Verlag Berlin Heidelberg 1993

dosed with adamantane through a calibrated microchannel plate doser. The adamantane (99.9%) was purified by heating to 70 C and continuously pumping the sample cell for 1 to 2 hours. Just before dosing with adamantane, its mass spectrum was also taken to ensure its purity. While holding the dosed sample temperature at 85 K, the surfaces were exposed to broadband SR. Once the surfaces had been exposed they were then annealed to 525 K.

The broadband SR exposures were carried out by tuning the monochromators to zero order. The low energy exposures in this study were carried out on both the 3-meter and 6-meter TGM beamlines. The high energy gratings on each of these beamlines were used and have working ranges of 40-140 and 70-200 eV respectively. The high energy exposures were performed on the the Mark II beamline which has a working range from 55-600 eV. The approximate photon flux densities of the zero order beam for the 6-meter TGM, the 3-meter TGM and the Mark II beamlines are 8.5, 5.5, and 0.45 x10^{13} photons/sec/mA/mm^2. It is important to note that only the high energy Mark II beamline transmits photons of sufficient energy to excite or ionize the C(1s) electron (hν>284).

3. Results and Discussion

Figure 1 compares the valence band (VB) spectra of the (a) freshly dosed, (b) the low energy exposed (5300 mA-min), and (c) the high energy exposed (880 mA-min) adamantane/Si surfaces. Figure 1d shows the valence band spectrum of a clean SiC. The nine well resolved spectral features in the dosed adamantane/Si surface have been assigned based on the gas phase photoelectron assignments.[4] The five most tightly bound orbitals are C(2s) states and are labled with their orbital assignments. The four remaining bands are assigned to C(2p) orbitals.

A comparison of the VB spectra of the low and high energy exposed adamantane/Si surfaces shows that the surfaces undergo strikingly different chemistry. Even with the much shorter exposure time and the much smaller photon flux density, the VB spectrum of the high energy exposed surface

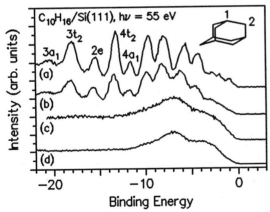

Figure 1. The VB spectra of the (a) unexposed, (b) low energy broadband SR exposed (5300 mA-min), and (c) the high energy broadband SR exposed (880 mA-min) adamantane/Si surfaces are shown. The surfaces were held at a temperature of 85 K during exposure and collection of the spectra. The valence band spectrum of a clean SiC surface is shown in Figure 1d. The inset is a stick figure diagram of the adamantane molecule. The numbered positions indicate the two distinct bonding configurations for the carbon atoms in the molecule.

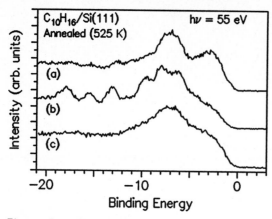

Figure 2. The VB spectra of the (a) unexposed, annealed, (b) the low energy exposed, annealed, and (c) the high energy exposed, annealed adamantane/Si(111) surfaces are compared. The surfaces were annealed to 525 K for approximately 10 seconds.

clearly shows that the adamantane has almost completely dissociated. In fact, the VB photoemission spectrum of the high energy exposed surface is very similar to the photoemission spectrum of the SiC surface. In contrast the adamantane adlayer has remained essentially intact during the low energy broadband SR exposure.

The effects of annealing the surfaces to 525 K are shown in Figure 2. Figure 2a shows the spectrum of the annealed, unexposed surface. Adamantane clearly dissociates on the Si surface when annealed. Annealing the high energy exposed adamantane/Si surface does not cause any significant changes in the already reacted surface (Figure 2c), but Figure 2b shows that the chemistry undergone by the low energy exposed surface is significantly different from the other two cases. The C(2s) states are still clearly present indicating that the molecular structure of the adamantane molecules is still intact; though, the C(2p) states do show an additional loss in structure. The overall intensity of the spectral features has also decreased which suggests that some desorption of the adlayer has occurred.

The most likely products that are formed when the adamantane dissociates on the Si(111) surface are Si-H and Si-C bonds. In order to determine if these assignments are reasonable, both the unexposed, annealed, and high energy exposed surfaces can be compared with the valence-band photoemission spectrum of the clean SiC surface which is shown in Figure 1d . While there are differences among the spectra, the similarities (specifically the -7 and the -10 eV features) suggest that the assignment of these features to Si-C bonds is reasonable. The observance of distinctly shifted (with respect to the photoabsorption spectrum) and structured edge jumps in the H[+] PSD spectra (not shown) taken around the Si(2p) edge suggest that Si-H bonds are also present.

Unlike the thermal and high photon energy induced reactions, the low energy exposed surface clearly reacts in a manner which leaves the adamantane molecules nearly intact and bound to the the Si surface. Even though the molecules remain intact, their symmetry appears to have decreased based on the broadening of the C(2p) features. One explanation for the observed changes is the formation of adamantyl radicals. The dissociation of a hydrogen atom from either the 1 or 2 position (see the inset in Figure 1) of the adamantane molecule will lower the molecule's symmetry and in turn remove the degeneracy of several of the molecule's

246

molecular orbitals. Using electron spin resonance (ESR) spectroscopy, Lloyd et al. demonstrated that the 1-adamantyl radical is the dominant product when adamantane is exposed to X-rays.[5] Based on the ESR results and our own findings, it is likely that the low energy exposures produce 1-adamantyl radicals which react with the dangling bond states of the Si surface, without dissociating further, to form adamantyl-Si species.

4. Conclusions

Our results demonstrate that the chemistry of adamantane on the Si(111)-7x7 surface is widely varying and can be selectively altered by exposing the surface to SR of different energies. Valence-level excitation results in the formation of adamantyl radicals, while C(1s), core-level excitation results in fragmentation and Si-C formation. These results suggest that by properly tuning the photon energy, it may be possible to selectively decorate the surface with potential nucleation sites.

Acknowledgements

The authors would like to thank the staff at the Synchrotron Radiation Center for their support.

References

1. F.F. Celii and J.E. Butler *Annu. Rev. Phys. Chem.* **42** (1991) 643, and references therein.

2. J.C. Angus and C.C. Hayman *Science* **241** (1988) 913, and references therein.

3. R.A. Rosenberg and C.-R. Wen *J.Vac. Sci. Technol. A* **6** (1988) 827.

4. R. Boschi, W. Schmidt, R.J. Suffolk, B.T. Wilkins, H.J. Lempka, and J.N.A. Ridyard *J. of Electr. Spectro. and Rel. Phenom.* **2** (1973) 377.

5. R.V. Lloyd, S. DiGregorio, L. DiMauro, and D.E. Wood *J. Phys. Chem.* **84** (1980) 2891.

Photolysis Enhancement by Excitation of Si(2p) Core Level Electrons in SiF$_4$/Ge(100)

S.P. Frigo[1], J.K. Simons[2], and R.A. Rosenberg[3]

[1]Department of Physics, Synchrotron Radiation Center,
University of Wisconsin-Madison, 3731 Schneider Drive,
Stoughton, WI 53589-3097, USA

[2]Department of Chemistry, Synchrotron Radiation Center,
University of Wisconsin-Madison, 3731 Schneider Drive,
Stoughton, WI 53589-3097, USA

[3]Advanced Photon Source, Argonne National Laboratory,
9700 South Cass Avenue, Argonne, IL 60439, USA

Abstract. Valence level photoemission was used to monitor the photolysis of SiF$_4$ adsorbed on Ge(100) at 30 K by resonant excitation of Si(2p) core electrons. The photolysis cross section was compared to the total electron yield at photon energies above and below the Si(2p) absorption threshold. It was found that the photolysis rate was 1-4 times the pre-edge value while the total yield increased at most by a factor of 1.2, thus showing that the photolysis is significantly enhanced by direct core excitation of the adsorbate.

1. Introduction

For many years researchers have been trying to discover if photoexcitation of adsorbates leads to significant chemical changes. Clear examples of these direct mechanisms have been elusive, and reasons range from (1) strong substrate coupling which leads to quenching of direct adsorbate excitation to (2) indirect processes whereby the underlying substrate absorbs most of the radiation and creates a pool of energetic electrons which then excite the adsorbate to form (a) dissociative precursor state(s).

It has been shown for a number of adsorption systems that photo-induced electrons can lead to a dissociative electron attachment (DEA) mechanism of bond cleavage. Using 5 eV laser radiation, Jo and White demonstrated DEA in the photolysis of CH$_3$Cl by pre-covering Pt(111) with D$_2$O and correlating the secondary photoelectron yield to the dissociation rate, determining that low energy ($\varepsilon < 0.4$ eV) electrons were most effective [1]. Wen and Rosenberg attributed the high cross section in the photolysis of CH$_3$F / Si(111) to a DEA mechanism, driven by substrate secondary photoelectrons [2]. In contrast, there is significant evidence of a direct excitation contribution in the photo-dissociation of COCl$_2$ / Pt(111) by low energy (4.4 eV) photons [3]. At higher energies, Rosenberg and Wen compared the desorption yield of neutral O$_2$ to the total electron yield (TEY) of YBa$_2$Cu$_3$O$_{7-x}$ about the Cu(3p) absorption edge to show that desorption was enhanced by direct excitation [4]. The reason that resonant core-level processes are intriguing is that they offer the possibility of selective bond breaking, thereby influencing the outcome of the final bonding configurations.

The equation used to infer the photolysis cross section (σ, cm^{-2}) from the time-dependent valence level (VL) photoemission (PES) is $n(t) = n(0) \, e^{-\sigma ft}$ where $n(t)$

Springer Series in Surface Sciences, Vol. 31
Desorption Induced by Electron Transitions DIET V
Editors: A.R. Burns · E.B. Stechel · D.R. Jennison   © Springer-Verlag Berlin Heidelberg 1993

is the surface number density (cm^{-2}) of adsorbate. Once the photon flux density (f, cm^{-2} s^{-1}) is known, the value for σ can be extracted from the decay constant of a fit to the PES signal (which is proportional to n) versus time.

The objective of this work is to determine if photoexcitation of adsorbate core-levels can lead to photolysis. Our approach is to compare σ, at photon energies above and below the threshold for adsorbate core-level excitation, to the TEY signal. An increase in σ greater than that in the TEY is evidence of direct photolysis following core-level excitation.

2. Experimental

The experiments were performed using radiation from the SRC 6m TGM beamline at the Synchrotron Radiation Center, University of Wisconsin-Madison. Experimental details have been given previously [2] so only pertinent matter will be presented here.

The TEY and PES were normalized by the restoring current to a 90% transmission Ni mesh. The energy-dependent response of the mesh was modeled using the Ni(3d) photoionization cross section [5] and used to obtain true normalization for the TEY. Incident photon flux was measured with the Ni mesh calibrated with an Au diode. The spot profile at 0-order, measured by scanning with an apertured photodiode in air, was used to determine the photon beam area. These measurements contain considerable systematic error and likely bias the calculated values of σ higher, we estimate, by as much as a factor of 2-5. The stated error in σ was calculated from estimated random error in the measurements of f and the statistical error from the curve fitting procedure.

The Ge(100) sample was cleaned by cycles of Ar$^+$ bombardment and annealing to ~ 600 C until VL PES indicated a well-ordered c(4x2) reconstruction [6]. Sample cooling to 30 $\pm$ 5 K was achieved using an open cycle liquid He cryostat. Gas exposures were performed using a calibrated doser employing a glass capillary array to collimate the beam. Absolute coverages were not determined, but for a sticking coefficient of unity, the dose (D) gives an upper bound. The chamber base pressure was 7 x 10^{-11} Torr and rose to 1 x 10^{-10} Torr during dosing.

A measurement consisted of taking a series of VL PES in time at a given photon energy. The photon energy was then changed and the sample moved to a new position for another series of measurements. A TEY measurement was made as well. Then σ was extracted from the decay constant after fitting the normalized intensity of a well-resolved feature in the PES to a single exponential function.

3. Results and Discussion

The initial adsorption and molecular orbital assignments of SiF$_4$ on Si(111) (7x7) at 30 K has been done at low photon flux densities where the system was stable to photolysis [7]. The present study monitors the time dependence of the peak which maps to the 5a$_1$ orbital feature in the PES of gas phase SiF$_4$.

Figure 1 shows a typical series of VL PES. The effects of photolysis are clearly evident in both the loss of structure and peak area. The inset to Fig. 1 shows the decay of the 89.5 eV normalized signal versus time. The fit is good and suggests the single exponential model is reasonable. Figure 2 is a comparison of the photon energy (ε) dependence of σ and the relative TEY increase (R$_T$), where R$_T(\varepsilon) \equiv$ TEY(ε) / TEY(100), for D = 2.0 x 10^{15} cm^{-2}. The selection of photon energies to measure σ follows from the TEY as it indicates the resonant transitions from the Si(2p) level to the various lower unoccupied molecular levels of SiF$_4$.

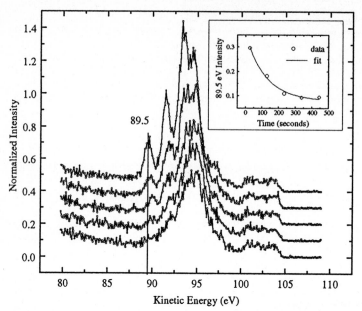

Figure 1. Sequence of VL PES for 109.0 eV photons and D = 0.5 x 10^{15} cm^{-2} SiF$_4$. The valence band maximum for the Ge substrate is at about 105 eV kinetic energy. Inset is an exponential fit to the 89.5 eV signal for the 5a$_1$ peak.

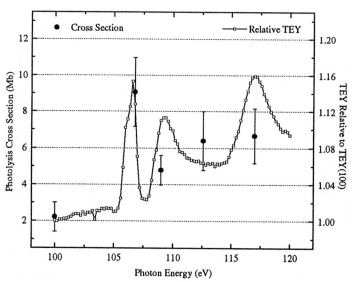

Figure 2. Composite of $\sigma(\epsilon)$ and R$_T(\epsilon)$ for D = 2.0 x 10^{15} cm^{-2}.

Table 1. Photolysis cross section and $R_\sigma(\varepsilon)$ / $R_T(\varepsilon)$ energy dependence

Photon Energy(eV)[a]	Dose[b] (10^{15} cm^{-2})				SiF$_4$[b]
	0.3	0.5	1.0	2.0	
100.0	12 (2)	14 (4)	14 (3)	2.2 (0.8)	7.5[c]
	1.0	1.0	1.0	1.0	1.0[c]
106.8	21 (3)	24 (8)	18 (4)	9.1 (1.9)	12.7
	1.6	1.6	1.1	3.7	1.7
109.0	39 (6)	18 (5)	21 (5)	4.8 (0.8)	11.2
	3.0	1.2	1.4	2.0	1.5
112.6	38 (6)	19 (4)	17 (6)	6.4 (1.6)	8.2
	2.8	1.3	1.2	2.7	1.1
117.0	59 (14)	18 (5)	24 (6)	6.7 (1.5)	11.2
	4.3	1.2	1.5	2.6	1.5

[a]Error in photon energy is ~ 0.2 eV [b]Error is ~ 20% [c]105 eV photon energy

Dose entries are σ (Mb) and R_σ / R_T (error is in parentheses). The SiF$_4$ entries are σ (Mb) and $R_\sigma(\varepsilon)$ for gas phase photoionization [10].

Table 1 shows the values of $\sigma(\varepsilon, D)$ and the ratio R_σ / R_T where the relative increase in σ is $R_\sigma(\varepsilon) \equiv \sigma(\varepsilon) / \sigma(100)$. The values of R_σ / R_T range from 1 to 4. If the photolysis were due to secondary electrons, then the ratio should be near unity. Therefore, for this system, DEA is not the predominant driving force in the photolysis.

In general, the absolute cross section decreases as the coverage increases. A possible explanation for this behavior is that as the coverage increases, the excitation responsible for the photolysis is quenched by interaction with neighboring molecules. Contrastingly, the behavior of R_σ / R_T is not monotonic, with the middle doses showing values near that of R_σ for gas phase SiF$_4$ but higher at the extremes. So, even though the cross section is diminishing overall due to quenching, the excitations which survive appear to be more effective at inducing photolysis. This behavior could be due to increased probability of a photolysis fragment (ion or neutral) interacting with a neighboring molecule or group of molecules and causing a reaction to occur. Such fragment-molecule interactions have been identified in several surface chemical reactions [8].

The value of σ obtained here is the overall cross section for the disappearance of the molecular specie; it doesn't yield any information on the paths the molecule may take: molecular desorption, or photodissociation where fragments then combine with the substrate atoms or leave the system as desorption products. It would be illustrative to know which channel(s) govern the formation of photolysis products and if there are any interadsorbate reactions. Time dependent measurements of the photoemission from the Si(2p) core level were also performed with 140 eV photons. The analysis of this data will yield insight on the formation of photolysis products [9].

4. Conclusion

From this study we can conclude the following: (1) SiF_4 / Ge(100) displays photolysis upon irradiation with photons in the range 100-117 eV; (2) The photolysis is significantly enhanced by excitation of adsorbate Si(2p) core electrons. This is the first experimental evidence to demonstrate that direct adsorbate core-level excitation can play a significant role in surface photochemistry.

Acknowledgments

The authors wish to thank the staff at the Synchrotron Radiation Center (SRC) for their enthusiastic support. The Ge samples were provided by A.J. Nelson. The SRC is supported by the National Science Foundation.

References

1   Sam K. Jo and J.M. White, Surf. Sci. **255,** 321 (1991).
2   C.-R. Wen and R.A. Rosenberg, Surf. Sci. Let. **218,** L483 (1989).
3   X.-Y. Zhu and J.M. White, J. Chem. Phys. **94,** 1555 (1991).
4   R.A. Rosenberg and C.-R. Wen, Phys. Rev. B **37,** 9852 (1988).
5   J.J. Yeh and I. Lindau, Atomic and Nuclear Data Tables **32** (1985).
6   S.D. Kevan, Phys. Rev. B **32,** 2344 (1985).
7   C.-R. Wen, S.P. Frigo, and R.A. Rosenberg, Surf. Sci. **249,** 117 (1991).
8   H. Kang, S.R. Kasi, and J.W. Rabalis, J. Chem. Phys. **88,** 5882 (1988).
9   S.P. Frigo, J.K. Simons, and R.A. Rosenberg, in preparation.
10 H. Friedrich, B. Pittel, P. Rabe, W.H.E. Schwartz, and B. Sonntag, J. Phys. B. **13,** 25 (1980).

Photon Stimulated Desorption of Fluorine from Semiconductor Surfaces

J.A. Yarmoff[1,2], D.K. Shuh[1], V. Chakarian[1], T.D. Durbin[1], K.A.H. German[1], and C.W. Lo[1]

[1]Department of Physics, University of California-Riverside,
Riverside, CA 92521, USA
[2]Materials Sciences Division, Lawrence Berkeley Laboratory,
Berkeley, CA 94720, USA

Abstract. Photon stimulated desorption (PSD) from XeF_2-exposed Si(111) and GaAs(100) is measured. The F^+ threshold from Si is at 27.5 eV, which corresponds to the transition from F 2s to the conduction band minimum (CBM). Features at the Si 2p edge, which distinguish the oxidation state of the bonding Si, are used to determine the structure of the XeF_2 etching reaction layer and to show how metal contaminants trap SiF_4 on the surface. For GaAs, the PSD threshold is at ~6 eV, which corresponds to the transition from F 2p to the CBM. Transitions from the 3d core levels of the bonding Ga and As atoms do not contribute to PSD.

1. Introduction

The interaction of fluorine with semiconductor surfaces is the fundamental chemical reaction in most processes employed in device manufacture. Fluorine containing molecules, such as CF_4, SF_6, etc. are routinely used in reactive ion etching of silicon. Metal fluorides, such as WF_6 and TaF_6, are used in chemical vapor deposition (CVD) processes. In all of these systems, however, there is little understanding of the basic surface chemical processes.

XeF_2 is routinely used as a fluorinating agent for studies of the basic interaction of atomic F with surfaces since it is easy to employ and readily dissociates at a surface. XeF_2 spontaneously etches Si at room temperature with the primary volatile reaction product being SiF_4 [1]. SXPS shows that a surface reaction layer is formed by the etching process. This reaction layer is composed of each of the fluorosilyl intermediate species, i.e. SiF, SiF_2, and SiF_3, and in some cases, trapped SiF_4 [2]. Although F does not spontaneously etch GaAs, because of the non-volatility of Ga fluorides, the interaction is important in CVD schemes that employ F-containing precursors. When a GaAs surface is reacted with XeF_2, mono- and higher Ga fluorides and As monofluoride are formed [3].

In this work, XeF_2-exposed Si(111) and GaAs(100) are studied with photon stimulated desorption (PSD) and soft x-ray photoemission spectroscopy (SXPS). The approach is to analyze the PSD data by first understanding the spectroscopy of the system and then using this information to determine the microscopic chemical mechanisms of the halogen-surface reactions. Both of these aims have been accomplished for Si, while for GaAs, the PSD spectroscopy is still not well understood.

2. Experimental Procedure

The experiments were performed at beamlines UV-8a and UV-8b at the National Synchrotron Light Source (NSLS), which are equipped with 3 and 6/10 m toroidal grating monochromators, respectively. The endstations consist of spectrometer, preparation, and dosing chambers connected by a UHV sample transfer system. The photoelectrons and desorbed ions were measured with an ellipsoidal mirror analyzer operating in an angle-integrating mode [4]. SXPS and PSD were collected as described previously [5]. For photon energies in the 25-50 eV range, a Mg filter was employed to attenuate higher order light.

Springer Series in Surface Sciences, Vol. 31
Desorption Induced by Electron Transitions DIET V
Editors: A.R. Burns · E.B. Stechel · D.R. Jennison   © Springer-Verlag Berlin Heidelberg 1993

XeF_2 reactions are sensitive to small amounts of metallic impurities. One source of impurities is vapor transport of volatile metal fluorides formed by reaction with the chamber walls, manipulators, etc. To reduce these effects, the dosing chamber was passivated with a large XeF_2 exposure prior to experiments. Also, the pressure in the dosing chamber was measured by a cold cathode gauge to preclude W-contamination from the hot filament of a conventional ion gauge.

Si wafers (n-type, P doped, 1.0-1.3 Ω-cm), cut within 0.25° of the (111) plane, were cleaned by ohmic heating to 1050°C to remove the native oxide and form the 7x7 reconstruction. Si 2p surface core level shifts (SCLS) were used to verify the cleanness and crystallinity of the samples. MBE grown, undoped GaAs(100) wafers were bombarded with 500 eV Ar^+ followed by annealing to ~550°C to produce clean, well-characterized surfaces. Both Si and GaAs wafers were exposed to XeF_2 at room temperature in a turbo-pumped dosing chamber, and transferred under UHV to the spectrometer chamber for measurement.

3. Results and Discussion

3.1 $XeF_2/Si(111)$

By varying the preparation technique, Si surfaces with different concentrations of each of the fluorosilyl species are obtained. Prior to a PSD measurement, SXPS is used to determine the relative concentrations of each of the fluorosilyls in the near-surface region. An exposure of Si(111)-7x7 to ~100 L of XeF_2 produces ~1 monolayer (ML) of fluorosilyl species consisting of SiF, SiF_2 and SiF_3 [6,8]. Annealing this surface to ~300°C removes the higher fluorides and forms a surface terminated solely by SiF [5]. An exposure on the order of 100,000 L is sufficient to reach steady state etching, after which the surface is covered with a reaction layer consisting of several ML's of SiF and SiF_3, and a small amount of SiF_2[2,7,8].

F 2s Edge. PSD spectra collected from Si(111) exposed to 100 L of XeF_2 and from that surface following annealing are shown in Figs. 1(a) and (b). The ESD threshold of F^+ from Si(100) has been previously measured at 27.5 eV [9]. The F^+ detected in ref. [9] originated from small amounts of F in the samples, and should therefore be compared to the PSD measurement from the annealed surface, which also represents a minimal coverage. Consistent with ref. [9], the PSD threshold is at 27.5 eV. In Fig. 1(c), SXPS of the F 2s level is shown with respect to the conduction band minimum (CBM). The edge in the annealed surface PSD spectrum corresponds closely to the SXPS spectrum, indicating that excitations from the F 2s to the CBM are responsible for the PSD threshold.

The spectroscopy of the unannealed surface is more complex because of the presence of quasi-molecular SiF_3 groups. As observed in PSD at the Si 2p edge, the electronic structure of adsorbed SiF_3 is sufficiently localized that 3s- and 3p-like σ^* molecular orbitals exist [5]. Structure at ~34.5 and ~40 eV in the PSD spectrum shown in Fig. 1(a) corresponds to the transition from F 2s to the 3s and 3p σ^* final states, respectively. In addition, there is another peak at ~28 eV in the unannealed surface PSD spectrum. This peak is not seen in the spectra collected from the annealed surface, and is therefore related to the higher fluorides. It cannot be due to a direct transition from the F 2s initial state to a final state in the solid, however, since this final state would be located a few eV below the valence band. The following scenario is presented to explain the origin of the 28 eV peak.

Houle [10] has shown that if above band gap laser radiation is coincident with a XeF_2 flux, then SiF_3 radicals are evolved from the surface by a photochemical process. These SiF_3 radicals are also emitted if the XeF_2 flux is turned off, provided that there has been a sufficient exposure to XeF_2 to form surface SiF_3. The energy of the photons incident on the surface during the PSD measurements is well above the band gap, so it is plausible to assume that, since SiF_3 is on the surface, neutral SiF_3 radicals will desorb as a consequence of the synchrotron radiation. Since the mechanism that produces the SiF_3 radicals involves charge carriers, and a large number of charge carriers are produced for each 28 eV photon adsorbed, the efficiency of SiF_3 production should be fairly high.

254

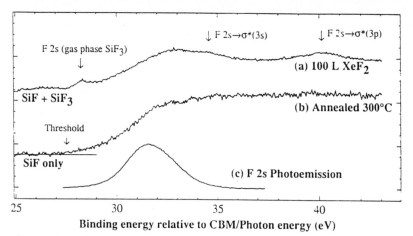

F 2s (gas phase SiF$_3$)
↓

↓ F 2s→σ*(3s)   ↓ F 2s→σ*(3p)

(a) 100 L XeF$_2$

SiF + SiF$_3$

(b) Annealed 300°C

Threshold
↓

SiF only

(c) F 2s Photoemission

25   30   35   40

Binding energy relative to CBM/Photon energy (eV)

Figure 1. PSD and SXPS at the F 2s edge for Si(111)-7x7 exposed to XeF$_2$. (a) PSD after a 100 L exposure, (b) PSD after annealing to 300°C, (c) F 2s SXPS.

After SiF$_3$ radicals desorb, they are ionized in the gas-phase. The translational energy of the photochemically produced SiF$_3$ is ~0.04 eV [10] in contrast to the ~2 eV F$^+$ ions produced by PSD [5]. Because the SiF$_3$ radicals have such a low translational energy, they remain in the vicinity of the surface for a sufficiently long time to be ionized directly by the photon beam. This has been verifed by a simple calculation of the cross section for such a process. A radical containing an unpaired electron has a transition from the F 2s to the lone electron orbital close in energy to the F 2s to CBM transition in the surface PSD process. A difference is that the gas phase transition is much sharper, i.e. like the 28 eV feature in Fig. 1(a). A process involving neutral PSD and subsequent ionization of gas phase products has an approximately square dependence upon the flux. This was tested by collecting a spectrum from an unannealed surface using a much smaller photon flux. The spectrum collected with a small flux showed no evidence of the 28 eV peak, consistent with a second order process.

Si 2p Edge. Figures 2(a)-(c) show Si 2p SXPS spectra after background subtraction. The spectrum in Fig. 2(a) has been fit with Gaussian-broadened Lorentzians to illustrate the chemically shifted components [6]. The peaks at ~100 eV binding energy are the two spin-orbit components of the bulk Si 2p level, split by 0.61 eV. Peaks representing SiF, SiF$_2$, and SiF$_3$ are located at ~1 eV higher binding energy per attached F. The dominant feature in Figs. 2(b) and (c) is the SiF$_3$ peak at ~103 eV, which is sufficiently broad that the spin-orbit splitting is not visible in the raw data. Note that Fig. 2(c) shows additional structure at ~104 eV due to the presence of SiF$_4$.

Figures 2(d)-(f) show Si 2p PSD spectra. The features, which are roughly indicated as (i)-(v) in the figure, correspond to the transitions from (i) bulk Si 2p to the CBM, (ii) SiF Si 2p to the CBM, (iii) SiF$_3$ Si 2p to the CBM, (iv) SiF$_3$ and/or SiF$_4$ Si 2p to the 3s-like σ* state and (v) SiF$_3$ and/or SiF$_4$ Si 2p to the 3p-like σ* state [5,11]. Transition (i) is the result of a secondary ESD process following the absorption of a photon by the substrate [5]. Transitions (ii) and (iii) are direct processes involving excitations to the CBM, and provide a measure of the relative proportions of SiF and SiF$_3$ at the surface. Transitions (iv) and (v) are associated with both SiF$_3$ and SiF$_4$, since they are to the σ* levels of the adsorbed molecule-like and molecular species. Note that the shape of the PSD spectrum at features (ii) and (iv) reflects the spin-orbit splitting of the 2p initial state.

Figure 2(e) shows the PSD spectrum collected from a surface that received a large enough XeF$_2$ exposure to reach steady-state etching. Although SXPS (Fig.

255

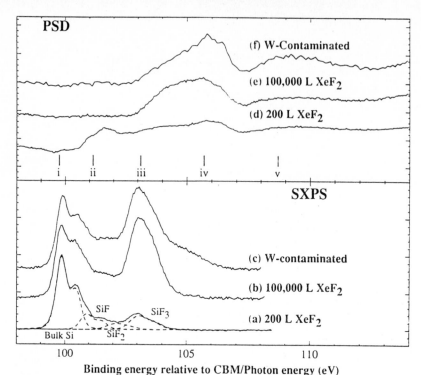

Figure 2. PSD and SXPS at the Si 2p edge for Si(111) exposed to XeF$_2$. (a) SXPS after a 200 L XeF$_2$ exposure, (b) SXPS after 100,000 L, (c) SXPS collected from a W-contaminated surface after an 80,000 L XeF$_2$ exposure. (d), (e), and (f) are PSD collected from the same surfaces as in (a), (b) and (c).

2(b)) shows the presence of SiF and SiF$_3$, the only PSD features observed are (iii), (iv), and (v), which each originate from an SiF$_3$ initial state [12]. The differences between the SXPS and PSD results are due to the higher surface sensitivity of PSD. This comparison shows that the reaction layer formed by XeF$_2$ etching has a tree-like structure, in which SiF$_3$ groups are the chain terminators at the surface [8].

Although SiF$_4$ molecules have been observed in the XeF$_2$ reaction layers, the means by which they are trapped is unknown [2]. After collecting a large number of SXPS spectra, a correlation was noted between the presence of SiF$_4$ and metallic contaminants. To investigate this, ~1 ML of W was intentionally deposited on Si via WF$_6$. After exposure of the W-contaminated surface to 80,000 L of XeF$_2$, an increased amount of SiF$_4$ was formed in the reaction layer, as seen in Fig. 2(c). In the PSD spectrum collected from that surface (Fig. 2(f)), transitions (iv) and (v) have increased. An increase in (iv) and (v), and not in (iii), indicates that SiF$_4$ is contributing to the PSD spectrum. Since SiF$_4$ is clearly visible in the PSD spectra, it is located near the top of the reaction layer, which demonstrates that the SiF$_4$ is not physically trapped. It is proposed that adsorbed M$^+$(SiF$_5$)$^-$ species give rise to the SiF$_4$ SXPS signal [8]. Stable pentavalent Si compounds, such as these, are well known.

3.2 XeF₂/GaAs(100)

Figure 3 shows SXPS and PSD collected after exposure to 100 L of XeF₂. These data are shown over a wide photon energy range to illustrate the overall behavior. There are four major features in the SXPS spectra. The valence band, at ~10 eV, is dominated by F 2p. The Ga 3d level, at ~20 eV, shows a small bulk and a large GaF₃ peak. The remaining two features are the F 2s and the As 3d levels. A close examination of the As 3d level shows that ~1 ML of AsF has formed [13].

The measured PSD threshold is at ~6 eV, which corresponds to the transition from the F 2p in the valence band to the CBM. In fact, the shapes of the PSD and the SPXS in this region are nearly identical. The thermodynamic threshold is ~20 eV, however, since the energy required to ionize F (17 eV), break the bond (~5 eV) and produce an ion with the measured 2 eV kinetic energy, less the workfunction (~4 eV), needs to be overcome. This leads to the conclusion that a multiphoton process must be involved. A possible scenario is that neutral F is desorbed by excitation of the F 2p followed by ionization in the gas phase. The gas-phase ionization must involve higher order light, since the F ionization potential is 17 eV. Further experiments aimed at determining the PSD mechanism are underway [13].

There is additional PSD structure at higher photon energies. The features in the 15-25 eV range are the result of second order light, and are identical to the first order features in the 30-50 eV range. Thus, within the limits of the measurement, there are no first order PSD features in the 15-25 eV range. At ~32 eV, the PSD shows a large feature that correlates with the F 2s to CBM transition. From 40 to 55 eV, there are additional features that have no simple correlation with particular initial states. Finally, at ~110 eV, there is a feature that correlates with the Ga 3p initial state. The structure in the 110 eV feature is the spin-orbit splitting of Ga 3p.

Correlations between the initial core levels and the PSD spectra are not as simple for GaAs as for Si. Despite having GaF₃ on the surface, for example, there is no PSD feature at the Ga 3d edge. Also, a close examination of the 40-60 eV range shows that none of the features overlap properly with As 3d to be direct transitions to the CBM. These features are the subject of future investigations [13].

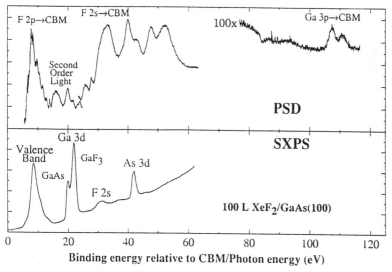

Figure 3. PSD and SXPS for a GaAs(100) surface exposed to 100 L of XeF₂.

4. Conclusions

PSD was collected from XeF$_2$-exposed Si(111) and GaAs(100). For Si, direct excitation of the bonding atoms' initial states induces PSD, and the threshold corresponds with F 2s. SiF$_4$ is seen in PSD collected from W-contaminated samples, suggesting that pentavalent Si compounds form at the surface. PSD from GaAs(100) is observed at F 2p, F 2s and Ga 3p. There is no simple correlation between the Ga and As 3d levels and the PSD spectra.

5. Acknowledgments

The authors thank F.A. Houle for useful discussions and GaAs samples. This work was conducted at the NSLS, which is supported by the Department of Energy (Divisions of Materials Science and Chemical Science, Basic Energy Sciences) under Contract No. DE-AC02-76CH0016. The authors also thank The Petroleum Research Fund, administered by the American Chemical Society, for partial support.

References

1. H.F. Winters and J.W. Coburn, Surf. Sci. Reports **14**, 161 (1992).
2. F.R. McFeely, J.F. Morar and F.J. Himpsel, Surf. Sci. **165**, 277 (1986).
3. A.B. McLean, L.J. Terminello and F.R. McFeely, Phys. Rev. B **40**, 11778 (1989).
4. D.E. Eastman, J.J. Donelon, N.C. Hein and F.J. Himpsel, Nucl. Instrum. Methods **172**, 327 (1980).
5. J.A. Yarmoff and S.A. Joyce, Phys. Rev. B **40**, 3143 (1989).
6. F.R. McFeely, J.F. Morar, N.D. Shinn, G. Landgren and F.J. Himpsel, Phys. Rev. B **30**, 764 (1984).
7. J.A. Yarmoff and F.R. McFeely, Phys. Rev. B **38**, 2057 (1988).
8. C.W. Lo, D.K. Shuh and J.A. Yarmoff, in preparation.
9. M.J. Bozack, M.J. Dresser, W.J. Choyke, P.A. Taylor and J.T. Yates, Jr., Surf. Sci. **184**, L332 (1987).
10. F.A. Houle, Phys. Rev. B **39**, 10120 (1989).
11. R.A. Rosenberg, J. Vac. Sci. Technol. A **2**, 1463 (1986).
12. J.A. Yarmoff, S.A. Joyce, C.W. Lo and J. Song, in *Desorption Induced by Electronic Transitions IV*, G. Betz and P. Varga, Eds. (Springer-Verlag, Berlin, 1990), p. 65; C.W. Lo, D.K. Shuh, V. Chakarian, K.A. German and J.A. Yarmoff, in *Proceedings of the Conference on Dry Processing*, Tokyo, 1990).
13. D.K. Shuh, T.D. Durbin and J.A. Yarmoff, in preparation.

ESDIAD Studies of Chlorine Adsorption at Silicon(100)

S.L. Bennett, C.L. Greenwood, and E.M. Williams

Interdisciplinary Research Centre in Surface Science and
Department of Electrical Engineering and Electronics,
University of Liverpool, P.O. Box 147, Liverpool L69 3BX, UK

Electron stimulated desorption ion angular distribution (ESDIAD) data are presented for Cl_2 adsorption at the Si(100) surface. The ion angular distribution and the total ion yield are found to be temperature dependent over the range studied (300 K to 670 K). These observations are interpreted in terms of the irreversible thermal conversion of tilted surface chloride species to normally orientated species bonded at asymmetric dimer sites.

1. Introduction

The use of halogen based etchants in the processing of semiconductor materials has prompted a great deal of interest in studies of chlorine chemisorption at silicon surfaces, including investigations directed at elucidating the geometry of the silicon-halogen bond. Early photoemission data of Rowe *et al.* [1] indicated adsorption sites at the dimer dangling bonds in an off-normal geometry. An electron energy loss study by Aoto *et al.* [2] again interpreted data in terms of an inclined substrate-adsorbate bond. In an application of the pseudo-intramolecular NEXAFS technique, Cl_2 adsorbed at the vicinal (100) surface at 500 K showed spectra similar in both the [011] and the [011] directions, and also similar to that obtained from the Si(111) (7 x 7)-Cl system, implying a Si-Cl bond along the surface normal [3]. Recently Yates *et al.* [4] have presented some preliminary findings of Cl^+ ESDIAD from Si(100) which indicate Si-Cl bonds inclined at ~30°.

2. Experimental details

Measurements were performed with a direct viewing ESDIAD/LEED system (similar to that used by Madey and co-workers [5]) which was incorporated into a UHV system together with facilities for Auger analysis and a time-of-flight ESD apparatus as described elsewhere [6].

Room temperature adsorption of chlorine from a Cd-doped AgCl electrochemical cell [7] was carried out on Si(100) samples off-cut by 4° towards the [011] direction. The samples were cleaned in vacuum by argon ion bombardment at 500 eV and resistive heating to 1000 K, to produce a clean well-ordered surface. Ion angular distributions were measured at room temperature and following annealing at a number of surface temperatures up to 670 K, measured by infrared pyrometry. To facilitate ion detection in ESDIAD, it was necessary to compress the ion beams towards to the surface normal by applying a sample bias of 80 volts. The determination of the true

Springer Series in Surface Sciences, Vol. 31
Desorption Induced by Electron Transitions DIET V
Editors: A.R. Burns · E.B. Stechel · D.R. Jennison   © Springer-Verlag Berlin Heidelberg 1993

ion desorption angle was carried out by computer modelling of the ion trajectory in the applied field conditions.

3. Results

Chlorine gas was adsorbed at room temperature, up to a Cl(LMM):Si(LVV) ratio of 1.1, corresponding to between 0.5 and 1 monolayer [3]. Electron stimulated desorption from the chlorine exposed surface at an electron beam energy 280 eV showed Cl^+ to be the majority species with a small signal (10%) of H^+ also present. An energy distribution curve obtained for the Cl^+ ions, taken approximately normal to the surface, is presented in fig. 1. Annealing to 670 K preserved the essential form of the curve, although the intensity of the main peak at 0.33 eV increased by a factor of two, while the shoulder around 2 eV remained unchanged.

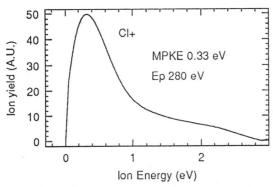

Figure 1. Cl^+ ion energy distribution at an electron energy of 280 eV.

The ESDIAD pattern recorded from a chlorine overlayer, adsorbed at room temperature, is presented as a raised contour plot in fig. 2a. Three maxima characterise the distribution, showing ion emission parallel to [011] and also [011], although, in the latter case, with significant intensity only in the down steps direction. The maxima occur at a polar angle of 40°±10°. The precision in quantitative measurement is restricted in this experiment by the low energy of the ions which causes significant angular dispersion in the compressive field. The ions are also subject to the usual distortion arising from the image charge and from reneutralisation. Following annealing at 670K the Cl(LMM):Si(LVV) ratio was reduced to ≤ 0.3. This was accompanied by a significant modification to the ESDIAD pattern, which is presented in fig. 2b. Here the degree of azimuthal structure is greatly reduced while the ion intensity close to the [100] direction is enhanced, leaving a distribution exhibiting a single maximum directed along the surface normal. The half width of the angular distribution in this case is estimated to be ~40°.

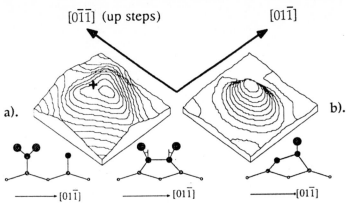

Figure 2. a) Cl$^+$ ESDIAD following Cl$_2$ adsorption at room temperature at 4° vicinal Si(100)2x1. b) Cl$^+$ESDIAD following annealing of a). to 670 K. Electron beam energy 280 eV (including 80 volt sample bias) at 10^{-7} A cm^{-2}. Both measurements taken at 300 K. The point corresponding to normal emission is marked by the symbol +. Models of possible bonding configurations are included (see text).

The angle integrated ion intensity following annealing was enhanced by a factor of around five despite, as already noted, a decrease in the total surface concentration of chlorine determined in AES. For annealing temperatures in excess of 670 K the pattern retained the form of fig. 2b, while decreasing rapidly in intensity with increasing temperature.

4. Discussion

The data presented here show that the angular distribution of Cl$^+$ desorbed from Cl adsorbed at room temperature at the Si(100) surface is strongly dependent upon the thermal history of the adlayer, consistent with previous TPD findings [8]. This behaviour contrasts with the thermal stability of the F$^+$ ion angular distribution observed at Si(100) [5].

Following room temperature adsorption, the observation of off-normal emission parallel to the dimer axis, accompanied by a shallow minimum along the [100] vector, is indicative of tilted surface species, probably coexisting with normally oriented groups. Emission of comparable intensity inclined from the normal is also observed along [011] in the down steps direction and perpendicular to the majority dimer axis. This observation may be due to emission from step sites at the vicinal surface, or alternatively due to etching at 300 K, as reported by Jackman *et al.* [8], which can be envisaged as acting within the terraces and with enhanced reneutralisation in the up steps direction.

The tilting of surface species would appear to originate either in single chlorine atoms bonded at both ends of symmetric silicon dimers, or from dichloride groups adsorbed at single silicon atoms. In contrast, annealing is envisaged to populate sites where Cl is bonded atop a buckled dimer as concluded in a recent NEXAFS investigation [3].

261

Modification of the angular distribution is accompanied by a marked enhancement of the total ion yield, which occurs against a decrease of the surface chlorine concentration as seen in AES. The Cl^+ ion energy distribution, however, shows little change in form. It would thus appear that thermally induced surface rearrangement induces the population of the monochloride state from which the ion yield is enhanced through final state lifetime effects.

Acknowledgements

The authors would like to thank P.L. Wincott and G. Thornton for the loan of the electrochemical cell. One of us (C.L.G.) would like to acknowledge the support of Hiden Analytical Ltd and the S.E.R.C. for the provision of a CASE studentship, and the British Vacuum Council for a travel bursary.

References

1. J. E. Rowe, G. Margaritondo and S. B. Christman, Phys. Rev. **B16**, 1581 (1977).
2. N. Aoto, E. Ikawa and Y. Kurogi, Surface Sci. **199** 408 (1988).
3. D. Purdie, C. A. Muryn, N. S. Prakash, K. G. Purcell, P. L. Wincott, G. Thornton and D. S-L. Law, J. Phys. Condens. Matter 3 7751 (1991).
4. J. T. Yates, Jr., M. D. Alvey, M. J. Dresser, M. A. Henderson, M. Kiskinova, R. D. Ramsier, A. Szabó, Science **255**, 1397 (1992).
5. A. L. Johnson, M. M. Walczak, and T.E.Madey, Langmuir, **4**, 277 (1988).
6. S. L. Bennett, C. L. Greenwood, E.M. Williams, Surface Sci. **251/252** 857 (1991).
7. N. D. Spencer, P. J. Goddard, P. W. Davies, M. Kitson and R. M. Lambert, J. Vac. Sci. Technol. A1 1554 (1983).
8. R. B. Jackman, H. Ebert and J. S. Foord, Surface Sci. **176** 183 (1986).

Probing GaSb(110) Surface and Interfaces by Photon Stimulated Desorption (PSD)*

Z. Hurych[1], P.S. Mangat[1], J. Peng[1], D. Crouch[1], D. Hochereau[2],
P. Soukiassian[2;1], L. Soonekindt[3], and J.J. Bonnet[3]

[1]Department of Physics, Northern Illinois University,
  DeKalb, IL 60115-2854, USA
[2]Commissariat à l'Energie Atomique, Service de Physique des Atomes
  et des Surfaces, Centre d'Etudes Nucléaires de Saclay, France; and
  Département de Physique, Université de Paris-Sud, Orsay, France
[3]Laboratoire d'Etudes des Surfaces, Interfaces et Composants,
  Université de Montpellier II, France

Abstract. We present photon stimulated desorption (PSD) studies of alkali
metals Na and Cs on a III-V semiconductor GaSb(110). It is shown that both
Na and Cs alter significantly the surface reactivity of the GaSb(110) as
indicated by the desorption of H⁺ which is observed in the presence of an
alkali metal overlayer but absent in the case of pure cleaved surfaces. The
different desorption sites of H⁺ on these surfaces are discussed together with
the difference between Na and Cs overlayers.

Alkali metals on semiconductor surfaces have been recently investigated for
a variety of reasons including the room temperature passivation of these
surfaces [1-6]. In contrast to all the other metal catalysts, alkali metals
can be removed from the surface of silicon after the catalytic oxidation or
nitridation by a rapid thermal annealing (RTA) at moderate temperatures
[2,6]. The alkali metal promoted oxidation of III-V compound semiconductors
has been studied for InP(110) [7,8], GaAs(110) [9-12] and GaP(110) [13]
surfaces. In the case of the Cs-InP(110) system, the oxidation rate was
enhanced as much as 13 orders of magnitude which is the largest promotion
effect ever observed for a solid [7]. The mechanism of electronic promotion
by the alkali atoms exhibits common features for all these systems:
i) weakening of the surface atoms back bonds, II) enhancement of the sticking
coefficient of the oxygen (nitrogen) molecule, iii) dissociation of oxygen
(nitrogen) by charge transfer into the antibonding orbitals of the molecule,
iv) migration of the oxygen (nitrogen) atoms below the surface into the bulk
and bonding with substrate atoms [13]. However, there are some considerable
differences between III-V and Si surfaces. In particular, while the alkali
metal/Si surfaces are nonreactive, it is observed that the catalytic
passivation of III-V surfaces by alkali metals occurs only if the interface
is a reactive one indicating the important role played by the surface
defects.

The above mentioned works include the study of alkali metals/semicon-
ductor interfaces and/or the oxidation/nitridation of these surfaces by the
technique of the photoelectron spectroscopy (PES). In view of successful
application of the technique of photon stimulated desorption (PSD) to the
studies of oxygen chemisorption and/or oxidation of Si surfaces [14] we have
decided to apply the PSD technique also to the alkali metals on III-V
semiconductors.

The experiments were performed at the Synchrotron Radiation Center of
University of Wisconsin. A 6m Toroidal Grating Monochrometer (TGM) was used
to provide monochromatic radiation in the photon energy range 25 to

* Supported by NSF grant No. DMR 88-07754 through NIU.

Springer Series in Surface Sciences, Vol. 31
Desorption Induced by Electron Transitions DIET V
Editors: A.R. Burns · E.B. Stechel · D.R. Jennison   © Springer-Verlag Berlin Heidelberg 1993

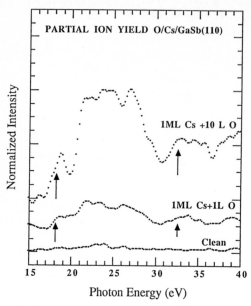

Fig. 1. Partial Ion Yield of H⁺ vs photon energy (in eV) for a clean GaSb, covered subsequently with 1 monolayer (ML) of Cs and 1L of O_2, and by an additional 10L of O_2.

130eV covering the core levels of the substrate and adsorbates. Two different gratings, overlapping at $h\nu$ = 70eV were used to cover the entire above spectral range. The pressure in the experimental chamber was better than 5 x 10^{-11} torr. Alkali metals were deposited from carefully outgassed SAES Chromate sources. The pressure increase during alkali metal deposition was always below 2 x 10^{-11} torr. The rest of the experimental details including ion detection and identification as H⁺, are the same as described previously [14,15].

The ion yield as observed by the PSD for GaSb surface is much below the yield in the case of Si surfaces, and there is no ion signal detected from freshly cleaved GaSb surfaces. Even upon deposition of one monolayer (1 ML) of Cs, there is still no PSD ion signal observed, despite the fact that Cs forms a reactive interface with GaSb. However, after additional exposure of the interface to 1 L of O_2 (1 L = 1 Langmuir = 10^{-6} torr sec), ion signal interpreted [14,15] as H⁺ is detected from surfaces with thresholds at Ga3d and Sb4d levels. This ion signal further increases with additional 10L of O_2 exposure as evident from Fig. 1. (The increase of H⁺ signal upon O_2 chemisorption is due to the decrease of the reneutralization of H⁺ as a result of the decrease of the surface free charge and has been reported previously for other semiconductor surfaces covered by alkali metal overlayers [14,15].) The desorption band in the range 18 to 30eV is interpreted to be a result of electronic transitions from the Ga3d level, indicating the applicability of the Knotek-Feibelman-like mechanism [16] and suggesting that the desorbed H+ comes from the Ga surface sites. This is certainly a very interesting observation in view of the fact that on clean GaSb surfaces the Ga has only empty lone p-pair orbitals which are unlikely to bond to hydrogen, and indicates that the Cs overlayer significantly modifies the surface electronic structure in agreement with previous observation that the Cs/GaSb interface is a reactive one. Another interesting observation is that the Ga assigned desorption band, in Fig. 1, starts with a sharp, resonant-like peak at ~18.5eV immediately following the Ga3d threshold.

264

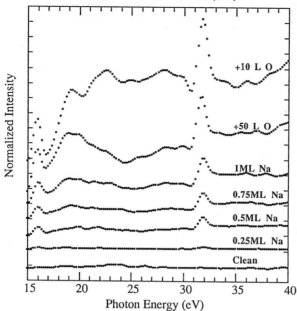

Fig. 2. Partial ion yield of H⁺ vs photon energy (in eV) for cleaved GaSb and for various Na coverages and oxygen exposures.

The second interesting structure in Fig. 1 is the desorption band at hν>31eV which is assigned to be a result of optical excitation of the Sb4d core level.

The situation for Na overlayers is indicated in Fig. 2. In contrast to Cs, even a submonolayer of Na on GaSb provides a detectable signal, and 1 ML of Na on freshly cleaved GaSb surface provides H⁺ signal of considerable magnitude. This signal is further increased by O_2 chemisorption as for other systems. However, in contrast with the Cs/GaSb interface, the Na/GaSb partial ion yield is different since it has thresholds at Ga3d and Na2p levels, but the broad structure at Sb4d level is missing completely. Furthermore, the Na2p desorption exhibits the similar sharp resonance at hν = 32eV as observed previously in the case of Na on Si surfaces and interpreted in terms of the Si-Na-H interface state [17,18].

The difference in PSD partial ion yields is interesting since both Cs and Na/GaSb interfaces are reactive ones, and both of them facilitate catalytic oxidation of GaSb. Therefore, the absence of a PSD signal for the Sb site in the case of Na/GaSb is especially interesting, indicating that there might be a difference in local bonding configuration between Na and Cs. Furthermore, in contrast to Na/Si interfaces, an additional coadsorption of Cs does not remove completely the H⁺ resonance at the Na2p threshold in the case of GaSb. A more detailed description and interpretation of this experimental work including comparison of the ion and partial photoelectron yields is beyond the scope of this article and will be published later [19].

In summary, the results indicate that the PSD technique could be a useful one to study the formation of the alkali metal/III-V semiconductor interface, and the catalytic passivation of these surfaces. Also, the results indicate that contrary to the case of Si, the surface electronic structure of the III-V semiconductors is considerably modified by the presence of alkali metals in agreement with the reactivity of these surfaces. The modification of the electronic structure by Na and Cs is observed by PSD to occur both at the Ga and the Sb sites.

Acknowledgement.

This work was supported by the U.S. National Science Foundation under contract No. DMR 88-07754 through Northern Illinois University and by the Commissariat à l'Energie Atomique--CEA. We are grateful to Dr. A. Gouskov for providing us the GaSb ingots used in this work. We thank the technical staff of the Synchrotron Radiation Center of the University of Wisconsin-Madison for expert and outstanding assistance.

References.

1. Physics and Chemistry of Alkali Metal Adsorption, Elsevier Science Publishers B.V., Amsterdam, H. P. Bonzel, A. M. Bradshaw and G. Ertl editors, Materials Science Monographs 57, 449 (1989) and references therein.
2. P. Soukiassian, M. H. Bakshi, H. I. Starnberg, Z. Hurych, T. M. Gentle and K. P. Schuette, Phys. Rev. Lett. 59, 1488 (1987).
3. H. I. Starnberg, P. Soukiassian, M. H. Bakshi and Z. Hurych, Surf. Sci. 224, 13 (1989).
4. P. Soukiassian, M. H. Bakshi, Z. D. Hurych and T. M. Gentle, Phys. Rev. B35, 4176 (1987).
5. C. A. Papageorgopoulos and M. Kamaratos, Surf. Sci. 221, 263 (1989).
6. G. Boishin, M. Tikhov and L. Surnev, Surf. Sci. 257, 190 (1991).
7. P. Soukiassian, M. H. Bakshi and Z. Hurych, J. Appl. Phys. 61, 2679 (1987).
8. J. A. Schaefer, F. Lodders, Th. Allinger, S. Nannarone, J. Anderson and G. J. Lapeyre, Surf. Sci. 211/212, 1075 (1989).
9. S. Valeri, M. Lolli and P. Sberveglieri, Surf. Sci. 238, 63 (1990).
10. H. Araghi-Kozaz, G. Brojerdi, M. Besançon, P. Dolle and J. Jupille, Surf. Sci. 251/252, 1091 (1991).
11. T. Kendelewicz, P. Soukiassian, M. H. Bakshi, Z. Hurych, I. Lindau and W. E. Spicer, Phys. Rev. B38, 7568 (1988).
12. R. Miranda, M. Prietsch, C. Laubschat, M. Domke, T. Mandel and G. Kaindl, Phys. Rev. B39, 10387 (1989).
13. P. Soukiassian and H. I. Starnberg in reference 1; H. I. Starnberg, P. Soukiassian and Z. Hurych, Phys. Rev. B39, 12775 (1989).
14. Z. Hurych, H. I. Starnberg and P. Soukiassian, Europhys. Lett. 8 (1989) 567.
15. Z. Hurych, P. Soukiassian, H. Starnberg, Physica Scripta 41, 955 (1990)
16. M. L. Knotek, P. I. Feibelman, Phys. Rev. Lett. 40, 964 (1978).
17. Z. Hurych, P. Soukiassian, S. Kapoor, P. S. Mangat, S. T. Kim, E. Bourdie, J. Peng. Submitted to Rapid Communications, Phys. Rev. B.
18. Z. Hurych, D. Crouch, J. Peng and P. S. Mangat, D. Hochereau and P. Soukiassian, in the Proceedings of this conference.
19. Z. Hurych, P. Mangat and P. Soukiassian, to be published.

Photon Stimulated H+ Ion Desorption Studies of Silicon Surfaces Covered by Alkali Metals*

Z. Hurych[1], D. Crouch[1], J. Peng[1], P.S. Mangat[1], D. Hochereau[2], and P. Soukiassian[2,1]

[1]Department of Physics, Northern Illinois University, DeKalb, IL 60115-2854, USA
[2]Commissariat à l'Energie Atomique, Service de Physique des Atomes et des Surfaces, Centre d'Etudes Nucléaires de Saclay, France; and Département de Physique, Université de Paris-Sud, Orsay, France

Abstract. We present the results of photon stimulated desorption (PSD) studies of O_2 chemisorption and catalytic oxidation of Si surfaces using Cs and Na overlayers. H+ is observed to be desorbed from both the Si and the alkali metal sites with comparable intensity suggesting the formation of Si and alkali metal hydrides. Of special interest is the observation in the case of Na overlayers of H+ yield resonance at hν ~32eV at the excitation of the Na2p level. This H+ resonance is unique to Na on Si surfaces and is actually quenched by coadsorption of other alkali metals. This ion resonance is interpreted as a result of the dielectric response of the near surface region (including the Si-Na-H interface state) to the departing H+ when the Na2p core level is excited.

Catalytic oxidation (nitridation) of semiconductor surfaces by alkali metals has been investigated by several techniques including photoelectron spectroscopy (PES), which has provided considerable insight into the changes of the electronic structure during catalytic passivation [1,2].

In an attempt to provide an additional insight into these catalytically activated reactions, the technique of the photon (electron)-stimulated desorption PSD(ESD) has also been employed [3,4]. Of particular interest is the ability of PSD and ESD to detect hydrogen which is not detectable by other techniques such as, e.g., Auger electron spectroscopy (AES). This is a very useful feature of PSD since the presence of hydrogen on the Si surface is known to affect, e.g., the process of Si oxidation, because hydrogen might compete with oxygen (and/or cesium) for Si surface sites. For this reason, several PSD studies of H+ have been performed on clean Si(100) and Si(111) surfaces exposed to H_2O [5,6].

PSD studies also have been recently performed for Cs overlayers on Si(100)2x1 and Si(111)2x1 surfaces [7,8] indicating that H+ is the only desorbed species observed from Si, Cs/Si and O_2/Cs/Si surfaces/interfaces. Recently studies have been extended to include Na overlayers on Si surfaces with similar results. In each case, the thresholds of H+ occur both at the Si2p and the alkali core levels indicating formation of alkali metal and Si hydrides at the surfaces. Of particular interest is an appearance of a sharp resonance in the partial ion yield (PIY) at hν~32 eV corresponding to the excitation of the Na 2p core level.

The experiments were performed at the Synchrotron Radiation Center of University of Wisconsin. A 6m Toroidal Grating Monochrometer (TGM) was used to provide monochromatic radiation in the photon energy range 25 to 130eV covering the core levels of the substrate and adsorbates. Two different gratings, overlapping at hν = 70eV were used to cover the entire above

* Supported by NSF grant No. DMR 88-07754 through NIU.

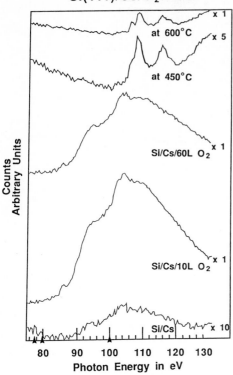

Si(111)/Cs/O₂/Heat

x 1

at 600°C

x 5

at 450°C

Si/Cs/60L O₂ x 1

Si/Cs/10L O₂ x 1

Si/Cs x 10

Counts
Arbitrary Units

80   90   100 110 120 130
Photon Energy in eV

Fig. 1   H⁺ ion yield vs photon energy for Si(100)2x1, Cs/Si, O₂/Cs/Si at room and increased temperatures.

spectral range. The pressure in the experimental chamber was better than 5 x 10^{-11} torr. Alkali metals were deposited from carefully outgassed SAES Chromate sources. The pressure increase during alkali metal deposition was always below 2 x 10^{-11} torr . The rest of the experimental details including ion detection and identification as H⁺, are the same as described previously [7,8].

The partial ion yield for the Cs overlayer on Si(100)2x1 and Si(111)2x1 surfaces is presented in Figs. 1 and 2 for both room temperature O_2 chemisorption as well as for the higher temperature range used for the catalytic oxidation. Both figures indicate that H⁺ desorption occurs both from Cs sites (threshold ~78eV; Cs4d level), and from Si sites (2p excitation at ~100eV). Also, it is noticeable that even a small O_2 exposure like 0.1L O_2 (1L = 1 Langmuir = 10^{-6} torr sec) of the Cs/Si system result in much stronger H⁺ yield due to the decreased ion reneutralization in agreement with previous reports [7,8]. At higher temperature (~ 400°C) the catalyst start to thermally desorb and the ion desorption is observed now only from the Si sites, and is shifted to ~104eV indicating formation of surface Si oxide complex including SiO_2. It is interesting to note that in the case of Si(111)2x1 surface, the PSD signal shows several other peaks following the 2p edge, due most likely to the inner well resonances.

Results of Si oxidation with Na overlayers are presented in Fig. 3 and yield similar overall results as in Figs. 1 and 2. The major new feature, however, is the strong resonant peak A at hν = 32eV corresponding to the

268

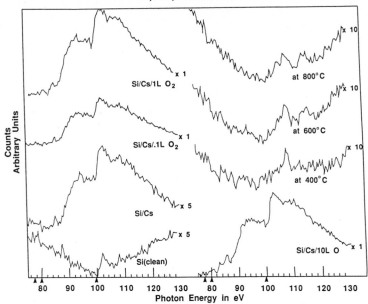

Fig. 2   H$^+$ ion yield vs photon energy for Si(111)2x1, Cs/Si, O$_2$/Cs/Si at
room and increased temperatures.

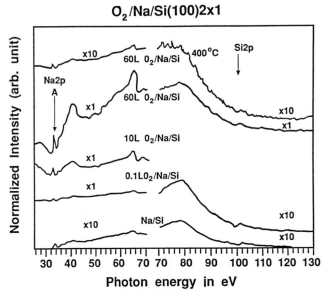

Fig. 3   H$^+$ ion yield vs photon energy hν in the range of 25 to 130eV for
1ML/Na on Si(100)2x1 at various oxygen exposures at room temperature
and at 400°C. The discontinuity at hν = 70eV reflects the use of
two different gratings necessary to cover the entire spectral
range.

269

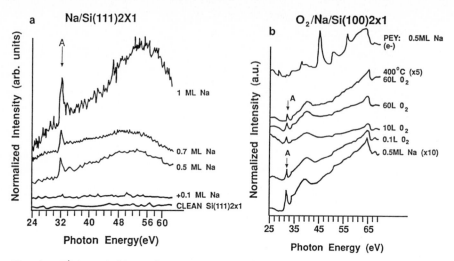

Fig. 4a H⁺ ion yield vs photon energy in the range 24 to 60eV for the Na/Si(111)2x1 system at various Na coverages.

Fig. 4b H⁺ ion yield vs photon energy in the range 25 to 70eV at various oxygen exposures of 0.5ML of Na/Si(100)2x1 at room temperature and at 400°C. For comparison the partial electron yield (PEY) for 0.5ML of Na on Si(100) is also displayed at the top of the figure.

excitation of the Na2p core level. The detailed behavior of this resonant peak A is very interesting, and is presented in Figs. 4a and 4b.

Figure 4a shows the H⁺ yield vs the photon energy hν, for the Na/Si(111)2x1 as a function of Na coverage. While there is an increase of the overall H⁺ signal with the coverage, the most noticeable feature here is a sharp peak A in H⁺ yield at hν = 32eV corresponding to the excitation of the Na 2p core level. The intensity of this peak increases initially with the Na coverage and saturates at about 1 monolayer (ML) coverage of Na. Figure 4b shows a similar sharp peak A in H⁺ yield for Na on Si(100)2x1 system, as well as for the case of additional oxygen chemisorption. It is obvious that while the position and shape of the peak A does not change with O₂ chemisorption, its intensity (as well as the overall magnitude of H⁺ yield), is increased by nearly one order of magnitude by such a low exposure as 0.6 L of O₂, while additional higher O₂ exposures cause no other significant changes of this peak.

Thus, Figs. 4a and 4b indicate that the H⁺ desorption starts at the Na2p core level threshold for both Si(111) and Si(100)2x1 surfaces. However, instead of a gradual increase of H⁺ at the core threshold, the H⁺ desorption is characterized by a very sharp peak A contrary to the case of H⁺ desorption from Cs/Si(100) [7,8]. The first explanation which could be reasonably provided for the peak at 32eV would be a sharp electronic transition including the Na 2p level and the s-p hybridized empty states near the conduction band minimum. However, this is <u>not</u> the case, as seen from Fig. 4b where such a peak is entirely absent in the optical absorption spectrum in the same photon energy range as indicated by the yield of secondary photo-electrons (upper curve labeled PEY in Fig. 4b) which shows just a gradual onset of absorption at hν ~32eV due to the sodium overlayer. Even at submonolayer coverages, the increase in the PEY is noticeable but does not show any pronounced resonance such as seen in the ion yield despite the fact that the signal for PEY is some five orders of magnitude higher than the ion

270

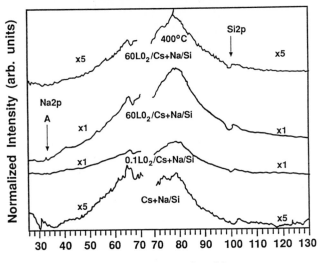

O₂/Cs+Na/Si(100)2x1

Fig. 5   H⁺ ion yield vs photon energy in the range 25 to 130eV for
coadsorbed Na and Cs/Si(100)2x1 followed by various oxygen exposure
at room temperature. The discontinuity at $h\nu$ = 70eV reflects the
use of two different gratings necessary to cover the entire spectral
range.

signal (see Fib. 4b).* This significant difference between the H⁺ yield and
the optical absorption (PEY) indicates that the origin of H⁺ desorption (peak
A) is not any optical excitation like, e.g., core excitons in the case of
alkali halides [9]. The PEY shows just a gradual onset at $h\nu \sim$ 32eV. Therefore, we conclude that the H⁺ resonance at 32eV is a result of the Si-Na-H
interface state characteristic of this system.

Figure 4b also shows the $h\nu$ dependence of H⁺ yield during the initial stage
of the catalytic oxidation at a temperature of 400°C when both Si³⁺ and the
stoichiometric SiO₂ oxide are formed. The sharp peak A as well as the overall
$h\nu$ features of the H⁺ yield remain unchanged.   This indicates that the
resonance characterized by the peak A is not affected by the growth of the
SiO₂ oxide which grows <u>beneath</u> the surface, and supports thus our assignment
of the peak A to the interface resonance state. As a further support of this
assignment, no such H⁺ resonant peak is observed in the PIY in case of Na
doped glasses for both clean surfaces as well as for additional Na overlayer
deposited onto such glasses [10].

In view of this very different behavior of H⁺ for Na as opposed to that
observed on Cs/Si previously [7,8], we have investigated further the behavior
of this H⁺ resonant peak near 32eV in the case of coadsorption of Na and Cs
(Fig. 5). As obvious from Fig. 5, the Na induced resonance peak A is barely
noticeable in the presence of coadsorbed Cs (even at low Cs coverages), and
has rather low intensity even when the overall H⁺ yield is subsequently
enhanced by O₂ chemisorption. The conclusion is that the intensity of the
resonant peak A is strongly suppressed by co-adsorption of Cs (and/or other
alkali metals), and additional subsequent deposition of Na onto the (Si+Na+
Cs) system does not restore the sharp resonant peak A.

---

*   The surface sensitivity of PEY to overlayers of Cs has also been shown
previously [7,8].

In conclusion, we show that the PSD technique can be used to monitor the alkali metal/Si interfaces and the alkali metal assisted electronic promotion of SiO_2. Interaction of hydrogen with these systems plays an important role as both Si and alkali metal hydrides are formed. The current results together with previous PSD studies support the existing model [11] of electronic promotion of Si oxidation including the non-local mechanism of catalytic oxidation and penetration of oxygen below the alkali metal overlayer.

Acknowledgement

This work was supported by NSF Grant No. DMR-88-07714 to Northern Illinois University and by the Northern Illinois University Graduate School Fund. We thank the staff of the Synchrotron Radiation Center, University of Wisconsin-Madison, for their technical assistance.

References

1. P. Soukiassian, T. M. Gentle, M. H. Bakshi and Z. Hurych, J. Appl. Phys. **60** (1986) 4339.
2. P. Soukiassian, M. H. Bakshi, H. I. Starnberg, Z. Hurych, T. M. Gentle and K. Schuette, Phys. Rev. Lett. **59** (1987) 1488.
3. T. E. Madey and R. Stockbauer, Methods Exp. Phys. **22** (1985) 465.
4. D. Menzel and R. Gomer, J. Chem. Phys. **41** (1964); P. A. Redhead, Can. J. Phys. **42** (1964) 886.
5. C.U.S. Larsson, A. S. Flodström, R. Nyholm, L. Incoccia and F. Senf, J. Vac. Sci. Technol. **A 5**, (1987) 3321.
6. R. A. Rosenberg, J. P. Love, V. Rehn, I. Owen and G. Thorton, J. Vac. Sci. Technol. **A 4**, (1986) 1451, and references therein; R. A. Rosenberg, C. R. Wen and D. C. Mancini, in Desorption Induced by Electronic Transitions, DIET III, edited by R. H. Stulen and M. L. Knotek (Springer-Verlag, Berlin) 1988, p. 220.
7. Z. Hurych, H. I. Starnberg and P. Soukiassian, Europhys. Lett. **8** (1989) 567.
8. Z. Hurych, P. Soukiassian, H. Starnberg, Physica Scripta **41**, 955 (1990).
9. P. H. Bunton, R. F. Haglund, Jr., D. Liu and N. H. Tolk, Surf. Sci. **243**, (1991) 227.
10. P. S. Mangat, D. Crouch, Z. Hurych, P. Soukiassian and H. Arribart, recent results.
11. H. I. Starnberg, P. Soukiassian and Z. Hurych, Phys. Rev. **B39**, (1989) 12775.

Part IX

Oxides

Low Energy Excitations and Desorption Dynamics from Oxide Surfaces

M. Menges[1], B. Baumeister[1], K. Al-Shamery[1], B. Adam[1], Th. Mull[1], H.-J. Freund[1], C. Fischer[2], D. Weide[2], and P. Andresen[2]

[1]Lehrstuhl für Physikalische Chemie I, Ruhr-Universiät Bochum,
 W-4630 Bochum, Germany
[2]Max-Planck-Institut für Strömungsforschung,
 Bunsenstr. 10, W-3400 Göttingen, Germany

1. Introduction

The interaction of photons with solid surfaces initiates processes which may be classified into various categories [1,2]. We will be concerned mainly with photochemical processes, including desorption of participating molecules. The measurement of the distribution of energy into translational and internal degrees of freedom possibly provides us with new insights into the mechanisms underlying the desorption after photoabsorption. In order to study the simplest cases first, various groups have studied photodesorption of NO and CO from metal and metaloxide surfaces [3-17]. A whole range of photon energies has been used so far. It appears that if we exclude photoinduced thermal desorption, the cross sections for photodesorption are orders of magnitude larger on weakly oxidized metal surfaces and in particular on oxide surfaces than on metal surfaces. Qualitatively, several effects are responsible for this difference in our view.

i) The electronic structure of the substrates is considerably different in the sense that a metal does not exhibit a band gap while an oxide often does. Energy that is dissipated into the substrate must exceed the gap energy in the case of an oxide unless there are defect states filling the gap. For a metal energy in any small quantity may be dissipated into the solid because excitation of electron-hole pairs of low energy is always possible. The probability of such excitations depends of course on the density of states at the Fermi energy. Metals with low density of states, such as Cu, Ag, Au, etc. should have a smaller probability for electron-hole-pair-creation.

ii) The degree of localization of the electronic charge distribution is typical for an oxide, while delocalization is a prototype metal property. This leads in the case of an oxide to a longer lifetime of the excited state. Consequently, the probability to escape the surface is larger for photodesorption from oxide surfaces due to the possibility to accumulate translational energy to leave the surface [13-15].

iii) A weak molecule-surface interaction will favour localization of the excitation on the molecule and thus increase the photodesorption probability.

Indeed the highest photodesorption cross-sections are observed for weakly bound molecules on oxide surfaces [13-17].The relatively large desorption cross sections allow us to simultaneously measure the energy distribution onto internal and translational degrees of freedom of the desorbing particles applying laser techniques. From the results of the studies we can deduce certain aspects of the mechanism of photodesorption through computer modelling of the complex experimental data.
 We review in the following our experimental setup as well as the results on some of the adsorbate systems.

2. Experimental setup

Fig. 1 shows a schematic diagram of our experimental setup. A desorption laser, in our case an ArF-excimer laser with 6.4 eV photon energy, is fired normal towards the surface for about

---

Springer Series in Surface Sciences, Vol. 31
Desorption Induced by Electron Transitions DIET V
Editors: A.R. Burns · E.B. Stechel · D.R. Jennison   © Springer-Verlag Berlin Heidelberg 1993

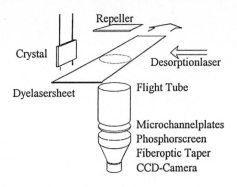

Fig.1. Schematic representation of the experimental setup.

Repeller

Crystal

Desorptionlaser

Dyelasersheet

Flight Tube

Microchannelplates
Phosphorscreen
Fiberoptic Taper
CCD-Camera

15 nsec. The molecules that escape from the surface are detected after a given time delay of the order of μsec with a resonant multiphoton ionisation (REMPI) detection system.

The REMPI signal is induced by an excimer laser pumped dye laser system fired parallel to the surface. The illuminated area may either form a line or be spread into a sheet perpendicular to the surface with a cylindrical lens system. In each point of the illuminated area the desorbing molecules undergo an excitation and ionisation (1+1) process. Via these processes the molecules resulting from a well defined rotational-vibrational state are converted into ions. These ions are than repelled into a flight tube behind the probe and desorption laser and are detected with a multichannel array. An image may be detected on the phosphorous screen behind the multichannel plates which is recorded by a videocamera and a personal computer. This image represents a distribution of time-of-flights because each line in the laser sheet is equivalent to a well defined distance from the surface. Simultaneously, one records the angular dependence of desorption along the lines parallel to the surface. One needs about three to four "shots" to record an image.

However, at the present time the signal-to-noise is limited in such images, which makes the identification of weak features difficult. We redose the surface after each laser shot by applying a background pressure of $5*10^{-8}$ Torr and take the difference between the REMPI signals with and without desorption which gives us the desorption signal. If we are interested in precise time-of-flight data we use the REMPI system without imaging and record the signal generated from a laser line by varying the time delay between the desorption and the probe laser [18].

Other groups have used imaging techniques to study desorption [10]. However, the laser sheet used in those studies was oriented parallel to the surface in order to determine the angular distribution of the desorbing molecules. In most other studies non-imaging detection has been employed and while we have mostly employed a desorption energy of 6.4 eV other groups have studied desorption at energies, which cover the region from 1.25 eV to 6.5 eV (see for example [1,2]).

3. Results and Discussion

3.1 The adsorbate systems NO on NiO (100) and on NiO(111)

The adsorption of NO and CO on well defined NiO surfaces has so far only been studied by very few groups [16,17,19] in contrast to NiO powder samples where numerous studies have been reported (see for example [20]).

We have used thin NiO films of (100) and (111) orientation grown on Ni(100) and Ni(111) single crystal metal surfaces [21,22]. With these samples all standard electron spectroscopic methods for adsorbate charactarization may be applied. The oxide surfaces exhibit reasonable long range order leading to LEED patterns and STM images compatible with oxide terraces of 100x150 Å width in the case of NiO(100).

On the terraces of NiO(100), NO adsorbs on top of the Ni ions with the NO axis tilted by 45° with respect to surface normal as revealed by NEXAFS. This is also supported by ab initio calculations [19]. HREELS shows that the vibrational frequency of the NO stretch is close to the gas phase value of NO which is typical for adsorption on oxide surfaces. The binding energy is 0.5 eV (peak desorption temperature T_p = 200 K) , which is in the regime of weak chemisorption. With XPS the coverage has been determined to be 0.25 on NiO (100). A detailed study has indicated that NO adsorbs on the terraces and not on the defect sides of the oxide surface [19].

On the polar NiO(111) surface it appears that adsorption of NO occurs on the Ni-terminated parts of the NiO(111) surface as opposed to the oxygen-terminated parts. The coverage is similar to NO on NiO (100). Even though a complete study of the adsorption behaviour for NO on NiO (111) has not yet been carried out, the TDS and XPS- results so far available are very similar to those found for NO adsorption on NiO(100).

3.2 Desorption dynamics NO/NiO(100)

Fig. 2 shows the determination of the desorption cross section both as measured on the surface via analysis of XP-spectra and in the gas phase via REMPI for the system NO/NiO(100) at T=100 K. In the left panel we determined the photodesorption cross section by XPS-measurements of the remaining coverage (N) after a certain photon exposure with respect to the initial coverage (No). The slope represents the photodesorption cross section. The value of $6*10^{-17}$ cm² is relatively large and within the experimental error identical with the one determined from the desorbed molecules in the gas phase (right panel) of $4*10^{-17}$ cm². The equivalence of cross sections determined on the surface and in the gas phase shows that the desorbing NO molecules are the majority species on the surface.

Fig. 3a) shows a set of velocity flux distributions for the desorbing NO molecules in one rotational state and three different vibrational states (v = 0,1,2). In Fig. 3b) we have plotted a set of velocity flux distributions for a fixed vibrational state and different rotational quanta from small to large values. Several issues are apparent:

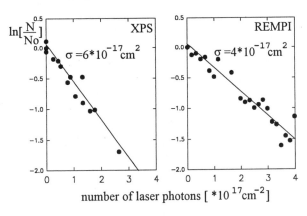

Fig. 2. Comparison of the desorption cross sections for the desorption of NO from NiO(100) measured via XPS (left side) and in the gas phase (right side).The logarithm of N/No is plotted against the number of photons.

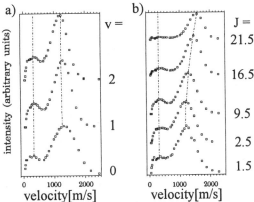

Fig. 3. Velocity flux distributions of desorbing NO from NiO(100).

a) Various vibrational states of one rotational state NO($^2\Pi_{3/2}$ J=11.5).
b) Various rotational quanta of the NO($^2\Pi_{3/2}$) for v =1.

277

i) The velocity distributions are bimodal (Fig.3).

ii) There is a coupling between rotational and translational motion for one of the maxima in the bimodal distributions (Fig.3.b).

iii) There is no obvious coupling of rotational and translational motion with vibration (Fig. 3.a). The molecules are vibrationally hot independent of whether they belong to the "slow" or the "fast" maximum of the bimodal distribution.

Neither the rotational nor the vibrational "temperatures" - if one wants to assign such a quantity - are directly connected with the surface temperature. The latter is about $T = 120$ K if we take the heating with the laser into account as estimated via the Bechtel-equation of heat flow [23]. The "slow" maximum in the velocity distribution exhibits the smallest "translational temperature" which is the region between 100 K $< T < 200$ K and thus close to the surface temperature. However, the "rotational temperature" is close to 400 K and the "vibrational temperature" around 1900 K. For the "fast" maximum the "translational temperature" varies between 1000 K $< T < 2000$ K as a function of the internal energy. For rotation and translation the values are comparable or even identical to those for the "slow" maximum. Clearly, it is not quite appropriate to use the term temperature in this connection the surface and the desorbing molecules are not equilibrated. The processes are of non-thermal origin.

3.3 Model Calculations

In order to explain the experimental observations we have carried out computer simulations of the desorption processes where we describe the excitation - deexcitation processes on the basis of quantum mechanical models and the propagation in the excited state as well as in the ground state potential quasiclassically [14,15]. We use the framework of the so called Menzel-Gomer-Readhead (MGR) model schematically shown in Fig. 4 [24,25].

Briefly, desorption is triggered by an electronic transition of the bound adsorbate-substrate complex. After the change of the potential energy curve has occurred the molecule propagates under the influence of the new potential energy curve for a time τ and may accumulate kinetic energy. After the time τ has elapsed the system relaxes into the ground state and transfers potential energy to the solid substrate via electron-hole-pair creation and/or phonon coupling. If the Franck-Condon concept is applied, the molecule keeps its kinetic energy, and if the accumulated kinetic energy is larger than the depth of the ground state well at the relaxation distance (Fig. 4), the molecule may escape the ground state well and desorption occurs.

To consider the dynamics of the molecule in the adsorbed state we use the model of the rigid rotor. This appears in the experimentally observed decoupling of the vibration and translation (see above). Fig. 5a illustrates typical motions of a molecule bound to the NiO(100) surface: In addition to the molecule-surface vibration a bending vibration of the molecule with respect to the surface may be excited. The vibration is fixed in amplitude to a region of solid angles by the angular dependence of the ground state potential (Fig. 5a). Due to these and possibly other motions a momentum k may be associated with the center of mass motion of the molecule in the ground state which is conserved during the electronic excitation of the molecule-surface complex [26]. With the assumption of an isotropic excited state potential, the electronic excitation leads to a rupture of the molecule-surface bond, i.e. the hindered rigid

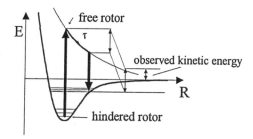

Fig. 4. MGR-model

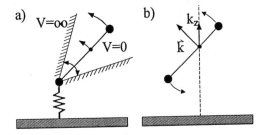

Fig. 5. Schematic representation of the models for the hindered rotor (a) and free rotor (b) as discussed in the text.

rotor in the adsorbed state becomes a free rigid rotor in the excited state [15]. The translational motion of the center of mass of the molecule may be characterized by the vector k (Fig. 5b). In case the electronic excitation occurs during the motion of the molecule away from the surface, k has a component pointing along the positive direction of the surface normal (k_z). In case it occurs while moving towards the surface, k has a component pointing along the negative direction of the surface normal.

The different signs of the k_z components have consequences for the desorption dynamics: For a repulsive excited potential, molecules with k_z components are moving away from the surface, while molecules with negative k_z are primarily moving towards the surface. After the initial momentum has been converted to potential energy, the molecules turn around and move outwards. Therefore, if the same lifetime in the excited state is chosen for all desorbing molecules, a molecule with an initially negative k_z component cannot acquire as much kinetic energy as a molecule with positive k_z. Consequently, relaxation to the ground state occurs at different positions on the potential energy surface. Thus the desorption dynamics will be different, which can lead for example to bimodal velocity (momentum) distributions. In other words, for a short lifetime of the excited state the relaxation occurs for both distributions at nearly the same position of the potentials and with nearly the same absolute values for k_z resulting in one peak of the momentum distribution after desorption. For longer lifetimes the relaxation of the two distributions occurs at different positions and with different absolute values of k_z, resulting in bimodal momentum distributions.

Generally, however, the observed behaviour strongly depends on the details (e.g. slope, existing minima, etc.) of the chosen excited state potential. We have sketched three typical situations in Fig. 6. The diagrams indicate that the molecules only probe a very limited region on the excited state potential where they are accelerated according to the slope of the potential. The observed velocity flux distribution is then strongly determined by the ground state potential. There are two issues to be adressed:

i) What is the nature of the repulsive potential energy curve?

ii) Why do we observe different populations in the two channels of the bimodal velocity flux distribution?

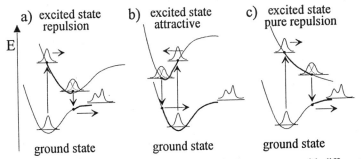

Fig. 6. Schematic drawing of three possible excitation processes with different excited states.

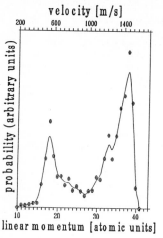

Fig. 7. Calculated velocity distribution

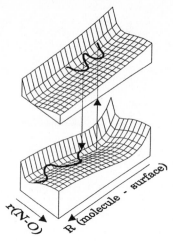

Fig. 8. Schematic representation of the potential energy surfaces.

i) Let us assume as the first elementary step the creation of an electron hole pair which initiates the electron to be charge transferred from the substrate to the molecule. In the literature there are some experimental indications that this is the situation encountered at the surface [12]. The excited state wavefunction of the adsorbed molecule would then look like a state of the NO^- ion. The remaining positive hole is screened on the oxide surface via the mobile electrons of the oxygen sublattice leading to an excited state potential with a minimum (Fig.6.a or b), the position of which depends on the details of the electronic interaction. However, upon excitation we assume that the molecules only probe the repulsive part of the excited state potential because of the short life time on the potential energy curve.

Therefore, we simulate this part of the potential curve by a repulsive 1/r-potential (Fig.6.c). We describe the rotational motion of the molecules on the excited state potential as a free rotor: if the interaction is basically electrostatic in nature it is reasonable to assume the potential to be isotropic. Therefore, in general the motion is thought to be more similar to a free than to a hindered rotor. Note, however, that the angular space probed by the molecule is small ($\Delta\Theta < 30°$) because the molecule has no time to fully rotate in the excited state. A model calculation where the parameters of the ground state potential [27] have been chosen according to existent experimental data leads to a typical time-of-flight distribution shown in Fig. 7. The lifetime enters as a parameter and was chosen as $\tau = 4 * 10^{-14}$ sec [15].

ii) The populations are governed by several factors, some of which are taken into account in the model calculations. Briefly, the probability to transfer from the excited state potential to the ground state will depend on the position on the excited state potential. Because the transfer occurs for the two components at different positions we expect differences in the populations. There is, however, another factor which has so far not been taken into account. The lifetimes of molecules initially moving inwards or outwards with respect to the surface may be different. This will lead to further changes in the relative populations. In fact, this latter aspect may turn out to be the more important one.

So far we have neglected the vibrations of the NO molecule in the adsorbate as well as in the desorbing particles. Our experimental results indicate that we can assume decoupling of rotation and translation from vibration. If we furthermore assume that the NO stretching vibration is decoupled from the NO-metal vibration we may find simple qualitative arguments to explain the observed vibrational excitations of the desorbing molecules. The situation is depicted in Fig. 8 where two potential energy surfaces (ground and excited states) are plotted as a function of the two independent degrees of freedom, namely the motion of the molecule away from the surface (R) and the NO stretching motion (r).

The excitation process may be described as follows: Before excitation the molecule is located in the global minimum of the lower curve. The photon excites the system via a Franck-Condon-like transition into the upper potential surface where the system only exhibits a minimum with respect to the NO stretching motion (r) and no minimum with respect to R. However, the position of the minimum as well as its shape differs from the situation in r-direction on the ground state potential energy surface. If we choose the potential well minimum to be located at larger distance, as would be the case if the nature of the excited state is NO⁻ like, the excitation leads to a non-eigenstate of the excited system in general. The propagation of this non-eigenstate may be described by solving the time dependent Schrödinger equation.

We assume to have significant quantum state populations at the surface just for v=0 in the ground state. Choosing the vibrational frequency from experimental data of various NO, or NO⁺ and NO⁻ states [28] we may determine via a solution of the time-dependent Schrödinger equation a relaxation time τ after which for example the experimentally determined vibrational populations of the desorbing molecules are reproduced. A value that fits the populations is of the order of several 10^{-14} sec and thus compatible with the lifetime determined independently from the time-of-flight distribution [15]. Of course, the two life times do not have to correlate necessarily. However, in our case the hopping of the electron from the assumed NO⁻ back into the substrate triggers the formation of the observed neutral NO. Therefore the similarity of the two lifetimes is in favour of the proposed mechanism.

3.4 Desorption Dynamics NO/NiO(111)

As the next step we would like to investigate the applicability of our model for NO/NiO(111). However, the adsorbate NO/NiO(111) behaves differently compared with NO/NiO(100). In particular Fig. 9 b) shows that NO does not only desorb from the surface but also undergoes a chemical transformation on the surface. Before exposing to UV-radiation we find via XPS a N1s doublet typical for NO adsorbed on NiO(100). The fine structure is most probably due to a shake-up structure of a single species and not to the presence of two chemically different species [19].

In contrast to NO/NiO(100) the exposure of NO/NiO(111) to the UV-laserlight induces changes of the relative intensity of the two signals and not just a decrease of the intensity and we observe the build-up of a signal at 398.5 eV binding energy (Fig. 9a).

Apparently NO dissociates under UV-laser irradiation on NiO(111) leading to the formation of atomic nitrogen and oxygen (not detected) on the surface. The oxygen may react with excess NO to form NO_2 leading to a signal overlapping with the lower binding NO feature of the NO N1s-spectrum (402.3 eV). The atomic nitrogen remains at the surface. The assignment of the proposed NO_2 spectrum is corroborated by experiments where XP-spectra are taken

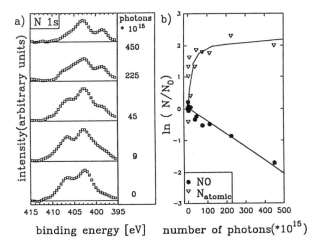

Fig. 9. Data for the laser induced desorption of NO from NiO(111).

a) A set of XPS-data for the N1s region after different numbers of photons exposed to the surface.

b) Logarithmic plot of ln(N/N₀) versus the number of photons. The filled circles refer to the signal at 406.7 eV (pure NO) and the open triangles belongs to the signal of atomic nitrogen at 398.5 eV.

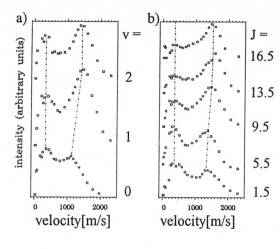

a)

intensity (arbitrary units)

velocity[m/s]

v =

2

1

0

0   1000   2000

b)

velocity[m/s]

J =

16.5

13.5

9.5

5.5

1.5

0   1000   2000

Fig. 10. Velocity distributions of desorbing NO from NiO(111).
a) Various vibrational states of one rotational state $NO(^2\Pi_{3/2}$ J=11.5).
b) Various rotational quanta of the $NO(^2\Pi_{3/2})$ for v =1.

after direct NO_2 adsorption on NiO(100) leading to a signal at 403.1 eV binding energy [29]. Therefore in our further discussion we will assign the reaction product as being NO_2. Further spectroscopic measurements in order to finally identify the reaction product have to be undertaken. Summarizing so far in contrast to NO/NiO(100), where only desorption can be observed, for NO/NiO(111) we also find a reaction channel for NO dissociation.

In Fig.9 b) ln (N/N_0) for the system NO/NiO(111) is plotted versus the number of photons exposed to the surface as obtained from an analysis of the XP-spectra. The plot reveals that the coverage of the atomic nitrogen saturates at the surface. However, the cross section observed in the gas phase is a result of the combined processes, namely the photoreaction and the photodesorption. We have investigated the velocity distributions for different vibrational and rotational quanta of the NO desorbing from NiO(111). These are shown in Fig. 10. Again we observe bimodal velocity distributions with velocity values similar to the NO desorbing from NiO(100), but the relative population of the "fast" and "slow" features are different for a given rotational state and varying vibrational state. The velocities only vary moderately. The rotational and vibrational temperatures are rather similar. The velocities of the "fast" channel for v=1 and v=2 are equal (about 1300 m/s), but the velocity in v=0 is smaller (1050 m/s). In addition we find strong coupling between translational and rotational motion in v=0 (900 - 1400 m/s), however the coupling is weak in the excited vibrational states (1300-1500 m/s).

The higher vibrationally and rotationally excited the desorbing NO-molecules are, the more similar the velocity flux distributions become to those for NO desorbing from NiO(100). The distributions show comparable positions of their maxima and of their populations. The velocity distributions are a result of two different channels (photodesorptive and photochemical). Further investigations are in progress in order to find out about the contributions of the two channels to the final velocity distributions.

We have also investigated the laser induced desorption of CO. On NiO(111) CO is adsorbed below 160 K in contrast to NiO(100) where it does not adsorb as revealed by XPS and TDS. Similar to the case of NO on NiO(100) CO desorbs without reaction with a cross section of the same order of magnitude.

Acknowledgements

We thank the Deutsche Forschungsgemeinschaft as well as the Ministerium für Wissenschaft und Forschung for financial support. H.-J. F. thanks the Fonds der Chemischen Industrie.

Literature:

1. X.L. Zhou, X.-Y. Zhu, J.M. White, Surf.Sci.Rep. **13**, 77(1991).
2. W. Ho, Desorption Induced by Electronic Transitions, DIET IV, Springer Ser.Surf.Sci. **19**, 48, Berlin, Springer-Verlag (1990).
3. A. Mödl, T. Gritsch, F. Budde, T.J. Chuang, G.Ertl, Phys.Rev.Lett. **57**, 384 (1986).
4. A.R. Burns, E.B. Stechel, E.R. Jennison, Phys.Rev.Lett. **58**, 250 (1987); E.B. Stechel, D.R. Jennison, A.R. Burns, Desorption Induced by Electronic Transitions, DIET III, Springer Ser.Surf.Sci.**13**, 137, Berlin, Springer-Verlag (1988); D.R. Jennison, E.B. Stechel, A.R. Burns, ibid., p.167
5. D. Weide, P. Andresen, H.-J. Freund, Chem.Phys.Lett. **136**, 106 (1987).
6. F. Budde, A.V. Hamza, P.M. Ferm, G. Ertl, D. Weide, P. Andresen, H.-J. Freund, Phys.Rev.Lett. **60**, 1518 (1988).
7. W.C. Natzle, D. Padowitz, S.J. Sibener, J.Chem.Phys. **88**, 7975 (1988).
8. E. Hasselbrink, S. Jakubith, S. Nettesheim, M. Wolf, A. Cassuto, G. Ertl, J.Chem.Phys. **92**, 3154 (1990).
9. M. Wolf, E. Hasselbrink, J.M. White, G. Ertl, J.Chem.Phys. **93**, 5327 (1990).
10. R. Schwarzwald, A. Mödl, T.J. Chuang, Surf.Sci. **242**, 437 (1991).
11. K. Mase, S. Mizuno, Y. Achiba, Y. Murata, Surf.Sci. **242**, 444 (1991).
12. J.W. Gadzuk, L.J. Richter, S.A. Buntin, D.S. King, R.R. Cavanagh, Surf.Sci. **235**, 317 (1990).
13. Th. Mull, H. Kuhlenbeck, G. Odörfer, R. Jaeger, C. Xu, B. Baumeister, M. Menges, G. Illing, H.-J. Freund, D. Weide, P. Andresen, Desorption Induced by Electronic Transitions, DIET IV, Springer Ser.Surf.Sci. **19**, 169, Berlin, Springer-Verlag (1990).
14. B. Baumeister, M. Menges, T. Mull, H.-J. Freund, D. Weide, P. Andresen, Proceedings of the Symposium on Surface Science, La Plagne, France, p. 147 (1990).
15. Th. Mull, B. Baumeister, M. Menges, H.-J. Freund, D. Weide, C. Fischer, P. Andresen, J.Chem.Phys., **96**, 7108 (1992).
16. J. Yoshinobu, T.H. Ballinger, Z. Xu, H.J. Jänsch, M. I. Zaki, J. Xu, J.T. Yates Jr., Surf.Sci **255**, 295 (1991).
17. M. Asscher, F.M. Zimmermann, L.L. Springsteen, P.L. Houston, W. Ho, submitted.
18. Th. Mull, Ph. D. Thesis, Bochum (1991).
19. H. Kuhlenbeck, G. Odörfer, R. Jaeger, G. Illing, M. Menges, Th. Mull, H.-J. Freund, M. Pöhlchen, V. Staemmler, S. Witzel, C. Scharfschwerdt, K. Wennemann, T. Liedtke, M. Neumann, Phys.Rev. **B 43**, 1969 (1991).
20. E. Escalona Platero, C. Coluccia, A. Zecchina, Langmuir **1**, 407 (1985).
21. M. Bäumer, D. Cappus, H. Kuhlenbeck, H.-J. Freund, G. Wilhelmi, A. Brodde, H. Neddermeyer, Surf.Sci. **253**, 116 (1991).
22. M. Bäumer , D. Cappus, G. Illing, H. Kuhlenbeck, H.-J. Freund, J.Vac.Sci.Technol. **A 10**, 1 (1992).
23. J.H. Bechtel, J.Appl.Phys. **46**, 1585 (1975).
24. D. Menzel, R. Gomer, J.Chem.Phys. **41**, 3311 (1964).
25. P.A. Redhead, Can.J.Phys. **42**, 886 (1964).
26. U. Landman, Israel J.Chem. **22**, 339 (1982).
27. J.E. Smedley, G.C. Corey, M.H. Alexander, J.Chem.Phys. **87**, 3218 (1987).
28. K.P. Huber, G. Herzberg, "Molecular spectra and molecular structure. Vol. 4: Constants of diatomic molecules", Van Nostrand Reinhold Company, New York 1979.
29. M. Menges, Diplomarbeit, Bochum 1990.

Electron Stimulated Desorption (ESD) of Ammonia on TiO$_2$(110): The Influence of Substrate Defect Structure

U. Diebold and T.E. Madey

Department of Physics and Laboratory for Surface Modification,
Rutgers, The State University of New Jersey, Piscataway, NJ 08855, USA

Abstract. Electron stimulated processes for NH$_3$ adsorbed on rutile TiO$_2$(110) have been studied by means of XPS and mass - resolved ESDIAD (electron stimulated desorption angular ion distribution). We have used three differently prepared TiO$_2$ surfaces to study the influence of the substrate defect structure on the interaction with NH$_3$: a stoichiometric nearly perfect surface, a high - temperature annealed slightly oxygen deficient surface and a sputtered highly oxygen deficient surface. In the limit of a *stoichiometric surface*, electron irradiation induces the *desorption* of NH$_3$ molecules. When the *highly oxygen deficient* TiO$_2$ surface is used as substrate, both desorption and electron stimulated *dissociation* of NH$_3$ take place with atomic nitrogen the final product of the dissociation process. These measurements provide direct evidence for the role of surface defects of an oxide substrate in electron stimulated reaction pathways.

1. Introduction

Electron Stimulated Desorption (ESD) from adsorbates on metal surfaces has been studied quite extensively during recent years, but little work has been done for adsorbates on insulators, e.g. transition metal oxides. In this work, we report on the electron stimulated desorption of ammonia adsorbed on three differently prepared TiO$_2$(110) surfaces with varying defect concentrations: A sputtered highly oxygen deficient surface, a slightly oxygen deficient surface with point defects, and a nearly perfect stoichiometric surface. The key observations are that the ESD process depends strongly on the defect concentration on the TiO$_2$ substrate, both qualitatively and quantitatively. The electron beam damage process was utilized in ESDIAD (electron stimulated desorption ion angular distribution) measurements to study the structure of the adsorbed molecules. Since the TiO$_2$ substrate itself is sensitive to electron irradiation [1], we have performed the ESDIAD measurements in a mass - resolved mode using a time of flight (TOF) technique to distinguish between the substrate and overlayer structure. To the best of our knowledge, these are the first TOF - ESDIAD measurements of a molecular overlayer on a substrate which is sensitive to electron irradiation, showing clearly the advantage of this technique.

2. Experimental

The experiments were performed in a UHV chamber which is described in some detail in [2, 3]. The TiO$_2$(110) sample was initially reduced by annealing to 1000 K to form n - type semiconducting TiO$_2$ [4], eliminating charging problems. During the ESD experiments, the sample was held at a temperature of ~ 160 K.

The experiments described below were performed on TiO$_2$(110) surfaces prepared in three different ways, for the preparation procedures see [3, 5]. (1) The *stoichiometric* surface is obtained by annealing a sputtered surface to 1000 K in O$_2$.

Springer Series in Surface Sciences, Vol. 31
Desorption Induced by Electron Transitions DIET V
Editors: A.R. Burns · E.B. Stechel · D.R. Jennison   © Springer-Verlag Berlin Heidelberg 1993

XPS shows sharp Ti *2p* peaks with no indication of reduced Ti states and LEED shows a (1 x 1) pattern. (2) The *slightly oxygen deficient* surface is obtained by annealing the stoichiometric surface to 1000 K for 3 minutes in UHV. A small amount of Ti^{3+} is observed using XPS; a (1 x 1) LEED pattern persists. (3) The *highly oxygen deficient* surface is obtained by sputtering the stoichiometric surface with 500 eV Ar^+ (total fluence ~ 1 x 10^{16} ions/cm^2). Several different Ti *2p* states are distinguishable in the XPS spectra and the surface has no long range order (no LEED pattern).

For ESD, electron bombardment of the front surface was done with emission from a W filament, with the sample biased to +325 V.

For the TOF - ESDIAD we use a setup similar to that described in [6]. The velocity resolution of this setup was sufficient to distinguish clearly between desorbed hydrogen ions and ions of higher mass; differentiation between mass 16 and 17 was not possible, however. The video and frame grabber system described in [2] was used to produce digitized ESDIAD images.

3. Electron stimulated desorption and dissociation

3.1. Results

In Figure 1, the XPS nitrogen *1s* region is shown following successive intervals of electron bombardment with 325 eV electrons. The *left* panel shows the spectra taken from NH$_3$ on a nearly perfect *stoichiometric* surface and the *right* panel on a *highly oxygen deficient sputtered* surface, respectively. The data are shown after

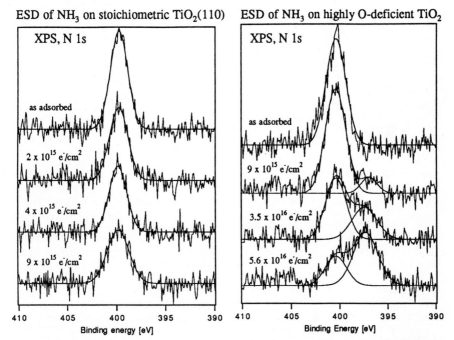

Figure 1 XPS spectra of the nitrogen *1s* region of NH$_3$ adsorbed on stoichiometric, nearly perfect TiO$_2$(110) (left panel) and sputtered, highly oxygen deficient TiO$_2$ (right panel). The curves were taken after irradiation with 325 eV electrons with the fluences indicated.

background subtraction and with Gaussians fitted to the curves. The starting points (top curves in each panel) for electron irradiation are the NH_3 saturation coverages at 160 K (0.19 and 0.16 ML, respectively). The N $1s$ peak with a binding energy of 399.8 eV and a peak width of 1.45 eV FWHM is assigned to molecularly adsorbed NH_3 [3, 7].

Irradiation of the NH_3 covered *stoichiometric* surface by 325 eV electrons causes mainly a decrease of the N$1s$ peak intensity. Assuming a simple first order desorption process, we determine the desorption cross section σ_{des} from the decrease of the N $1s$ peak area as 1×10^{-16} cm^2. During electron bombardment, the peak width increases somewhat, from 1.45 eV in the beginning to 1.7 eV after 1×10^{16} electrons/cm^2. We conclude that simultaneously with electron induced desorption of intact NH_3, dissociation to NH_x species occurs to some extent, but we were not able to distinguish extra XPS peaks.

The main process on the stoichiometric surface is desorption of N containing species and is very different from the case of *highly oxygen deficient sputtered* TiO_2 (right panel of fig. 1). After some electron irradiation, a peak with lower binding energy (E_B = 397.2 eV) appears. The binding energy is very close to 397.4 eV, observed for the N$1s$ level from stoichiometric and substoichiometric TiN [8]. We assign the lower E_B peak to nitrogen from fully dissociated ammonia that interacts strongly with surface titanium atoms. The complete dissociation of ammonia probably occurs via several steps involving the production of intermediate NH_x species (x=1, 2).

Electron irradiation of NH_3 on the highly oxygen deficient sputtered surface decreases the N coverage as well. The total intensity of the N$1s$ region decreases much slower as compared to the stoichiometric surface. (Note the higher electron fluences used for the right panel of fig. 1.) Using several approximations [3], the main one being that the concentration of NH_x (x = 1, 2) is small, we determined a dissociation cross section, σ_{diss}, of $\sim1.5\times10^{-17}$ cm^2 for the rate limiting step in the electron mediated reaction $NH_3 \rightarrow NH_2 \rightarrow NH \rightarrow N$. The cross section for desorption of ammonia, σ_{des}, is $\sim0.9\times10^{-17}$ cm^2, ten times smaller than for the stoichiometric surface. We have shown that prolonged electron bombardment of the highly defective, NH_3 saturated surface leads to the formation of a highly nitrided surface layer [3].

3.2. Discussion

The different behavior of NH_3 on sputtered and stoichiometric surfaces is striking. On a *thermally treated slightly oxygen deficient* surface, there was some evidence for formation of atomic nitrogen by electron bombardment, but the effect was too small to be quantified. Nevertheless this provides evidence that reduced surface Ti states are responsible for catalyzing the production of atomic nitrogen due to electron bombardment. Oxygen vacancies on a $TiO_2(110)$ surface induce mainly Ti 3d derived states in the band gap [9]. It is well known from studies of ESD processes that the lifetime of an excited state of an adsorbate will mainly govern the probability for desorption. Relaxation processes are concurrent with the conversion of electronic excitation into nuclear movement, which takes place within $\sim 10^{-14} - 10^{-13}$ sec. The cross section for loss of ammonia from a stoichiometric surface is surprisingly high, 1×10^{-16} cm^2, approaching values of gas phase ionization cross sections. It is ten times higher than that of a highly oxygen deficient surface where Ti atoms in different oxidation states are available. In other words, electronic excitation of NH_3 on TiO_2 induced by electron bombardment will decay faster the more metallic the TiO_2 surface is.

Hayes and Evans [10] compared ion yields induced by electron bombardment of a variety of organosilane molecules adsorbed on TiO_2 and SiO_2. They found higher positive ion yields in the case of a SiO_2 substrate, where reneutralization of ionized adsorbates is less likely due to the wider band gap as compared to TiO_2. In our case, not only is the absolute value of the cross section changed, but also an additional decay channel (dissociation) is opened on a highly oxygen deficient surface with excess titanium, making the ESD behavior more similar to that found on metal surfaces [11].

4. Mass resolved ESDIAD

The saturation coverages of NH_3 on both a stoichiometric, nearly perfect and a sputtered, highly oxygen deficient surface are quite similar [3]. However, the different sensitivity to electronic excitation and different reaction channels could also be caused by different adsorption geometries of the NH_3 molecule. For ion desorption, the influence of defect sites and surface steps has been observed on metal surfaces [12]. Trying to separate geometric and electronic effects on the NH_3/TiO_2 surfaces, we have initiated mass resolved TOF - ESDIAD measurements.

Fig 2 shows two ESDIAD images taken after a dose of 1L (1 x 10^{-6} torr.sec) NH_3 on the sputtered TiO_2 surface at 160 K. The right panel shows the ESDIAD image for lower mass (hydrogen ions) and for the left image for the higher mass region (oxygen ions). The differences in the two images immediately shows how valuable the TOF technique is to study this system: The hydrogen signal shows the halo typical for a nitrogen bonded NH_3 molecule standing upright on the surface, with the N end bonded to a substrate Ti atom (see insert in fig. 2). Either no rotational barrier in the N-H bond or azimuthal disorder of this bond gives rise to a uniform halo similar to that observed previously for NH_3 on metal surfaces [11, 13, 14]. The non-circular shape and the increased intensity on the left side of the image in fig. 2 are due to inhomogeneities of the electric fields and of the detection probability of the analyzer, respectively. The halo can be observed for low coverages only.

The maximum intensity of the oxygen signal is in the center of the image, indicating forward emission [15]. Without mass resolution we would not have been able to observe the hydrogen halo, since superposition of the two images shows a broad, forward peaked intensity distribution.

The observation of the halo in the spatially resolved hydrogen signal corroborates our assumption that the bonding of the ammonia molecule on highly oxygen deficient sputtered TiO_2 is very similar to that on metal surfaces and that the electronic changes in the substrate with defect densities causes the main difference observed in the ESD processes.

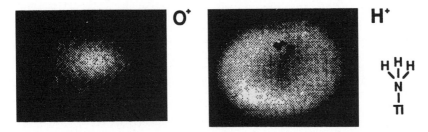

Figure 2 Mass resolved TOF - ESDIAD images of 1L NH_3 adsorbed on a highly oxygen deficient, sputtered TiO_2 surface. The images were taken with 200 eV primary electron energy and at a sample temperature of 160K. The right panel shows H^+ ions, the left panel the higher mass region of the desorbed ions (mainly O^+).

5. Conclusion

We have shown that the ESD behavior of adsorbed ammonia molecules is strongly influenced by the substrate defect structure of a $TiO_2(110)$ substrate. On a stoichiometric nearly perfect TiO_2 surface, mainly desorption of molecularly adsorbed ammonia is observed during electron bombardment with 325 eV electrons, with a cross section of $\sim 1 \times 10^{-16}$ cm^2. On a highly defective sputtered TiO_2 surface, the cross section for desorption is reduced drastically to $\sim 0.9 \times 10^{-17}$ cm^2 and electron irradiation causes mainly dissociation. By performing mass resolved ESDIAD measurements, we are able to distinguish between substrate and overlayer structure. The desorbed H^+ ions show a halo like structure that is characteristic for molecularly adsorbed ammonia. We believe that the differences in the electronic structure of the substrate, induced by defects, are the main reason for the change in the ESD reaction pathways.

Acknowledgment This work was supported in part by the National Science Foundation, grant DMR-8907533. The authors want to thank Dr. A.L. Johnson for valuable assistance with the TOF - ESDIAD measurements.

References

[1]   P.J. Feibelman and M.L. Knotek, Phys.Rev. B **18**, 6531 (1978)
[2]   B.L. Maschhoff, J.-M. Pan and T.E. Madey, Surf.Sci. **259**, 190 (1991)
[3]   U. Diebold and T.E. Madey, J.Vac.Sci.Techn. (in press)
[4]   D.C. Cronemeyer, Phys.Rev. **87**, 876 (1952)
[5]   J.-M. Pan, U. Diebold, B.L. Maschhoff and T.E. Madey, J. Vac. Sci. Techn. (in press)
[6]   S.A. Joyce, A.L. Johnson and T.E. Madey, J.Vac.Sci.Techn. **A7**, 2221 (1989)
[7]   E. Román and J.L.de Segovia, Surf.Sci. **251/252**, 742 (1991)
[8]   H. Höchst, R.D. Bringans, P.Steiner and T. Wolf, Phys.Rev.B **25**, 7183 (1982)
[9]   S. Munnix and M. Schmeits, Phys.Rev.B. **31**, 3369 (1985)
[10] T.R. Hayes and J.F. Evans, Surf.Sci. **159**, 466 (1985)
[11] C. Klauber, M.D. Alvey and J.T. Yates Jr., Surf.Sci. **154**, 139 (1985)
[12] T.E. Madey, M. Polak, A.L. Johnson and M.M. Walczak, in *Desorption Induced by Electronic Transitions,* edited by R. H. Stulen and M. L. Knotek, Springer, Berlin (1988)
[13] F.P. Netzer and T.E. Madey, Phys.Rev.Lett. **47**, 928 (1981)
[14] C. Benndorf and T.E. Madey, Surf.Sci. **135**, 164 (1983)
[15] R.L. Kurtz, Surf.Sci. **177**, 526 (1986)

Electron Stimulated Desorption of O^+ from $TiO_2(110)$–SO_2

M.C. Torquemada[1], J.L. de Segovia[1], E. Román[1], G. Thornton[2],
E.M. Williams[2], and S.L. Bennett[2]

[1]Laboratorio de Física de Superficies, Instituto de Ciencia de Materiales,
  CSIC, Serrano 144, E-28006 Madrid, Spain
[2]Interdisciplinary Research Centre in Surface Science,
  University of Liverpool, Liverpool L69 3BX, UK

Abstract. Adsorption of SO_2 on stoichiometric $TiO_2(110)$ at 153K has been studied by electron stimulated desorption (ESD). At this temperature the coverage of SO_2 estimated from complementary Auger studies is about 0.27 ML. Following SO_2 dosing, the ESD O^+ ion yield is enhanced for electron energies above 155 eV and 225 eV, and is identified with excitation of the S 2p and 2s core levels. The clean surface has a O^+ threshold at 35 eV corresponding to excitation of the Ti 3p level. Changes in ion yield with substrate temperature up to 450 K are described within a model consistent with a recent NEXAFS study which revealed the surface reaction sequence: $SO_2 \rightarrow SO_3^{2-} \rightarrow SO_4^{2-}$.

1. Introduction

Processes of gas adsorption at well defined oxide surfaces remain poorly understood in comparison with equivalent reactions at metal surfaces. There is the particular question as to the role of defects such as steps or vacancies which had been thought to play an important part. However, recent studies of TiO_2 indicate that their contribution is not a necessary prerequisite to gas uptake and chemisorption (1,2). TiO_2 has been at the center of interest in electron stimulated desorption (ESD) as a model system for the Knotek and Feibelman desorption mechanism (3), yet comparatively little attention has been devoted to ESD with gases deliberately introduced to the surface. Apart from the intrinsic interest in observations of ESD with sulphur-containing molecules at the surface of TiO_2, there is the background of technological activity centered on gas sensors and catalyst poisoning effects which serves to motivate the study.

For SO_2 on $TiO_2(110)$ the findings concerning room temperature reaction are somewhat controversial. However, the evidence from recent NEXAFS investigations at low temperature (105 K) supports the formation of a chemisorbed SO_2 phase, which upon heating transforms to a sulphate-like species via a sulphite intermediate (4). In this paper we present the preliminary findings of an ESD study of the interface formed by SO_2 gas dosing of stoichiometric TiO_2 at 153 K. The immediate aims in the study were: (1) to characterise the response of ESD ion desorption at the interface formed at low temperature. (2) to investigate changes in the ion yield with heat treatment, in order to compare the transformation between the bonding modes deduced in the earlier NEXAFS work (4). As far as we are aware this represents the first ESD study of a S-containing species.

2. Experimental

Experiments were carried out in an instrument equipped with a single pass CMA, a quadrupole mass spectrometer (QMS), a gas doser and an Ar^+ ion gun. Ultimate pressure after bakeout was in the 10^{-11} mbar region. The QMS was arranged to accept surface ions emitted within $\pm 5°$ of the surface normal with the electron beam at $45° \pm 2°$ to the normal. Electron currents for ESD and AES were always maintained at 10 nA ($\sim 1.2\ 10^{-7}$ A/cm^2) to minimise electron damage. The kinetic energy of the ESD ions was determined by shifting the mid-axis potential of the QMS to introduce a retarding field. The ion kinetic energy distribution is obtained by differentiating this integral spectrum.

The $TiO_2(110)$ sample was bulk reduced in-situ to prevent charging effects, and then cleaned by Ar$^+$ ion bombardment (500 eV and 1 μA). The stoichiometry was re-established by heating the sample at 800 K in oxygen gas. This gave a Ti(LMM) / O(KLL) Auger ratio of 1.7 in agreement with previous work (5). It was also seen that the ratio of the inter to intra-atomic transition at 417 eV corresponded with TiO_2 stoichiometry (6,7). The sample preparation procedure adopted here is known to yield an ordered 1x1 surface (4). The temperature at the sample was measured with a thermocouple inserted between the sample and its tantalum-foil support plate, which in turn was fixed to copper support rods serving both as current leads for resistive heating and as coupling rods to a liquid nitrogen reservoir for cooling. The presence of H on the sample was monitored by the ESD spectrum of surface ions, with a residual contamination of <1% of mass 16 (O$^+$) during adsorption experiments.

3. Results and discussion

The O(KLL)/Ti(LMM) Auger ratio after saturation dosing the 153 K sample with SO_2 increases from the clean surface value of 1.70 to 1.93. The S(LVV) structure at 150 eV is consistent with non-dissociative adsorption. Surface coverage was determined on the basis that a S/Ti ratio of 0.31 corresponded to a coverage of 0.4 ML as obtained at 105 K in the previous NEXAFS experiments (4). From the S/Ti ratio of 0.2 obtained in present adsorption experiments at 153 K, a coverage of 0.27 ML is obtained which is in agreement with the coverage at the same temperature obtained in the NEXAFS experiments.

Figure 1 shows the ESD O$^+$ ion kinetic energy distribution for the clean surface and the SO_2 saturated surface at 153 K. The clean surface spectrum is characterized by

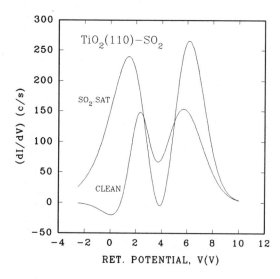

Figure 1: Ion kinetic energy distribution of ESD O$^+$ ions at $E_p = 250$ eV from $TiO_2(110)$ at 153 K and after SO_2 saturation at 153 K.
Supplementary experiments with the gas phase production of O$^+$ ions from SO_2 carried out in the same equipment indicated a most probably kinetic energy at -0.5 eV retarding potential (scale is uncorrected).

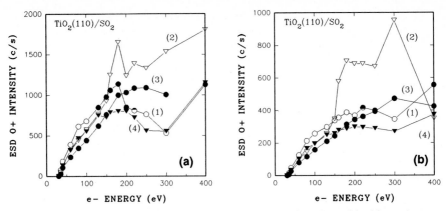

Figure 2: ESD O⁺ ion yield from TiO$_2$(110)-SO$_2$ as a function of incident electron energy. (a) At 0 V retarding potential.(b) At 5 V retarding potential. (1) Clean surface at room temperature.(2) After SO$_2$ saturation at 153 K.(3) Heated to room temperature. (4) Heated to 450 K.

a bimodal energy distribution with values of most probable ion kinetic energies of 2.5 eV and 5.9 eV. After SO$_2$ saturation the form of the ion energy distribution, also bimodal, shows significant changes. The origin of the bimodal energy distribution of the clean and SO$_2$-dosed surfaces are currently being investigated. They could arise from either two bonding sites or two desorption mechanisms. The discrepancy between the clean surface result and that reported earlier (7) could arise from a difference in electron beam energy, E$_p$, suggesting a secondary electron origin for the low energy component.

Figure 2(a) shows a family of ion yield curves at 0 V retarding potential to detect all emitted ions seen in Fig. 1. Spectra are shown for the clean surface, after SO$_2$ dosing and after further annealing at the indicated temperatures. There is a common threshold for all the curves at about 35 eV, corresponding Ti 3p core level excitation, in agreement with previous results (3). It has been argued that the O⁺ signal derives mainly from emission of bridging O-atoms (7). After SO$_2$ saturation there is seen to be a second threshold at about 155 eV incident electron energy corresponding to excitation of the S 2p core level. The O⁺ yield from the saturated surface increases again after 240 eV electron energy consistent with excitation of the S 2s core level. At energies below 150 eV the O⁺ intensity is lower than that from the clean surface. When the layer is annealed to room temperature and heated to 450 K a systematic decrease of the O⁺ intensities is observed, although the indication of an onset at 150 eV does not completely disappear. When only the more energetic ions are monitored the ion yield curves shown in Fig.2 (b) are obtained. The threshold features found in the earlier data with zero retardation are now less prominent and the response is generally broader in form in all cases of surface treatments, the differences can possibly be interpreted in terms of a secondary electron process as noted earlier. After heating the surface to room temperature, the SO$_2$ coverage, as revealed by AES, decreased to 0.17 ML. Further heating to 450 K did not reveal a significant decrease of the coverage. This is consistent with the NEXAFS data which suggest that mainly SO$_3^{2-}$ species should be formed at 153 K , converting at higher temperatures to a stable sulphate species, with some loss of SO$_x$. Fig.3 shows the ESD O⁺ production model from the clean and SO$_2$ dosed surface.

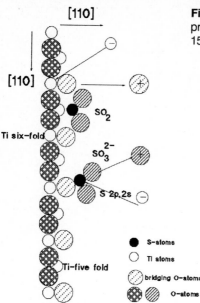

Figure 3. ESD model of O^+ production from TiO_2-SO_2 at 153 K.

[110]

[110]

Ti six-fold

SO_2

SO_3^{2-}

$S\ 2p,2s$

Ti-five fold

● S-atoms
○ Ti atoms
bridging O-atoms
O-atoms

Although we can not rule out a significant contribution to the O^+ yield from SO_2, we infer from the NEXAFS data that the principal contribution at 153 K arises from the SO_3^{2-} reaction intermediate. That the O^+ ion yield decreases rapidly with anneal temperature suggests that the SO_4^{2-} product of the $TiO_2(110)$-SO_2 reaction has a low ESD cross-section. However, the low yield could also be due to an emission-angle effect. The lowering of the ion yield with anneal temperature for $E_p < 150$ eV most likely reflects occupation of the O-bridge sites by SO_2 (4). Indeed, a much greater effect may have been anticipated on the basis of the site coverage (4), again suggesting a secondary electron contribution to the O^+ yield.

This work has been performed under the auspices of the "Acciones Integradas" Hispano- Británicas. The Spanish part has been supported by the CICYT num. MAT89-0541.

REFERENCES

[1]. R.L. Kurtz, R. Stockbauer, T.E. Madey, E. Román and J.L. de Segovia, Surface Sci.218(1989)178.
[2]. C.A. Muryn, P.J. Hardman, J.J. Grouch, G. N Raiker and G. Thorton, Surface Sci. 251/252(1991)747.
[3]. P.J. Feibelman, M.L. Knotek, Phys. Rev. B18(1978)6531.
[4]. D. R. Warburton, D. Purdie, C.A. Muryn, K. Prabhakaran, Surface Sci. in press.
[5]. Y.W. Chung, W.J. Lo and G.A. Somorjai, Surface Sci. 64(1977)588.
[6]. E. Román, M Sánchez-Avedillo and J.L. de Segovia, Appl. Phys. A35(1984)35.
[7]. R.E. Walkup and R.L Kurtz, Diet III,ed. by R. H. Stulen and M. L. Knotek, Springer Series in Surface Science, Springer-Verlag,(1988), page 160.

Part X

Alkali Halides

Surface Deformation and Change in Reflectivity During Electron Irradiation: Results for the CaF$_2$(111) Surface

M. Reichling, R. Wiebel, and E. Matthias

Institut für Atom- und Festkörperphysik, Freie Universität Berlin,
Arnimallee 14, W-1000 Berlin 33, Germany

Abstract. Photothermal techniques have been applied to monitor surface alterations induced by an electron beam during ESD measurements. A probe HeNe laser beam was used to measure both the periodic surface deformation and the change in reflectivity of a CaF$_2$(111) surface induced by a chopped electron beam with a current density of $500\mu A/cm^2$ in the energy range of 1.5-2.7 keV. At low modulation frequencies (10Hz) the surface deformation was dominant which most likely signals the thermoelastic response of the surface, although lattice expansion due to accumulation of F-centers may contribute. At higher modulation frequencies (10kHz) the change in reflectivity prevailed which can be attributed to either free carrier effects or refractive index changes due to defect formation.

1. Introduction

In DIET the state of the surface is expected to play a major role for the emission of particles. Therefore, in situ monitoring of the surface condition during ESD seems highly desirable and can, be carried out employing thermoreflectance and thermoelastic displacement techniques. Following electron bombardment of dielectric single crystals, a *change in reflectivity* is expected by charging of the surface layer, absorption by localized defects, and by lattice destruction. *Surface deformation* is expected when heating becomes significant or due to color center accumulation near the surface. The ultimate goal is to correlate these quantities with the desorption yield of neutrals and ions.

The index of refraction of materials is perturbed by free carriers due to their intrinsic polarizability [1]. This effect can be utilized as a non-invasive probe for optical charge sensing. Forget and Fournier [2], for example, studied carrier transport in semiconductors with the modulated reflectance method using above-bandgap laser light for the excitation of the carriers. In these experiments two contributions to the modulated reflectance signal had to be considered: (1) the change in refractive index directly resulting from the modulated carrier density [3], and (2) a thermomodulation due to the heat waves generated by the periodically absorbed energy [4]. Both contributions can be separated by applying elaborate theoretical models [5,6]. Studies on thermal and carrier transport in semiconductors have been extended to low energy electron excitation using mirage detection [7] or surface displacement [8] as a probe.

The purpose of the present work is to demonstrate that the methods of modulated reflectance and surface displacement can also be used for charge and heat transport studies in irradiated insulators.

The lifetime of free electrons injected into the insulator lattice is very short and their kinetic energy is immediately transfered to electron/hole-pairs, a process often followed by production of various types of lattice defects [9,10]. The creation of these defects leads to an increased optical absorption of the crystal in specific spectral regions (coloration) and, therefore, to a change in the refractive index. Hence, models for the photothermal response of dielectric crystals have to be extended by including defect-induced changes of the refractive index. For semiconductors where the free carrier motion can be described by a Drude type theory it has been shown that the change in surface reflectivity is proportional to the density of the charge carriers and the surface temperature rise [11]. For insulators, the modulated surface reflectance must be calculated by applying Kramers-Kronig analysis to the well known absorption bands of the lattice

Springer Series in Surface Sciences, Vol. 31
Desorption Induced by Electron Transitions DIET V
Editors: A.R. Burns · E.B. Stechel · D.R. Jennison   © Springer-Verlag Berlin Heidelberg 1993

defects. It can also be shown that in this case the reflectivity is proportional to the defect density. The simultaneous application of the displacement technique, which is solely sensitive to thermal diffusion or lattice expansion due to defects, provides additional information about heat and charge transport phenomena in irradiated insulator crystals.

2. Experimental

Here we report about introductory measurements of the reflectivity and displacement of a $CaF_2(111)$ surface exposed to a chopped electron beam. The experiments were carried out using a standard photothermal displacement setup as described in the literature [12] and sketched in Fig.1. In our case the pump was typically a beam of about 1keV electrons focused onto the sample surface (focal diameter: $\approx 1mm^2$, beam current: $\approx 5\mu A$) and chopped by applying a square wave voltage to the deflection plates of the electron source. The sample, a CaF_2 single crystal with a cleaved and polished (111)-surface, was mounted in a UHV-chamber maintaining a pressure of less than 10^{-9}mbar during the experiments. Modulated surface reflectivity and displacement variations were monitored by a HeNe-probe laser beam reflected at the irradiated spot and detected by a quadrant detector. Detector signals were processed with a Lock-In amplifier synchronized to the electron beam modulation frequency. To obtain the reflectivity, signals from all quadrants of the detector were added, while for displacement measurements signals of two opposite quadrants were subtracted since the measured signal is proportional to the angular deflection (typically μrad and less). When the relative position between pump and probe beam is scanned, the deflection signal as a function of relative beam position is proportional to the spatial derivative of the surface deformation. Hence, the signal amplitude is zero in the center and shows maxima in the slopes of the deformation profile as illustrated in the upper left insert of Fig. 1. Two-dimensional (2-D) scanning of the electron beam reveals a map of the displacement gradient as shown in the left part of Fig. 2.

Such gradient profiles can be integrated to obtain the magnitude of the deformation in the center. Note that this technique only records periodic variations caused by the periodic electron irradiation but does not observe static phenomena like the slow average temperature rise of the sample that occurs during the experiment.

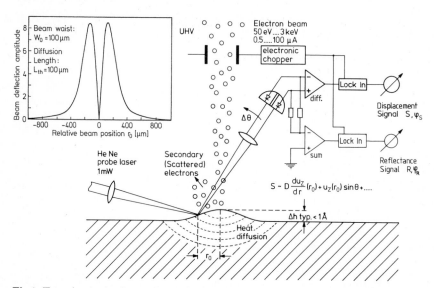

Fig.1: Experimental scheme for measuring modulated reflectance and surface displacement during electron irradiation.

3. Results and Discussion

Figs. 2(a) and (b) show 2-D scans at a modulation frequency of 1kHz illustrating the typical phenomena. Both signals, reflectivity and displacement, decrease in amplitude with increasing frequency; however, the frequency dependence is stronger for the displacement signal. Hence, in the high frequency region the picture is dominated by the circular reflectivity pattern while the displacement profile with its characteristic double-peak structure has nearly vanished. More detailed results are shown in Fig. 2(c) where the integrated amplitudes for reflectivity and displacement are plotted as a function of the chopper frequency.

We observe that the modulated reflectance signal is about one order of magnitude larger than the displacement signal. This observation contrasts with previously reported results on semiconductors [4] and metals [8] where the displacement signal was found to be three times larger than the reflectivity signal. Since the thermal expansion coefficient of CaF_2 is comparable to values for metals and larger than those for semiconductors, and in view of the fact that the dielectric crystal has a much smaller thermal conductivity (enhancing both the displacement signal and the thermally induced reflectivity change), the result in Fig. 2 cannot be explained by a thermal process. Hence, we conclude that there are strong nonthermal contributions to the reflectivity signal. A more quantitative analysis of results like the ones shown in Fig. 2(c) requires extended model calculations which are currently in progress. At the end these data will provide the changes in surface temperature and charge density during electron irradiation. It is planned to extend frequency dependent measurements to higher frequencies where the defect diffusion length reaches about the same order of magnitude as the thermal length (provided the sample temperature is set correctly). In this frequency regime the phase of the reflectivity signal should reveal information about carrier and defect diffusion.

The final goal of our experiments is to establish a connection between the desorption rates of particles and the photothermal response of the surface of dielectric crystals. A preliminary result along this line is shown in Fig. 3. The amount of neutral fluorine atoms emitted from the CaF_2-surface during electron bombardment was measured using a quadrupole mass filter. Simultaneously, the modulated reflectance was recorded. We observe that the modulated reflectance signal rises with increasing electron beam energy, while the desorption signal mainly remains constant with a slight tendency towards an increased yield for higher energy.

The reflectivity change in an electron energy dependent measurement clearly indicates that defect generation plays an important role since an increased electron energy does not result in more primary electrons per unit time but in an enhanced electron/hole-pair production. Obviously, the photothermal experiment is sufficiently sensitive to measure the defect creation at a level well below the onset of a measurable desorption yield for the fluorine atoms.

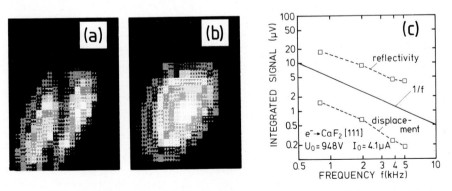

Fig. 2: 2-D displacement (a) and reflectivity (b) profiles for 1kHz. E= 948eV, I=4.1μA (c): Integrated modulated reflectance and displacement signals from CaF_2(111) during electron irradiation as a function of the electron chopper frequency.

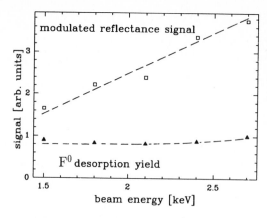

Fig. 3: Comparison of simultaneously recorded results for modulated reflectance and neutral fluorine desorption from $CaF_2(111)$ during electron bombardment as a function of the incident electron energy.

4. Acknowlegement

This work was supported by the Deutsche Forschungsgemeinschaft, Sfb 337.

5. References

1. A.F. Gibson, Proc. R. Soc. London **B66**, 588 (1953)
2. B.C. Forget, and D. Fournier, 7th International Topic Meeting on Photoacoustic and Photothermal Phenomena, Doorwerth 1991, proceedings in press
3. D.E. Aspnes and A. Frova, Solid State Comm. **7**, 155 (1969)
4. A. Rosencwaig, J. Opsal, W.L. Smith, and D.L. Willenborg, Appl. Phys. Lett. **46**(11), 1013 (1985)
5. D. Fournier, C. Boccara, A. Skumanich, and N.M. Amer, J. Appl. Phys. **59**(3), 787 (1986)
6. F.A. McDonald, D. Guidotti, and T.M. DelGiudice, in D.O. Thompson and D.E. Chimenti (eds.): Review of Progress in Quantitative Nondestructive Evaluation Vol. 6B, Plenum, New York 1987, p. 1361
7. J.C. Murphy, J.W. Maclachlan Spicer, R.B. Givens, L.C. Aamodt, and G. Chang, in J.C. Murphy et al. (eds.): Photoacoustic and Photothermal Phenomena II, Springer, Heidelberg 1990, p. 249
8. M. Reichling, A. de la Calle, and E. Matthias, in J.C. Murphy et al. (eds.): Photoacoustic and Photothermal Phenomena II, Springer, Heidelberg 1990, p. 241
9. R.T. Williams, M.N. Kabler, W. Hayes, and J.P. Scott, Phys. Rev. **B14**(2), 725 (1976)
10. K. Tanimura, T. Katoh, and N. Itoh, Phys. Rev. **B40**(2), 1282 (1989)
11. R.E. Wagner and A. Mandelis, J. Phys. Chem. Solids **52**(9), 1061 (1991)
12. M.A. Olmstead, N.M. Amer, S. Kohn, D. Fournier, C. Boccara, Appl. Phys. **A32**, 141 (1983)

The Role of Hot Hole Migration in Electronic Sputtering of Alkali Halides

Z. Postawa[1], J. Kolodziej[1], P. Czuba[1], P. Piatkowski[1], A. Poradzisz[1], M. Szymonski[1], and J. Fine[2]

[1]Institute of Physics, Jagellonian University,
ul. Reymonta 4, PL-30-059 Krakow, Poland
[2]Surface and Microanalysis Science Division, National Institute
of Standards and Technology, Gaithersburg, MD 20899, USA

Abstract. The electron stimulated desorption (ESD) of alkali and halogen atoms from single crystal alkali halide samples has been investigated to determine the dependence of the sputtering yield on primary electron energy. Kinetic energy distributions of emitted neutral particles have been measured for the (100)KBr surface; these data provide strong experimental evidence that long-range carrier migration is necessary to account for the observed primary energy dependency of the nonthermal halogen atom emission. Results are discussed using the recently proposed model of hot hole diffusion.

1. Introduction

The concept of hot holes has been recently proposed to explain various experimental findings on defect production and electronic sputtering/desorption (ESD) of alkali halides [1-5]. It appears that the holes created in the valence band of an insulator can have a rather wide distribution of initial kinetic energies, corresponding to the valence band density of states [2]. For example, in KBr the width of the valence band is about 2.6 eV [6] so that, on average, the hot holes could have an excess energy of 1.3 eV. This high initial energy acquired in primary excitation makes the hot holes very mobile: they can migrate over considerable distances in alkali halides thus providing an efficient transport mechanism for the excitation energy to go from the bulk to the surface [1,2].

In a recent series of experiments, we have measured the angular-resolved and mass-selected kinetic energy distributions for ESD emitted alkali and halogen atoms on stoichiometric, single crystal surfaces of alkali halides [5,7,8]. It has been found that, in addition to the emission of particles with thermal energies, directional ejection of nonthermal halogen atoms takes place predominantly along the ⟨100⟩ crystallographic direction of the (100) and (110)KBr, and (100)KCl crystals investigated. Such nonthermal emission was not observed, however, to be collimated along the ⟨110⟩ direction as was previously anticipated from the "Pooley-Hersh" model[9,10]. The time-of-flight distributions of the nonthermal atoms were found to be very narrow as would be expected for the decay of electronic states at the surface. These findings clearly indicated the need for a new theoretical approach to

Springer Series in Surface Sciences, Vol. 31
Desorption Induced by Electron Transitions DIET V
Editors: A.R. Burns · E.B. Stechel · D.R. Jennison   © Springer-Verlag Berlin Heidelberg 1993

explain this dynamic ejection process in alkali halides. In recent publications we have proposed such a model which invokes the non-radiative decay of a highly excited ("hot") hole localized at the surface of the crystal [5,8].

The purpose of the present work is not to discuss the details of the surface localization and decay of hot holes, but to perform an experimental test to determine if and to what extent long range diffusion of hot carriers is required for a realistic description of nonthermal emission in the ESD of alkali halides. We believe that the depth of the primary deposited energy can affect critically the balance between the nonthermal and the thermal halogen emission and provide the required experimental test.

2. Experimental

The measurements were made in an UHV chamber with a base pressure of 2×10^{-10} Torr. The details of the apparatus and the experimental procedure have been described previously [5]. Typically, a (100)KBr crystal was bombarded with a well controlled electron beam of about 30 $\mu A/cm^2$. The electron impact angle was 45° and only particles emitted along the normal to the surface were detected. Since the surface composition for KBr remained stoichiometric for temperatures above 90 °C [5], all experimental data reported here were taken above this temperature.

3. Results and discussion

A typical set of time-of-flight distribution data of Br atoms electronically desorbed from a (100) surface of KBr crystal is presented in Fig. 1. In the energy range from 0.2-2.3 keV the distributions show a double structure corresponding to the two energy components of the desorbing particles. The broad peak with a temperature dependent maximum around kT (where T denotes the macroscopic temperature of the sample) is due to thermally emitted atoms with Maxwellian energy distributions. The sharp peak, however, with a temperature independent position of the maximum at 0.4 ms (0.3 eV) and a cutoff below 0.2 ms (1.2 eV), corresponds to the nonthermal Br component.

It is quite clear that the shape of the TOF spectrum varies with the energy of the primary electrons. The nonthermal signal increases steeply at low electron energies, reaches a maximum around 900 eV, and then drops (Fig. 2). On the other hand, the thermal yield at first decreases with the electron energy up to about 2500 eV and then it starts to rise. It is relatively easy to explain qualitatively the shape of the nonthermal data. At low energies, the penetration range of the primary electrons is smaller than the "effective migration range" of the carriers responsible for the emission of nonthermal particles. At this point the nature of the carriers is not important. It can be, for instance, a carrier produced by an incident electron which can be localized on the surface and emit nonthermal halogen. In this energy range, the number of emitted halogens will increase with the energy of the incident electrons because more and more

300

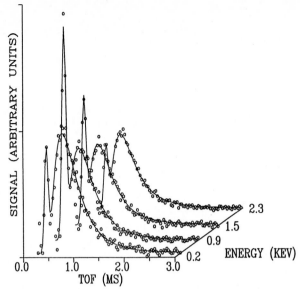

Fig. 1 A set of angular-resolved time-of-flight distributions for Br atoms leaving a (100) surface of KBr. The data were collected along the normal to the surface. Only particles emitted in the solid angle of 8×10^{-4} sr were recorded. The energy of primary electron beam was varied between 0.2 and 2.3 keV. These distributions were normalized to reproduce the angular dependence of the total intensity.

energy is deposited in the volume of the crystal contributing to the desorption. However, once the energy at which the range of the primary electrons exceeds the "effective range" of carrier migration L is reached, the fraction of primary electron energy deposited in the "active" subsurface region will decrease and the desorption yield should drop. To address this problem more quantitatively we incorporated the hole diffusion theory of Ref. 1 and 11 into our previous model [12]. The range of penetrating electrons in KBr was taken from the measurements done by I. Bronstein and A. Procenko [13]. We assumed that the hole diffusion transforms the spatial distribution of energy deposited in the crystal by primary beam H(x) into a new spatial distribution for self trapped holes P(x). Good agreement between the experimental results and the theory can be obtained if L=100 Å. This value is quite reasonable for a hot hole diffusion range. On the other hand, it is very unlikely that such a long focussed replacement collision sequence can develop in the crystal at 150 °C as required by the Pooley-Hersh mechanism. The value of L does not depend on the state of the surface, i.e. whether the surface is reflecting or absorbing. Only the total signal is sensitive to these boundary conditions. It is worth to mention that the "effective range" L is the only parameter of the model. Therefore, it seems that a measurement of this type is a simple

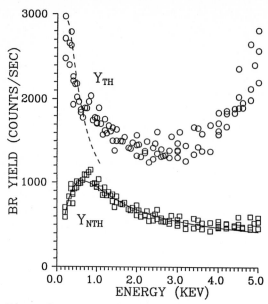

Fig. 2 Energy dependence of the thermal and nonthermal partial yields for Br atom desorption. The measured behavior can be explained if the diffusion range of hot holes is 100Å (solid line) and the diffusion range of H centers is taken to be 20Å (broken line). Ranges of primary electrons were taken from Ref. 13.

and straightforward method to estimate the hot hole diffusion length.

An analogous formalism can be used to describe the thermal desorption. The depth distribution of the primary defects (F and H pair centers) is rather difficult to predict because part of H(x) is already redistributed by hot holes. As a first approximation, one can assume that the resulting distribution is a linear combination of H(x) and P(x). If P(x) dominates one would expect the thermal Br yield to depend on energy in a similar way to the nonthermal signal; this is not observed in our measurements (Fig. 2). A satisfactory fit can only be obtained for the low energy part of the experimental data; P(x) has to be small compared to H(x) and the diffusion length of H centers is taken to be 20Å. At higher electron energies this description completely breaks down. The model predicts that the thermal yield should decrease monotonically with energy, whereas the experimental data show a strong increase of the emission of thermal halogen atoms above 2500 eV. The reason for such behavior is not known at this moment. It is possible that Auger like processes can efficiently contribute to the desorption in this energy range. These processes are disregarded in the present approach. In KBr, for instance, the LMM transition (1396 eV) is very strong [14] and can be a new channel of energy absorption and a source of additional, energetic electrons. As a result, the energy of the primary electrons will be deposited closer to the surface. Because the mean diffusion range of H centers is short, even a small change in the amount of energy deposited in the vicinity of the surface will be very important for the thermal desorption process. The same effect is much less important for the nonthermal Br atoms, since this component is predominantly controlled by the long range diffusion of hot holes.

In conclusion, we find that the yield dependence of emitted nonthermal Br atoms on primary electron energy can only be explained if migration range of carriers responsible for this emission is assumed to be long. This concept is consistent with a hot hole diffusion model previously proposed for nonthermal halogen atom ejection.

4. Acknowledgments

Research founded by the Polish-American Maria Sklodowska-Curie Fund II, grant no. MEN/NIST-89-6 and partially supported by the Polish Research Council, grant no. PB 2603/2.

5. References

1. V.N. Kadchenko and M. Elango, Phys. Status Solidi (a) **46**, 315 (1978).
2. M.A. Elango, V.N. Kadchenko, A.M. Saar, and A.P. Zhurakovski, J. Luminescence, **14**, 375 (1976).
3. R.T. Williams, Radiat. Eff. Defects Solids, **109**, 175 (1989).
4. T.A. Green, G.M. Loubrier, P.M. Richerds, L.T. Hudson, P.M. Savundararaj, R.G. Albridge, A.V. Barnes, and N.H. Tolk, in *Desorption Induced by Electronic Transitions*, edited by G. Betz and P. Varga, Springer-Verlag, Berlin, 1990), p 281.
5. M. Szymonski, J. Kolodziej, P. Czuba, P. Piatkowski, and A. Poradzisz, Phys. Rev. Lett., **67(14)**, 1906 (1991).
6. P. Kowalczyk, F.R. McFeely, L. Ley, R.A. Pollack, and D.A. Schirley, Phys. Rev. B **9**, 3573 (1974).
7. Z. Postawa and M. Szymonski, Phys. Rev. B **39**, 12950 (1989).
8. J. Kolodziej, P. Czuba, P. Piatkowski, Z. Postawa, M. Szymonski, and J. Fine, Nucl. Instrum, Methods B, in press.
9. P.D. Townsend, R. Browning, G.J. Garlant, J.C. Kelly, A. Mahjoobi, A.L. Michael, and M. Saidoh, Radiat. Eff., **30**, 55 (1976).
10. D. Pooley , Proc. Phys. Soc., **87**, 245 (1966).
11. C.T. Reimann, W.L. Brown, and R.E. Johnson, Phys. Rev. B **37**, 1455 (1988).
12. Z. Postawa, P. Czuba, A. Poradzisz, and M. Szymonski, Radiat. Eff. Defects Solids, **109**, 189 (1989).
13. I.M. Bronstein and A.H. Procenko, Radiotekh. Elektron., **4**, 805 (1970).
14. D.R. Frankl, *Handbook of Auger Electron Spectroscopy*, (Physical Electronics Industries, Inc., Eden Praire, p. 253).

New Mechanism for the Desorption of Excited Atoms by Photon Bombardment of Alkali Halide Crystals

D. Liu[1], R.G. Albridge[1], A.V. Barnes[1], P.H. Bunton[1], C.S. Ewig[1], N.H. Tolk[1], and M. Szymonski[2]

[1]Department of Physics and Astronomy, and Department of Chemistry, Vanderbilt University, Nashville, TN 37235, USA
[2]Institute of Physics, Jagellonian University, ul. Reymonta 4, PL-30-059 Krakow, Poland

We present a new model for desorption of excited alkali atoms from alkali halide crystal surfaces following photon bombardment. The mechanisms responsible for the production of excited alkali atoms by photon-stimulated desorption (PSD) and electron-stimulated desorption (ESD) have been the subjects of much controversy. Two models have been invoked most often to account for the production of the excited atoms. One model involves ion neutralization, and is analogous to the Knotek-Feibelman mechanism for ion desorption [1]. According to this mechanism, the interatomic Auger decay of a core hole results in the formation of two valence holes on a single surface anion resulting in ion desorption. Subsequent neutralization of the desorbing ion into an excited state then accounts for the observed excited atom yield. The second model assumes that the excited atoms are produced by secondary-electron excitation of desorbed ground state atoms in the gas phase [2]. Recent work [3] shows that valence excitations alone can lead to excited alkali atom desorption; thus core-hole creation, an essential component of the Knotek-Feibelman mechanism, is not always required for the production of excited alkali atoms. Previous work [3-5] also shows that the excited alkali atom yield of PSD decreases markedly with increasing temperature while both the total electron yield and ground state alkali atom yield increase, in contradiction to the second model involving secondary electron excitation. Thus, both models have been found to exhibit deficiencies when applied to PSD.

The new mechanism we propose is based on one or more exoergic surface reactions between alkali dimers (M_2) and halogens (X or X_2). The most direct possible reaction [6-9] is $M_2 + X \rightarrow M^* + MX$ [10]. This reaction was first suggested to explain the observation of sodium D-lines when sodium vapor was mixed with chlorine, bromine, or iodine gas [6]. In general, the reaction occurs with most of the possible combinations of alkali vapors and halogen gases. Subsequently this reaction was studied directly by molecular-beam experiments [7 and references therein] and its feasibility was demonstrated by semiempirical theoretical calculations [8]. The gas-phase reaction cross section for producing excited alkali atoms was found by theoretical calculations [9] and by crossed beam studies [7] to be very large, 10 – 100 $Å^2$. Figure 1, based on the calculations in Struve's paper [8], depicts a section of the energy surface of collinear Na_2Cl. The reaction proceeds through a transition state in which an electron from the sodium dimer shifts to the chlorine atom positioning the excited molecular complex, $(Na-Na^+Cl^-)^*$, at a point on the potential energy surface corresponding to neutral sodium in a repulsive state. The

Springer Series in Surface Sciences, Vol. 31
Desorption Induced by Electron Transitions DIET V
Editors: A.R. Burns · E.B. Stechel · D.R. Jennison   © Springer-Verlag Berlin Heidelberg 1993

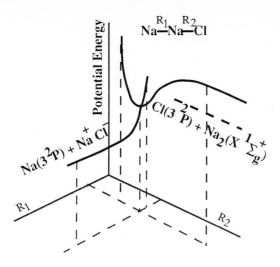

Na—Na—Cl

Figure 1: A schematic representation of a section of the semiempirical energy surface for the collinear $Na_2 + Cl \rightarrow Na^* + NaCl$ reaction based on figure 1 of Struve's paper (ref.8). The vertical axis represents energy, R_2 is the distance between a chlorine atom and the nearer sodium atom of the dimer, and R_1 the distance between the two sodium atoms.

system may then evolve through this position to form a free sodium atom in its first excited state and an ionically bonded NaCl molecule. The available reaction energy is no greater than 3.50 ± 0.04 eV for Na_2+Cl and 3.30 ± 0.13 eV for Li_2+F [11]. This limitation rules out the population of those higher excited states that decay in the visible, unless there is a contribution from reagent collision energy.

The proposed model depends upon the presence of alkali dimers and halide atoms on the surface. It is well accepted that in PSD or ESD from alkali halide crystals at elevated temperatures the majority desorbing species are ground state halogen atoms and alkali atoms [12]. The desorption of these ground state halogen and alkali atoms is thought to be mediated by self-trapped excitons created either on the surface or in the bulk [13 and references therein]. The incident photon or electron creates one or more holes in occupied valence bands in the near surface bulk. Energetic "hot" holes may migrate within a picosecond before coming to rest and forming self-trapped excitons. Self-trapped excitons in the bulk may then decay forming an F-center/H-center pair. Subsequent thermally assisted migration of the H-center can produce temperature-equilibrated halogen at the surface, some of which thermally desorb leaving excess metal. Nonthermal halogen emission may arise from the decay of surface self-trapped excitons again resulting in excess metal. Depending on the degree of metalization, which in turn depends on irradiation history and temperature, the accumulated neutralized alkali atoms may then form dimers, trimers, and larger clusters on the surface [14,15]. At some accumulated irradiation dosage, the number of dimers should reach a maximum. Thus during photon or electron bombardment of alkali halide crystals, defect mediated processes create at the surface the necessary reactants for the postulated reaction.

The PSD experiments reported here were performed at the Synchrotron Radiation Center of the University of Wisconsin with the Vanderbilt/SRC 6-meter toroidal-grating monochromator. Excited atoms were detected by monitoring their characteristic optical emission above the surface with a MacPherson 218

305

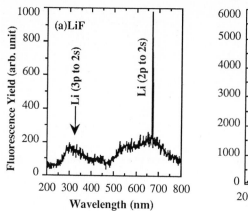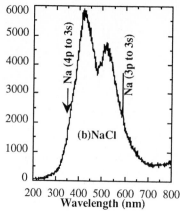

Figures 2a and 2b: Optical fluorescence spectra of LiF and NaCl crystals, respectively, due to photon bombardment.

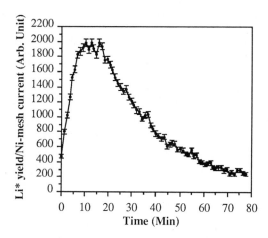

Figure 3: Effect of irradiation time on the yield of excited lithium atoms. The photon energy was 62.2 eV and the yield is normalized to the Ni mesh current.

monochromator and an EMI S-20 photomultiplier tube. After cleaving, the alkali halide single crystals were exposed to air for less than 30 minutes before the chamber was evacuated. The crystals were then annealed six to eight hours at 350° C.

Shown in figures 2a and 2b are optical spectra of LiF and NaCl crystals under photon bombardment. Sharp peaks superimposed on the continuous bulk luminescence correspond to atomic transitions from the 2p to the 2s state for lithium atoms and the 3p to the 3s state for sodium atoms, respectively. Transitions from higher excited states, such as from the 3p to the 2s state for lithium atoms (323.3 nm) and from the 4p to the 3s state for sodium (330.2 nm) are not seen in our spectra, but will be searched for in future experiments. The absence of the higher states is consistent with the prediction of our surface-reaction model: the reaction energy available is not sufficient to populate higher states provided the reactants on the surface have negligible collision energy.

Figure 3 shows the excited lithium atom yield plotted as a function of irradiation time during 62.2 eV photon bombardment of LiF. The dependence of the yield on the irradiation time, shown in Fig. 3, is consistent with the hypothesis that the yield reflects the instantaneous alkali dimer concentration which in turn depends largely on the degree of metalization of the surface. The nonzero yield at the beginning of the measurement is due to the presence of metal on the surface, even for the annealed samples [14]. Initially there are relatively few alkali dimers on the surface; thus even though the halogen yield is large at this stage, the excited atom yield is small. As the metalization increases, there is an optimum stage where many alkali dimers are present while the halogen atom concentration remains large. At this stage the excited alkali atom yield is at a maximum as indicated by the peak in figure 3. With further irradiation the yield decreases because of a decrease in the concentration of dimers due to the formation of larger clusters, and because the metal clusters block halogen atom desorption.

We postulate that our proposed mechanism is also a contributor to *electron*-stimulated desorption of excited alkali atoms. Other likely contributors include excitation of desorbed ground-state neutral alkalis by primary beam electrons and also to some extent by secondary electrons [2] In support of this postulate, for the case of LiF we have previously measured the temperature-dependent yield of excited lithium atoms during electron bombardment [4]. The yield was found to be appreciable at room temperature (where desorption of thermal ground-state atoms is low) but increased by two orders of magnitude with temperature up to 400° C (where radiation induced desorption of ground state lithium atoms is increased by *four* orders of magnitude)[4]. In contrast, the yield from photon-stimulated desorption is also appreciable at room temperature but *decreases* markedly with increasing temperature [3]. This indicates that (a) for PSD, there is little contribution from gas-phase excitation by secondary electrons, and (b) for ESD at low temperatures, where gas-phase excitation does not contribute because the desorption yield of ground-state alkalis is low, our proposed mechanism is the principal contributor.

In conclusion, we suggest a new model for the production of excited alkali atoms by PSD. This model is based on experimental measurements of photon-stimulated desorption and the previously known gas-phase chemiluminescent reaction between alkali dimers and halogen atoms. Radiation-induced metalization and halogen atom formation provide the two necessary reactants for the surface reaction.

The authors wish to thank the staff at the University of Wisconsin's Synchrotron Radiation Center for their excellent support and M. Riehl-Chudoba, T. A. Green, N. Itoh and J. C. Tully for helpful discussions. This work was supported in part by the Air Force Office of Scientific Research under contracts F49620-86-C-0125 and F49620-88-C-0080, grant AFOSR-90-0030, and by the Office of Naval Research under contract N00014-87-C-0146.

References:

1. M. L. Knotek and P. J. Feibelman, Phys. Rev. Lett. 40, 965 (1978); Phys. Rev. B 18, 6531 (1978).
2. R. E. Walkup, Ph. Avouris, and A. P. Ghosh, Phys. Rev. Lett. 57, 2227 (1986).
3. P. H. Bunton, R. F. Haglund, Jr., D. Liu, and N. H. Tolk, Phys. Rev. B 45, 4566, (1992).

4. E. Taglauer, N. Tolk, R. Riedel, E. Colavita, G. Margaritondo, N. Grershenfeld, N. Stoffel, J. A. Kelber,   G. Loubriel, A. S. Bommanavar, M. Bakshi, and Z. Huric, Surf. Sci. 169, 267 (1986).

5. P. Wurz, E. Wolfrum, W. Husinsky, G. Betz, L. Hudson, and N. Tolk, Rad. Effects 109, 203 (1989).

6. Alkali Halide Vapors, Structure, Spectra, and Reaction Dynamics, edited by P. Davidovits and D. L. McFadden, Acadenic Press, Inc. 1979, Chapter 11, 361.

7. P. Arrowsmith, S. H. P. Bly, P. E. Charters, P. Chevrier, and J. C. Polanyi, J. Phys. Chem. 93, 4716 (1989).

8. W. S. Struve, J. Mol. Phys. 25, 777 (1973).

9. J. R. Krenos and J. C. Tully, J. Chem. Phys. 62, 420 (1975)

10. Other reactions include the vibrational-to-electronic energy transfer reaction $M_2 + X \rightarrow MX^\dagger + M^O \rightarrow$ $MX + M^*$, where $MX^\dagger$ stands for a vibrationally excited molecule; and the reaction $M_2^+ X_2 \rightarrow M^* +$ $MX + X$. We consider both reactions to be possible contributors to our proposed surface mechanism but less likely based on gas phase studies (see references 12-17).

11. M. W. Chase, Jr., C. A. Davies, J. R. Downey, Jr., D. J. Frurip, R. A. McDonald, and A. N. Syverud, J. Phys. Chem. Ref. Data, Suppl. No.1, Vol. 14, 1985.

12. T. A. Green, G. M. Loubriel, P. M. Richards, L. T. Hudson, P. M. Savundararaj, R. G. Albridge, A. V. Barnes, and N. H. Tolk, Desorption Induced by Electronic Transitions, DIET IV, edited by G. Betz and P. Varga, Springer Series in Surface Sciences, Vol. 19, Springer-Verlag Berlin, Heidelberg 1990, 281.

13. M. Szymonski, J. Kolodziej, P. Czuba, P. Piatkowski, A. Poradziaz, and N. H. Tolk, Nucl. Instr. Meth. B58, 485 (1991).

14. P. Wurz and C. H. Becker, Surf. Sci. 224, 559 (1989).

15. Qun Dou and D. W. Lynch, Surf. Sci. 219, L623 (1989).

Simultaneous Measurements of Optical Absorption and Electron Stimulated Li Desorption on LiF Crystals

N. Seifert[1], H. Ye[1,], D. Liu[1], R. Albridge[1], A. Barnes[1], N. Tolk[1], W. Husinsky[2], and G. Betz[2]*

[1]Center for Molecular and Atomic Studies at Surfaces, Department of Physics and Astronomy, Vanderbilt University, Nashville, TN 37235, USA

[2]Institut für Allgemeine Physik, Technische Universität Wien, Wiedner Hauptstraße 8–10, A-1040 Wien, Austria

*Permanent address: Department of Nuclear Science, Fudan University, Shanghai, 200433, P.R. China

It has been known for some time that ionizing radiation incident on alkali halide crystals lead to the formation of F- and H-centers [1]. When an F-center reaches the surface of the crystal, it results in the neutralization of an alkali ion. If the temperature of the crystal is high enough the neutral alkali atoms desorb thermally from the surface [2]. It has been observed in many experiments that large doses of energetic neutrons, ions, or X-rays lead to the formation of colloids (alkali agglomerations) in the bulk of the crystals [3,4]. However, agglomeration processes at or close to the surface of the crystals have only been investigated recently [5,6]. Electron energy loss spectroscopy, and Auger electron spectroscopy investigations have supplied evidence for the formation of alkali islands on the surface of alkali halides during electron bombardment [7]. In previous publications [5,6] the desorption kinetics of lithium atom delayed emission (i.e. emission following the cessation of electron bombardment) has been investigated. The results show the occurrence of a prompt and a delayed decay. The delayed decay takes seconds, whereas the prompt decay is faster than a few msec. Under certain circumstances the Li desorption rate even increases after the cessation of electron bombardment ("delayed maximum"). In all these publications conclusions have been drawn about the processes occuring during and after electron bombardment by monitoring the ground state yield of lithium desorbing from the surface. Clearly, this method alone is not capable of differentiating in which manner F-center clusters, lithium islands on the surface, or lithium clusters individually contribute to the observed desorption phenomena. Simultaneous correlated transmission optical absorption spectroscopy (which provides information about defect densities in the crystal) combined with measurements of the ESD of alkali atoms from alkali halides provides a new approach to this problem.

The goal of our investigations as reported in this paper is to correlate the ESD of neutral lithium atoms with microscopic processes happening immediately under or directly on the surface of the crystals using in situ transmission optical absorption spectroscopy. The position of the F-, F_2- (or M-), F_3- (or R) and F_4- (or N-) center bands and colloidal bands are well known in LiF crystals [3,8].

In our experiment, LiF <100> surfaces were bombarded with electrons of 400 - 600 eV over a range of target temperatures

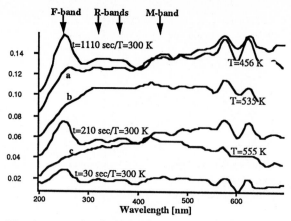

Fig.1: Optical density for different bombarding times and temperatures is plotted as a function of wavelength. One can clearly see the occurrence of the F-band at 250 nm and a very broad band which extends over the investigated wavelength range. Heating the sample to 456 K (a,b,c) and higher results in a decrease of the F-band intensity.

(27° - 410 °C). Typical current densities used were about 150 $\mu A/cm^2$. Ground-state lithium atoms were detected with a quadrupole mass analyzer following electron beam post ionization. For our transmission optical absorption experiments, we used a deuterium lamp as a light source. The UHV system maintained a base pressure of less than 1×10^{-9} Torr. The crystals were cleaved in air and cleaned in the vacuum system by prolonged heating at 400 °C.

At room temperature, no lithium desorption was observed during electron bombardment indicating that although lithium metal may be forming at the surface due to bombardment, the vapor pressure of lithium at room temperature is too low for lithium thermal desorption. After bombardment with electrons we recorded the time dependent transmission optical absorption spectrum (figure 1). One clearly sees the development of the F-band at approximately 250 nm after 30 sec of bombardment. After 1110 sec of bombardment we heated the sample and stopped the electron bombardment (figure 1: a, b, c). The F-band intensity clearly decreases for increasing temperature.

A broad absorption band was observed throughout the whole investigated temperature regime. The only observed differences are that the half width of the band clearly decreased, and that the F-band was not observed for temperatures higher than about 300 °C. For high current densities the broad band has a maximum around 500 nm at temperatures as high as 410 °C.

Figure 2 shows simultaneous measurements of the optical density at a fixed wavelength and the desorption yield of lithium atoms . Note that the change in optical density is very similar independent of the wavelength in a range from 250 to 700 nm. The lithium yield saturates after a few seconds, whereas the optical density increases steadily. After the electron beam has been turned off, the lithium signal features the characteristic prompt and slow decays, whereas the optical density decreases continuously. At higher temperatures similar phenomena are observed.

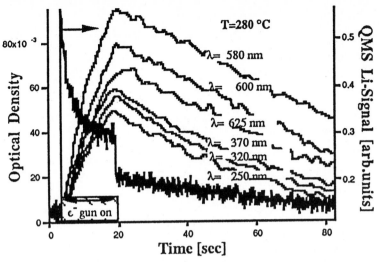

Fig.2: Smoothed optical data for different wavelengths and the correlated Li-Yield are plotted as a function of irradiation time

The occurrence of the very broad absorption band (several hundred nm in width) (fig.1) cannot be related to electron centers such as F-centers or small F-centers agglomerates, since the half-widths of the absorption bands of these centers are usually an order of magnitude smaller than our observed broad band half-widths [8]. Broad band absorptions of the kind that we have observed arise from the presence of metal phase agglomerates in the bulk or on the surface of optical materials [4,9]. This suggests that the broad absorption band is caused by the formation of lithium particles in the crystal, or lithium islands on the surface of the LiF crystals. Radchenko [10] calculated the extinction coefficient of lithium colloids *in* LiF based on classical Mie theory. His theoretical results would agree with our experimental data only for 50 nm or larger lithium colloids. This is not very realistic, since the penetration depth of 400 eV electrons is of the order of a few nm and that diffusion does not play a significant role at room temperature [11].

Rasigni et al. [9] performed very detailed studies of the transmission of lithium droplets of various sizes deposited *on the surface* of a quartz substrate. Their results show marked similarity to our data. That is, half-widths and positions of their transmission minima (transmission optical absorption maxima) measured for lithium droplets of about 20 nm radii are comparable to our values.

Our correlation experiments contrasting transmission optical absorption data and lithium desorption yield also fit very well into this physical picture which assumes that surface concentrations of lithium metal play a major role. In figure 2 one sees that the delayed emission of lithium is correlated to the change in optical density at wavelengths in the region of the broad band. Because single lithium metal atoms are only weakly bound to the surface, their vapor pressure is much higher than lithium atoms bound to lithium atoms in droplets.

Diffusion of F-centers cannot be the rate limiting factor since at temperatures as high as 600 K diffusion is very fast [2]. This suggests that the disintegration of surface lithium islands is the rate limiting step which accounts for delayed emission. The rate limiting step for the prompt decay is some combination of the diffusion of F-centers to the surface, and the time of evaporation for the lithium atom. The fact that no prompt change can be observed in the optical density after the electron gun has been turned off indicates that very small lithium clusters are not detected by our optical absorption detection system. The slow disintegration of the lithium islands due to evaporation of lithium atoms is indicated by the decrease of the optical density after the electron beam has been turned off.

Also of interest is the observation that the optical density continues to increase during bombardment whereas the lithium desorption rate approaches saturation (figure 2). This can easily be explained in the light of the lithium droplet model considering that larger droplets have a lower vapor pressure, which results in a decrease of the desorption rate although the radii of the clusters and consequently the measured optical density are still increasing.

Our measurements show that the delayed maximum in the desorption rate of lithium atoms does not correlate with the changes in optical density at any wavelength. This suggests that metallic clusters (broad absorption band observed at 410 °C) of Li do not contribute directly to the occurrence of the delayed maximum. This agrees with recent investigations and conclusions drawn by Seifert et al. [5]. But the data also suggest that changes in the F-center, M-center etc. concentration during and after bombardment at high temperatures are probably below our detection limit since it is well known that these centers are formed and decay (with different rate constants) in the investigated temperature regime.

The authors gratefully acknowledge financial support in part by the Air Force Office of Scientific Research under contracts F49620-86-C-0125 and F49620-88-C-0080, grant AFOSR-90-0030, and by the Office of Naval Research under contract N00014-87-C-0146 and grant N 00014-91-J-4040.

References

[1] R.T. Williams, Semicond. Insul. **5**, 457 (1983)
[2] G.M. Loubriel, T.A. Green, and P.M. Richards, R.G. Albridge, D.W. Cherry, R.K.Cole, R.F. Haglund Jr., L.T. Hudson, M.H. Mendenhall, D.M. Newns, P.M. Savundaraj, K.F. Snowdon, and N.H. Tolk, Phys. Rev. Lett. **57**, 1781 (1986).
[3] Many relevant references can be found in: A.E. Hughes, and S.C. Jain in *Metal colloids in ionic crystals*, Advances in Physics, 717 (1979)
[4] N.G. Politov and L.F. Vorozheikina, Soviet Phys. - Solid State **12**, 2, 277 (1970)
[5] N. Seifert, D. Liu, R.G. Albridge, A.V. Barnes, N. Tolk, W. Husinsky, G. Betz, accepted for publication in Phys. Rev. B
[6] G. Betz, J.Sarnthein, P.Wurz, and W.Husinsky, Nucl.Instr. Methods **B 48**, 593 (1990)
[7] Qun Dou, D.W. Lynch, Surface Science **219** (1989) L623 - L627
[8] Y. Farge and M.P. Fontana, Defects in Crystalline Solids, Vol **11**, North Holland (1979)
[9] M.Rasigni, and G. Rasigni, J.Opt.Soc. Am. **67**, 4, 510 (1977)
[10] I.S.Radchenko, Soviet Physics-Solid State **11**, 7, 1476 (1970)
[11] P. Wurz and C.H. Becker, Surf.Sci. **224**, 559 (1989)

Rotational Distributions Following DIET: CN* from Alkali-Rich Surfaces

J. Xu[1], L. Hulett, Jr.[1], A. Barnes[2], R. Albridge[2], and N. Tolk[2]

[1]Oak Ridge National Laboratory, P.O. Box 2008, MS 6142,
Oak Ridge, TN 37831-6142, USA
[2]Department of Physics and Astronomy, Vanderbilt University,
Nashville, TN 37235, USA

Abstract: We studied rotational distributions of excited CN ejected due to electron- and photon-stimulated desorption (ESD and PSD) from alkali-metal and alkali-halide surfaces. Measured rotational spectra exhibit temperature-independent non-Boltzmann features that are systematically correlated to the particular alkali-metal component present on the surface. The temperature-independent, substrate-dependent rotational distributions indicate that the molecular rotation following ESD and PSD is determined by direct electronic bond breaking

1. Introduction

Measurements of internal energy distributions of molecules constitute a powerful tool in characterizing the dynamics of molecular interactions with surfaces. In thermal desorption experiments, rotational distributions are observed to be Boltzmann-like and to depend on temperature.[1] In surface-molecule scattering experiments, rotational distributions of scattered molecules are found to be distinctly non-Boltzmann, to vary only weakly with surface temperature and to depend strongly on the incident energy and angle.[2] In a recent ESD experiment, the rotational distributions of the ground-state NO molecule desorbed from a metal substrate were observed to be non-Boltzmann.[3] In addition, a number of laser-induced desorption experiments have been carried out also on the NO-metal system which also showed non-Boltzmann rotational distributions.[4-6] A theoretical model by Burns, Stechel and Jennison,[7] based on the transition from hindered to free rotors originally discussed by Gadzuk and coworkers,[8] has been proposed to describe qualitatively the ESD NO results. Another model has been proposed by Hasselbrink[9] to explain rotational distribution arising from laser-induced desorption in terms of excited potential surfaces. In these ESD and PSD experiments, there were no measurements of the dependence of the rotational distributions on substrate, temperature, and incident energy. The data presented here suggest that these dependences are in fact important in elucidating the underlying mechanisms by which molecules acquire rotation following desorption induced by electronic transitions (DIET).

2. Experimental

The detailed description of the experimental setup appeared in ref 10. The work was carried out in an UHV system with a base pressure of 1.0×10^{-10} Torr. Single photons emitted from desorbed excited CN which decay in free space in front of the samples were imaged onto the entrance slit of a McPherson 0.3-m monochromater, and detected by a cooled photomultiplier. Electron beams (60 to 900 eV, 30 to 500 μA, and 2 mm in diameter) were produced by a gun. The photon measurements (8 to 190 eV) were carried out at the University of Wisconsin Synchrotron Radiation Center (SRC) using the Vanderbilt/SRC joint beamline. The samples were in thermal contact with a closed-cycle helium cooler which can vary the sample temperature from 60 K to 800 K. The alkali-halide crystals were cleaved in air, mounted with their (100) surfaces facing the beam and heated under UHV conditions for cleaning. The alkali-metal samples were prepared by evaporating thick alkali-metal layers onto glass slides.

3. Results

Bombardment of CN-adsorbed alkali-metal and alkali-halide surfaces with a low-energy electron beam and synchrotron photons produces optical emission in the 360-420 nm region,

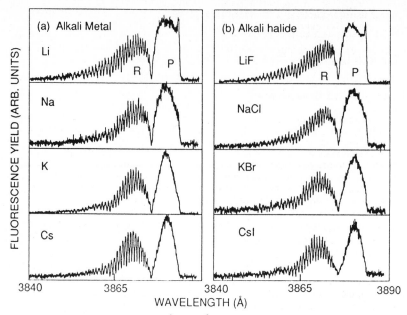

Figure 1. P branches and R branches of $B^2\Sigma^+ \rightarrow X^2\Sigma^+$ transition of CN* desorbed, respectively, (a) from alkali-halide surfaces under 400-eV electron bombardment at 60 K.

attributed to the $B^2\Sigma^+ \rightarrow X^2\Sigma^+$ electronic transition in the desorbed CN radical.[11] In our previous experiments,[10] we found that the desorbing CN molecules are generated through the following sequence: (1) pre-irradiation of alkali-halide crystal produces alkali metal-rich surfaces, (2) surface CN molecules are formed when the alkali-rich surface is exposed to CO_2 and N_2, and (3) electron or photon bombardment induces the desorption of excited CN from the surface. This procedure to enhance the formation of surface CN adsorbates is similar to that reported by Watanaba and co-workers.[12] On alkali metal surfaces, similar results were obtained except that pre-irradiation by electrons was not required.

Figures 1(a) and 1(b) show the R and P branches of the B→X transition of CN* desorbed from alkali-metal and alkali-halide surfaces under 400-eV electron bombardment at a temperature of 60 K. The array of peak intensities represents the population distribution of the CN $B^2\Sigma^+$(v=0) rotational states. There are two prominent features of the overall distributions which should be noted: the width (at half maximum) of each of the envelopes, and the position of the maximum of the envelope in each of the distributions. The widths as defined above change with the alkali component of the substrate; the data obtained for the low-atomic-number alkali metal case such as lithium show a distinctly wider population distribution than do the data for the high-atomic-number metals such as cesium. The position of the distribution maxima are observed also to shift with the alkali-metal component of the substrate: the peak position associated with the low atomic number alkali metal surface components are located at lower J values than those from the high atomic number metals. The population of the band head in the P branch mirrors the populations of the high-J states (J > 21) shown explicitly in the R branch. This is seen from the P-branch associated with lithium substrates which shows a more intense head than that from the high atomic-number metals due to the enhanced population of higher J-states in the lithium case. We have analyzed the spectra by a least squares fit procedure to determine the v=0 rotational populations and a negligibly small overlapping 1-1 contribution. The fitted results confirm that: (1) the rotational distributions are non-thermal since the Boltzmann plots are not straight lines, (2) rotational distributions arising from alkali and alkali halide surfaces systematically depend on the type of alkali-metal.

314

Measurements of rotational distributions were carried out over a wide range of surface temperatures and with energy-resolved electron and photon beams. The distributions demonstrate that rotational distributions of desorbed CN* in ESD are found to be independent of surface temperature in the range 60-300 K. The distributions are also found to be independent of the beam energy.

4. Discussions and Conclusions

Some controversial theoretical studies have been attempted to understand rotation following electronic interactions. Some explain the rotational distribution by ground-state potential surfaces,[7,8] while others explain the distribution by the repulsive-state potential surface.[9] We suggest a quantum description of desorption and rotation in which angular momentum can arise both from the initial wavefunction in the ground state and from the time evolution of the wave functions in the repulsive state as the molecule is ejected from the surface. In figure 2 we illustrate schematically the constraints imposed by a solid surface upon the rotation of a molecular adsorbate and relaxation of this constraint as the adsorbate leaves the surface.

At zero temperature, the potential surface in the ground state completely determines the initial wavefunction. The potential energy, $V(Z, \theta)$, is a function of the distance between the surface and the center of mass of a CN molecule (Z) and the angle that the CN bond makes with the surface normal (θ). (We assume CN standing on the surface with a cylindrical symmetry.) Initially in the ground state, the Z-dependent potential restricts the CN molecule moving away from surface and the θ-dependent potential (given here as a square potential) hinders the volume in which the molecule rotates. For a strongly hindering potential in the ground state, the angular part of the initial wavefunction, $\phi(\theta, \varphi)$, is highly localized.

Electron or photon bombardment of the surface leads to an excitation from the bonding state to a repulsive state. The potential energy surface in the repulsive state determines the time evolution of the initial wavefunction. Recently, Hasselbrink's works explain the final rotational distribution in terms of the potential energy surface in the repulsive state. In these classical models, the anisotropy of the potential results in a net torque exerted on the molecule. Such a torque will influence the molecular rotation. We believe that both the ground state potential surface and the repulsive-state potential surface affect the molecular rotation of desorbed molecules.

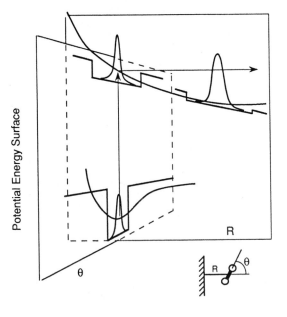

Figure 2. Scheme of the ground-state and excited-state potential surfaces of a CN-alkali system. The square wells illustrate the potentials varying with angle, and the curves illustrate the potentials varying with the reaction coordinate.

315

The above discussions clerify that the final rotational state populations of desorbing CN* are determined by the substrates since the interaction between CN and the substrate determines both the ground-state and excited-state potential surfaces. This prediction is consistent with the quantitative experimental observation in this work that the rotational distribution of desorbing CN* is systematically correlated to the alkali component of the substrate.

In conclusion, we have measured the rotational distributions of desorbing CN* for LiF, NaF, NaCl, KCl, KBr, CsI, lithium, sodium, potassium, and cesium metal surfaces, in the surface temperature range of 60 K to 300 K, at 100 eV and 400 eV electron energies. The results show that rotational distributions of excited CN arising from ESD and PSD exhibit a non-thermal feature that are systematically correlated to the alkali component of the substrate, independent of surface temperature, incident energy, and halide component of the substrate. The temperature-independent, substrate-dependent rotational spectra indicate that the molecular rotation following ESD and PSD is determined by direct electronic bond breaking.

Acknowledgments

Authors thank P. Nordlander for helpful discussions. This research was sponsored in part by the Air Force Office of Scientific Research (AFOSR) under Contract AFOSR-90-0030, and by NASA under Contract NAS8-37744.

References

1. D. A. Mantell, R. R. Cavanagh, and D. S. King, J. Chem. Phys. **84**, 1531 (1986).
2. A. W. Kleyn, A. C. Luntz, and D. J. Auerbach, Phys. Rev. Lett. **47**, 1169 (1980).
3. A. R. Burns, Phys. Rev. Lett. **55**, 525 (1985).
4. D. Weide, P. Audresen, and H.-J. Freund, Chem. Phys. Lett. **136**, 106 (1987).
5. S. Buntin, L. Richter, R. Cavanagh, D. King, Phys. Rev. Lett. **61**, 1321 (1988).
6. F. Budde, A. V. Hamza, P. M. Ferm, G. Ertl, D. Weide, P. Andresen, and H.-J. Freund, Phys. Rev. Lett. **60**, 1518 (1988).
7. A. R. Burns, E. B. Stechel, and D. R. Jennison, Phys. Rev. Lett. **58**, 250 (1987).
8. J. W. Gadzuk, Uzi Landman, E. J. Kuster, C. L. Cleveland, and R. N. Barnett, Phys. Rev. Lett. **49**, 426 (1982); J. Electron. Spectrosc. Relat. Phenom. **30**, 103 (1983).
9. Eckart Hasselbrink, Chem. Phys. Lett. **170**, 329 (1990).
10. Jun Xu, R. Albridge, A. Barnes, M. Riehl-Chudoba, Akira Ueda, N. Tolk, D. Russell, and P. Wang, Surface Sci. **262**, 77 (1992).
11. Jun Xu, Marcus Mendenhall, and Joel Tellinghuisen, J. Chem. Phys. **93**, 5281 (1990).
12. H. Nakagawa, T. Deguchi, H. Matsumoto, T. Miyanaga, M. Fujita, K. Fukui, E.Ishiguro, I. H. Munro, T. Kato, and M. Watanabe, J. Phys. Soc. Japan **58**, 2605 (1989).

Auger Studies of Damage and Charging in NaCl

D.P. Russell, W. Durrer, J. Ramirez, and P.W. Wang

Department of Physics, The University of Texas at El Paso,
El Paso, TX 79968, USA

Abstract. In this study we observe that the positions of Auger signals shift from slightly below (~10 eV) the tabulated values to well above (>150 eV). The observed shifts can be understood by a simple model of surface charging based on the change in the secondary electron emission of the damaged region.

1. Introduction

We use Auger spectroscopy to study radiation damage and charging due to electron radiation. Several studies [1-4] have shown that the primary effect of electron radiation is the metallization of the NaCl crystal. In this study we observe that the positions of Auger signals shift from slightly below (~10 eV) the tabulated values to well above (>150 eV). These energy shifts are interpreted as an overall charging of the surface. The primary quantitative experimental result is that the transition from slight positive charging to strong negative charging occurs at a definite electron fluence of 5.4 coulombs/cm^2. This electron fluence is comparable to those occurring in ESD studies on room temperature alkali halides. The observed charging can be understood by a simple model based on the change in the secondary electron emission of the damaged region.

2. Experimental System

These studies were performed in a PHI-560 system with several surface-sensitive spectrocospies including, AES, XPS and SIMS. Only AES was used for these experiments. The experimental set-up consisted of a NaCl sample mounted on a carousel(or hot/cold probe) with the (100) face exposed. The electron beam was incident on the sample at 45° from normal. The Auger electrons were detected with a double pass CMA. The same electron beam was simultaneously used for the Auger measurement and to produce the damage in the crystal.

Throughout the study the beam energy was 3.0 keV. The beam current was typically between 0.1 μA and 0.7 μA. The electron gun was run both in spot mode and raster mode. The spot mode allowed for higher current densities, but gave poorer estimates of the spot size for use in computing current densities. The raster mode allowed for more accurate measurement of the irradiated area. The irradiated area was measured by viewing an image of secondary electron emission from the sample and moving the sample stage with a micrometer.

Springer Series in Surface Sciences, Vol. 31
Desorption Induced by Electron Transitions DIET V
Editors: A.R. Burns · E.B. Stechel · D.R. Jennison   © Springer-Verlag Berlin Heidelberg 1993

3. Experimental Results

The Auger spectra obtained for a negligibly damaged surface is shown in Fig. 1(a). An example of the Auger spectra after the strong charging is shown in Fig. 1(b). The two primary features in the second spectra are the peak at ~150eV and the broad feature coming to a head at ~330 eV. We associate the peak with the secondary electrons emitted from the sample and the broad region with the inelastic Auger

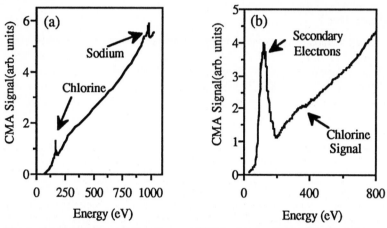

Fig. 1. (a) Auger survey scan for a negligibly damaged NaCl(100) surface.
(b) Auger survey after strong negative charging has occurred.

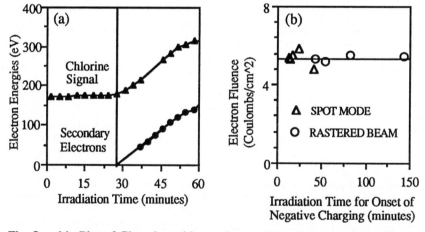

Fig. 2. (a) Plot of Cl peak position and secondary electron peak position as a function of irradiation time. Note the onset of negative charging. (b) Plot of the electron fluence(current density * time) at onset of negative charging for current densities from 600 μA to 8000 μA/cm². Due to uncertainty in determining the irradiation area in spot mode, the spot mode data has been normalized to the raster beam results.

signal from the chlorine(the elastic peak cannot be discerned). The unusual energies for these features are taken to be a signature of strong negative charging. It is difficult to get an impression of the time evolution of the data from a few static pictures. Figure 2a shows the time dependence of the chlorine and secondary electron signals throughout a typical run. Note the onset of the negative charging.

Once we established the features and our interpretation of the phenomena, we wanted to identify an experimental parameter that would predict the onset of the strong negative charging. We did this by repeating the experiment at different currents and both in spot mode and raster mode in order to change the irradiated area. The results of this study clearly showed that for lower current densities longer irradiation times were required to produce strong negative charging. The onset of the strong negative charging is seen to be associated with the total electron fluence. Figure 2b shows a plot of the electron fluence at the onset of negative charging versus irradiation time. Scatter in the raster data is much smaller than in the spot mode data due to the reduction in the uncertainty in the irradiated area and minimization of edge effects. This establishes a connection between the electron fluence and strong negative charging.

4. Analysis and Discussion

In this section we present an explanation of the qualitative behavior observed in this study. This model is based on the established results of the high secondary electron emission efficiency of alkali halides and the metallization of the alkali halide surface under electron radiation. The primary feature we wish to explain is the potential at the surface of the irradiated volume. Our experiment indicates that the potential is initially slightly positive (<10volts) with respect to ground but as the irradiation continues the potential goes through zero and becomes highly negative(~150 volts).

These observations can be understood by a novel application of an analysis used by von Seggern [5] to explain electron beam stability on Teflon. In order to do this we consider two situations. Neglecting ground current, we consider an electron beam incident on an insulating surface which has a secondary electron emission coefficient greater than one. In this case irradiating the sample depletes the electron concentration on the surface. This charges the surface positively until the attractive field is sufficiently strong to capture enough secondary electrons to stop the depletion. This leaves the system with a stable beam and a surface that has a slightly positive (~10 volts) potential. Next, we consider the case in which the beam is incident on an insulating surface with a secondary electron emission efficiency less than one. In this case, the emission of secondary electrons is insufficient to compensate for the influx of primary electrons and the surface charges to a negative potential sufficient to deflect the beam. In the case of Teflon, von Seggern went from the stable beam to the unstable beam by varying the beam energy from an energy for which the secondary electron emission coefficient is greater than one to less than one. In our situation the irradiation damage appears to do this automatically. For a 3 keV beam the secondary electron emission efficiency of NaCl is determined to be approximately 5 to 8 electrons emitted for each incident electron, while for a pure sodium metal it is less than 0.82. Thus as the irradiated volume becomes predominantly metal the secondary electron emission efficiency decreases to below one. Thus we propose the initial positive charging of the surface occurs to a point at which sufficient secondary electrons are recaptured to prohibit further charge build up. Then, at the onset of strong negative charging the secondary

319

electron emission efficiency drops below unity and capturing secondary electrons can no longer maintain charge stability.

5. Conclusions

From this work we conclude that for sodium chloride irradiated with a 3keV electron beam at room temperature strong negative charging will occur when the incident electron fluence reaches 5.4 coulombs/cm^2. This phenomenon appears to be associated with the change in secondary electron emission efficiency as the surface region becomes metallized.

Acknowledgements

One of us J. R. would like to thank the Materials Research Center of Excellence (M.R.C.E.) for support during this project. We also thank Brian Davies for all his assistance in getting this work started.

References

1. L. S. Cota Araiza and B. D. Powell, Surf. Sci. **51** 504 (1975)
2. M. Szymonski, J. Ruthowski, A. Poradzisz, Z. Postawa and B. Jorgensen, in **DIET II**, eds. W. Brenig and D. Menzel, (Springer Verlag, 1985) p 160
3. Q. Dou, D. W. Lynch and A. J. Bevolo, Surf. Sci. **219** L623 (1989)
4. P. Wurz and C.H. Becker, Surf Sci **224** 559 (1989)
5. H. von Seggern, I.E.E.E. Trans. on Nucl. Sci., **NS-32** 1503 (1985)
6. N. Whetten, **CRC Handbook of Physics and Chemistry 53 Ed**, CRC Press, 1972) p 203.

Part XI

Cryogenic Overlayers

DIET from Cryogenic Layers:
Subject and Tool of Investigation

P. Feulner

Physik-Department E 20, Technische Universität München,
W-8046 Garching, Germany

1. Introduction

This paper is devoted to two aspects of DIET (ESD and PSD) from Van der
Waals bound layers: i) the efforts to elucidate the microscopic details of
the desorption reaction, and ii) the use of DIET as a tool for the
investigation of electronic excitations and properties of electronically
excited states. "Simple" Van der Waals solids, namely rare gas and
molecular solids, have been used for the experiments. Although the studies
of desorption reactions from these samples are somehow "academic", they
nevertheless offer the chance of a detailed understanding of the different
steps involved in DIET by comparing theoretical and experimental results.
The examples selected for this paper are: 1) ESD of neutral particles from
Ar films of various thicknesses; 2) PSD of ions from Ar films; 3) high
resolution PSD of ions from solid N_2. Topic 1) demonstrates how comparison
of experiment and calculation can pinpoint certain desorption mechanisms,
particularly when dependencies on film thickness and dopants are included
in the investigation; 2) will show how DIET experiments hitherto exclusively give access to a distinct type of electronic excitation in rare gas
solids (RGS), and topic 3) will finally demonstrate the aid of high
spectral resolution for the detection of the nature and symmetry of
electronic excitations.

The experimental data reported here have been obtained with two
different set-ups. The kinetic energy distributions of neutral particles
desorbed by electron impact (ESD) have been recorded with a time of flight
(TOF) apparatus described in [1]. The PSD results have been measured in a
second UHV apparatus equipped with a quadrupole mass spectrometer (QMS), a
partial electron yield (PY) detector, and a hemispherical electron energy
analyser. Here, the axis of the sample manipulator was situated in the
direction of the synchrotron light. The surface plane of the sample was
tilted versus the light by 7 degrees, e.g., data were always collected
under grazing incidence condition for enhanced surface sensitivity. All the
detectors were rotatable around the sample. In this way, the polarization
angle δ and that of detection Φ could be varied independently: δ by
rotating the sample around its manipulator axis, and Φ by rotating the
analysers around the light beam [2].

2. Results

2.1. ESD of neutral atoms desorbed from Ar films.

Ample experimental and theoretical work has been devoted to DIET from mono-
and multilayers of rare gases on metals. As a function of film thickness,
different mechanisms and, therefore, different translational energy
distributions are expected. DIET from monolayers has been described in the
Antoniewicz picture as well as by the wave packet squeezing model [3]. The
shape of the kinetic energy distribution $N(E_k)$ obtained experimentally for
<u>monolayers</u> (fig.1) is compatible with both models; the experimental yield

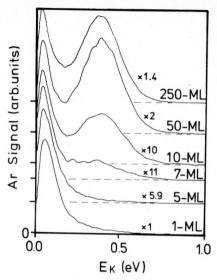

Fig.1: Kinetic energy distributions of Ar atoms desorbed by electrons of 200eV from Ar films of various thicknesses θ (given in units of one monolayer ML). For monolayers, ESD is due to the mechanisms described in ref.3. For thick films (θ≈50 ML), the molecular mechanism dominates. DIET via the cavity mechanism is important for films of intermediate thickness (see text).

data (the desorption cross section is ≈ 60% of the primary ionisation cross section) favour the latter one [3]. The correlation of emission angle and kinetic energy should allow to discriminate between the two models [3]; these data are presently not yet available.

DIET of neutrals from Ar multilayers occurs mainly through two competitive mechanisms: i) the cavity mechanism, i.e., expulsion of an electronically excited atom [4] or excimer [5] from the surface by repulsive interaction with the lattice, and ii) the molecular mechanism, i.e. the dissociation of a molecular self-trapped exciton (Ar_2^*) upon electronic relaxation [1,6]. For both processes distinct predictions exist about the kinetic energy distributions of the ejecta. Atoms as well as dimers desorbed via the cavity process should have kinetic energies not more than ≈100 meV; lattice imperfections are expected to affect the repulsion seriously [4,5]. Because desorption occurs within picoseconds, ejected dimers should be vibrationally hot. The two atoms originating from their radiative decay in the gas phase should exhibit far less kinetic energy than expected from vibrationally cold dimers (about 0.5 eV/atom [1,6]). The existence of the cavity mechanism has convincingly been shown by luminescence experiments [7]. DIET via the molecular mechanism on the other hand is slow; it requires the vibrational relaxation of the Ar_2^* by phonon emission. The time scale is rather tenths of nanoseconds than picoseconds. Upon electronic relaxation, bimodal kinetic energy distributions are expected (and have previously been experimentally detected [1,6]), namely direct desorption of fast fragments with E_k ≈ 0.4 eV, and atoms sputtered by those fast fragments directed into the solid; the latter contribution should lead to a maximum in $N(E_k)$ at E_k ≈ U/2 (U: surface binding energy; for Ar: U = 78 meV [8]). The $N(E_k)$ traces obtained for films of various thickness (fig.1) are in good agreement with these predictions. Thick layers show clearly the bimodal energy distribution with maxima at 43 meV and 385 meV expected for the molecular mechanism. The position of the "direct" peak is insensitive to sample perfection, which according to [4,5] makes contributions from the cavity process for those samples unlikely. The latter process is important only for films of medium thickness, for which the quenching of the electronic excitation by the vicinity of the metal surface is faster than the vibrational relaxation. Impurities in the layer also very effectively suppress the molecular

mechanism for which a long lifetime of the electronic excitation is a
necessity [1] (the delayed decrease of the yield when going from the mono-
to the multilayer must mean that desorption by interaction with the surface
is important not only for the first layer). In summary, DIET of neutrals
from solid rare gas films is an example where joint efforts of theory and
experiment succesfully revealed detailed information on the desorption
process.

2.2. PSD of ions from rare gas solids.

PSD [9] and ESD [10] data show that DIET of <u>singly</u> charged ions from RGS
can proceed via primary excitation of singly charged, electronically
excited ions X^{+*}, as well as doubly charged ions X^{2+}; DIET of <u>doubly</u>
charged ions however stems from primarily triply charged particles X^{3+}
[10]. Beside these one-atom excitations, thresholds have been found in ESD
of Ar^+ and Ar_2^* at about the double surface and bulk outer valence exciton
energies [10]. They have been assigned to biexcitonic excitations, i.e.,
formation of excitonic molecules, Ar_2^{**}, which desorb via the cavity
mechanism (no such thresholds have been obtained for Kr where desorption
via the cavity mechanism is not expected) and autoionize in the gas phase
[10]. PSD can give even more detailed information on these interesting
excitations in RGS (fig.2), which up to now have not been detected by other
techniques, probably because of their low cross section, which requires the
selective "projecting out" by DIET. Obviously, the primary interpretation
derived from ESD has to be revised. Excitonic molecules excited by a single
photon should show up as narrow features in the ion yield. However, such
maxima are visible only within the band Ia between 23.1 and 23.5 eV
(fig.2); they are stronger for Ar^+ than for Ar_2^+ emission. The other
absorption features between these excitonic molecule excitations and the
narrow maxima related to $3s^{-1}np^{+1}$ excitons around 27.5 eV (band III in
fig.2) are quite different. They do not show fine structure, and their
width depends strongly on film thickness. We have assigned them to coupled
exciton pairs whose momenta add to that of the photon, i.e., electronic
polarons [11]. From the different dependence on film thickness of the
features Ib and II we conclude that II is due merely to an excitation in
the bulk, whereas Ib probably contains contributions from the surface. It
is not yet clear how the range of momentum depends on layer thickness, and

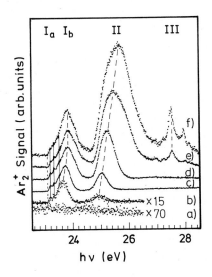

Fig.2: Ar_2^+-PSD from Ar films of
various thicknesses θ. The bands Ia,
Ib, II, and III are due to excitonic
molecule, electronic polaron, and Ar
$3s^{-1}$ excitonic excitations, respectively. For Ar^+ emission similar
spectra are obtained, though with
enhanced Ia intensity.

325

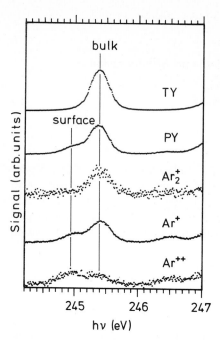

Fig.3: Total electron yield (TY) and partial electron yield (PY), compared with Ar_2^+, Ar^+, and Ar^{2+} PSD from Ar multilayers for resonant Ar $2p_{3/2}^{-1}4s^{+1}$ excitation (from ref.12). Surface and bulk contributions are clearly discernible (see text).

how the coupling of these obvioulsy non-localised excitations to desorption occurs. The importance of a cavity-like mechanism (and the formation of excitonic molecules as intermediate steps) appears likely from the fact that, as for ESD, no DIET by those excitations was seen for Kr. Further theoretical support is urgently needed.

Let me give another example for the versatility of DIET as a tool for the study of RGS. It again concerns the discrimination between bulk and surface excitations, however for primary core excitations. Fig.3 displays partial and total electron yield signals, as well as the Ar_2^+, Ar^+, and Ar^{2+} ion signals from solid Ar for $2p_{3/2}^{-1}4s^{+1}$ excitation (from ref.12). Bulk and surface excitations are clearly discernible; they are separated by about 0.4 eV. The relative amounts of surface and bulk contributions to the ion signals are 1.14:1 for Ar^{2+}, but only 0.3:1 for Ar^+. For Ar_2^+, the bulk contribution is essentially zero. This indicates that the exciton pairs which are the main source for emission of ionic dimers are mainly due to XESD, whereas desorption of Ar^+ and Ar^{2+} obviously proceeds via true PSD. Believing the ESD results that Ar^{2+} stems from $Ar^{3+}Ar \rightarrow Ar^{2+}Ar^+$, and Ar^+ from $Ar^{2+}Ar \rightarrow Ar^+Ar^+$ charge exchange reactions, an intrinsic asymmetry is obvious. The doubly charged ion will only desorb if it forms the end of the pair which is directed to the surface, whereas for Ar^+ emission the primary orientation is irrelevant. Moreover, if the time of residence of Ar^{2+} in the matrix is long, further charge exchange to singly charged ions can occur. This explains why 60% of desorbing Ar^{2+} ions stem from excitations in the topmost layer. This species, therefore, is a good probe for surface excitations. On the other hand, the large abundance of bulk signal in the Ar^+ yield demonstrates that photon absorption events in at least the 4 to 5 topmost layers are contributing to this species (note that the appearance of the dipole forbidden $2p_{3/2}^{-1}4p$ excitation at 246.5 eV is also indicative of excitations at the surface).

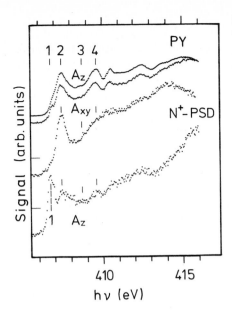

Fig.4: Partial electron yield (PY) compared with N^+ PSD from solid N_2 for resonant excitation of N1s core electrons to Rydberg levels (1: 3sσ; 2: 3pπ; 3: 4sσ; 4: 5sσ [16]). States with σ symmetry are projected out by A_z light, such with π symmetry by A_{xy} polarization in PSD.

2.3. High resolution PSD from solid nitrogen.

The capability of high resolution techniques in PSD have been introduced by the paper of D.Menzel [13], and by ref.14. It was shown that clear discrimination between bonding and antibonding core excited states can be derived from the appearance of vibrational structure, and that dynamic effects can be investigated by comparing the occupancy of vibrational levels seen in electron and ion yield. High resolution, polarization resolved PSD can also be used to pinpoint the electronic symmetry particularly of Rydberg states. This is demonstrated in fig.4 for N^+ emission by N1s$\rightarrow$(Rydberg state) excitations of solid N_2 [14,15]. The symmetry of these Rydberg states is known from gas phase studies [16-18]. The data of fig.4 show that A_z light projects out the excitations with σ symmetry, and A_{xy} light those with π symmetry, as expected. On the other hand, no clear polarization dependence is visible for the electron yield. According to this test system N_2, it is obvious that PSD of ions can be a very simple method for the assignment of electronic excitations of unknown symmetry.

3. Acknowledgement

All the measurements as well as the evaluation of the data have been performed in collaboration with E.Hudel, G.Remmers, R.Scheuerer, M.Scheuer, E.Steinacker, and W.Wurth. I also thank them and D.Menzel for the joint development and shaping of the concepts presented here. This work was funded by the German Ministry of Research and Development, BMFT, through project 05 466 CAB, and by the Deutsche Forschungsgemeinschaft through SFB 338 C4.

327

References

[1] E.Hudel, E.Steinacker, and P.Feulner, Phys. Rev. B44 (1991) p.8972, and references therein.
[2] R.Scheuerer, K.Eberle, and P.Feulner (to be published).
[3] Z.W.Gortel, this conference.
[4] S.Cui, R.E.Johnson, and P.T.Cummings, Phys. Rev. B39 (1989) p.9580.
[5] S.Cui, R.E.Johnson, C.T.Reimann, and J.W.Boring, Phys. Rev. B39 (1989) p.12345.
[6] D.J.O'Shaughnessy, J.W.Boring, S.Cui, and R.E.Johnson, Phys. Rev. Lett. 61 (1988) p.1635.
[7] F.Coletti, J.M.Debever, and G.Zimmerer, J.Physique Lett. 45 (1984) p.L-467.
[8] H.Schlichting, Ph-D thesis, Technische Universität München 1990.
[9] G.Dujardin, L.Hellner, M.J.Besnard-Ramage, and R. Azria, Phys. Rev. Lett. 64 (1989) p.1289.
[10] Y.Baba, G.Dujardin, P.Feulner, and D.Menzel, Phys. Rev. Lett. 66 (1991) p.3269.
[11] T.Schwabenthan, R.Scheuerer, E.Hudel, and P.Feulner, Solid State Commun. 80 (1991) p.773.
[12] E.Hudel, diploma thesis, Technische Universität München 1990
[13] D.Menzel, this conference.
[14] P.Feulner, R.Scheuerer, M.Scheuer, G.Remmers, W.Wurth, and D.Menzel, Appl. Phys. (in press).
[15] R.Scheuerer, P.Wiethoff, W.Wurth, P.Feulner, and D.Menzel (to be published).
[16] C.T.Chen, Y.Ma, and F.Sette, Phys. Rev. A40 (1989) p.6737.
[17] M.Nakamura, M.Sasanuma, S.Sato, M.Watanabe, H.Yamashita, Y.Iguchi, A.Ejiri, S.Nakai, S.Yamaguchi, T.Sagawa, Y.Nakai, and T.Oshio, Phys. Rev. 178 (1969) p.80.
[18] D.M.Hanson, this conference.

Low-Energy ESD of Metastable Atoms from Ar, Kr, Xe(111) Condensed Films

A. Mann, G. Leclerc, and L. Sanche

MRC Group in the Radiation Sciences and Canadian Centre of Excellence
in Molecular and Interfacial Dynamics, Faculté de Médecine,
Université de Sherbrooke, Sherbrooke, Québec, Canada J1H 5N4

Abstract. Thin condensed films of rare-gas atoms (Ar, Kr, Xe), grown on a Pt(111) crystal, have been exposed to monochromatic low-energy electrons (E = 5 - 25 eV, ΔE = 60 meV). Desorbed metastable atoms are abundant for Ar films, and several desorption components are identified based on the information on flight time and desorption angle. With Kr and Xe, a significant metastable atom signal is obtained only by covering a Kr film with a monolayer of Xe. Various desorption mechanisms are discussed.

1. Introduction

Rare-gas solids are fundamental for studying DIET from van-der-Waals bound systems. The excitonic states which are regarded as precursor to the desorption have been investigated for a number of years and are now well understood [1,2]. There exist a number of radiative decay channels some of which are accompanied by desorption of neutral ground-state atoms. But an excited atom (or dimer) may also be ejected into vacuum prior to the radiative decay. These species probably carry more "unrelaxed" information in the sense that they allow to study the dynamics during the relaxation of the lattice around a trapped exciton.

In the present contribution, we report briefly on the desorption of metastable atoms from pure Ar films as well as from Kr films covered by a monolayer of Xe. The available experimental tools to study the relevant mechanisms are variation of the incident energy, as well as flight-time and desorption-angle resolved detection. Details will be published elsewhere [3,4].

2. Apparatus

The apparatus used to obtain the present results has been described recently [3,5]. Essentially, it consists of an electron monochromator, a cryogenically cooled Pt(111) target held at a temperature of 20 K, and a system capable of detecting metastable particles and photons. The electron monochromator provides a collimated beam with an intensity of 1 nA and an energy resolution of 60 meV.

The detection system is a LEED-type configuration with four grids, a block of three micro-channel plates, and a position-sensitive anode. In the present context, the grids are biased to reject charged particles, and only photons and metastable particles with an excitation energy above about 7 eV are counted. The pulsed electron beam mode allows the determination of the velocity of the metastable particles. The detector covers an angular range of θ = ±35° with the normal of the target surface being tilted by 18° with respect to the normal of the detection plane. Within that detection range, the angular distribution of the signal can be calculated from the particles' impact positions.

Springer Series in Surface Sciences, Vol. 31
Desorption Induced by Electron Transitions DIET V
Editors: A.R. Burns · E.B. Stechel · D.R. Jennison   © Springer-Verlag Berlin Heidelberg 1993

3. Desorption from Ar films

The desorption of metastable Ar* atoms from 100 ML Ar(111) films has been studied in a combined time-of-flight and desorption-angle experiment at a constant impact energy of 14.5 eV corresponding to the maximum in the excitation function [5]. After a 5 μs electron pulse, desorption patterns have been collected in time-windows 25 μs wide. Additionally, a high-resolution time-of-flight (TOF) spectrum with steps of 5 μs has been recorded for near-normal desorption ($\theta < 6°$). Typical spectra are depicted in Fig. 1; a detailed presentation will be published elsewhere [3]. Owing to a distinct behavior in one or both of the studied attributes, five desorption components have been identified and labelled M1 to M5. In a first approach, Gaussian functions have been fitted to both the kinetic-energy distribution and the normalized desorption-angle distribution. The desorption components are sketched in Fig. 2.

In solid Ar, the lattice is repelled around a localized Ar* atom and a cavity is formed [6]. Coletti et al. [7] proposed that at the surface such an interaction would lead quite naturally to the desorption of the Ar* atom, since the repelling forces are not balanced. Cui et al. [8] performed molecular-dynamics studies and concluded that the kinetic energy of a desorbed Ar* atom crucially depends on the number of nearest neighbors in the surface plane. Regarding the decreasing kinetic energy and the increasing angular width of the observed components M3, M4, and M5, we suggest that they are originating from surface sites with six, five, and four nearest neighbors, respectively. M1 and M2 do not fit into this scheme and will be discussed below.

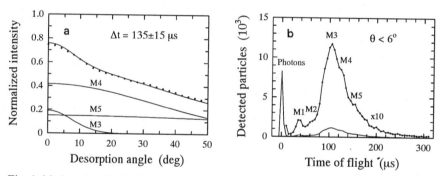

Fig. 1. (a) Angular distribution of metastable Ar* atoms in the time window 135±15μs, normalized to the photon signal. (b) Time-of-flight spectrum of metastable Ar* atoms for desorption angles $\theta < 6°$.

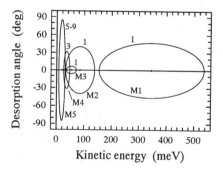

Fig. 2. Kinetic-energy and desorption-angle distributions for the five detected components in the Ar* desorption signal. The ellipses correspond to a signal decrease by a factor exp(-1/2) in the Gaussian fit functions. The numbers indicate the relative intensity of each contribution.

4. Desorption from monolayer Xe / multilayer Kr films

Pure Kr films show only little metastable particle signal, and pure Xe films none at all with the present sensitivity. However, from a 35 ML Kr film covered by 1 ML Xe, a considerable desorption signal is observed (Fig. 3). Since Kr and Xe rare-gas solids are expected to show no repulsive cavity around an excited atom but rather an attractive interaction, other mechanisms have to be considered.

The threshold for the metastable particle signal is found in the vicinity of the threshold for the creation of a Xe exciton in a Kr matrix ($E_{exc} + V_0 = 9.01$ eV - 0.2 eV [1]). This value probably depends on the specific concentration and location of Xe and Kr atoms around the excited atom and is regarded only as a rough estimate to describe the observed onset. Initially exciting a Xe atom in the neighborhood of Kr atoms, the heteronuclear excited complex Kr-Xe* may be formed. The estimated energy of this initial state is higher than both the energy of a relaxed m-STE (molecular self-trapped electron) in the crystal (7.8 eV [6]) and the energy of a desorbed Xe* ($E^* + D_C = 8.45$ eV [1]). As indicated in Fig. 4, motion in both "directions" is energetically possible. Inward motion will result in a vibrationally relaxed heteronuclear m-STE which eventually decays with luminescence to the repulsive ground-state. Outward motion of the Xe* leads to desorption, and according to the surplus energy and the Xe/Kr mass ratio an estimate of 220 meV for the kinetic energy of the desorbed Xe* atom is obtained. This scenario defines the threshold for the desorption of an excited Xe atom from the Xe/Kr film.

As seen in Fig. 4, a similar process is possible for pure Ar films, but not for pure Kr and pure Xe films. We expect the Ar* desorption component M2 with a mean kinetic energy of 85 meV to be related to this mechanism. While the cavity expulsion assumes an average interaction potential with all neighbors, the presently proposed mechanism relies on the preservation of a repulsive dimer interaction with a certain

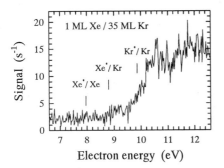

Fig. 3. Incident electron energy dependence of the metastable particle signal from a 35 ML Kr film on Pt (111) covered by a 1 ML Xe film. The thresholds for exciting bulk excitons in pure Xe, Xe in Kr, and pure Kr are indicated.

Fig. 4. Potential-energy values for excitations in rare-gas crystals. The distance to the nearest neighbor indicates schematically three locations of the excited atom. Initially, an atom in an undistorted surface is excited presumably to the lowest exciton state. Inward motion is possible in all cases giving rise to a dimer-type self-trapped exciton, while outward motion (metastable particle desorption) occurs only for Ar*/Ar and Xe*/Kr.

331

neighbor similar to the attractive interaction leading to m-STE states. Consequently, the kinetic energy of the desorbed particle should be higher in the latter case. For Ar films, where both processes are possible, the signal attributed to dimer repulsion is one order of magnitude lower than the signal from cavity expulsion [3].

The third energy value in Fig. 3 corresponds to the excitation of a Kr bulk exciton and is correlated with the onset of the region of maximum signal intensity for the metastable particle signal in Fig. 2. A possible mechanism seems to be the formation of a Kr^*-Xe complex, where the excitation is initially with the Kr atom. During relaxation, multiple curve crossings with repulsive Kr-Xe^* potential energy curves occur, and the transition can easily lead to desorption of the Xe^* atom:

$$Kr^*\text{-Xe (attr)} \rightarrow Kr\text{-}Xe^* \text{ (rep)} \rightarrow Kr \text{ (fast)} + Xe^* \text{ (fast)}.$$

The total surplus energy for this process is 1.7 eV. However, since the dissociation energy depends on the amount of energy released into lattice vibration by the time the transition occurs, we expect a broad kinetic-energy distribution centered around 0.4 eV.

A similar process can be described for pure Ar films and is postulated to explain the observed fast desorption component M1 (mean kinetic energy: 345 meV). In this case, potential energy is transferred to kinetic energy via attractive-repulsive curve crossings according to the scheme

$$Ar_2^{**} \text{ (attr)} \rightarrow Ar_2^* \text{ (rep)} \rightarrow Ar \text{ (fast)} + Ar^* \text{ (fast)}.$$

If the Ar_2^{**} state is formed by an n=2 exciton [9], the total surplus energy amounts to about 2 eV to be shared between the lattice and the repelling atoms giving rise to desorbing Ar^* atoms with high kinetic energy and substantial energy spread.

Acknowledgment

This work was sponsored by the Medical Research Council of Canada.

References

1. N. Schwentner, E.-E. Koch, and J. Jortner: *Electronic Excitations in Condensed Rare Gases* (Springer, Berlin 1985)
2. G. Zimmerer: in *Excited-State Spectroscopy in Solids*, eds. U. M. Grassano and N. Terzi (North-Holland, Amsterdam 1987)
3. G. Leclerc, A. D. Bass, A. Mann, and L. Sanche: Phys. Rev. B (in press)
4. A. Mann, G. Leclerc, and L. Sanche: to be published
5. G. Leclerc, A. D. Bass, M. Michaud, and L. Sanche: J. Electr. Spectros. Relat. Phenom. **52**, 725 (1990)
6. I. Ya. Fugol': Adv. Phys. **27**, 1 (1978)
7. F. Coletti, J. M. Debever, and G. Zimmerer: J. Phys. Lett. (Paris) **45**, 467 (1984)
8. S. Cui, R. E. Johnson, and P. T. Cummings: Phys. Rev. B **39**, 9580 (1989)
9. I. Arakawa and M. Sakurai: in *Desorption Induced by Electronic Transitions DIET IV*, eds. G. Betz and P. Varga (Springer, Berlin 1990)

Photo-Stimulated Desorption of Ne Metastables from Thin Ne Layers on Solid Ar, Kr, and Xe

D.E. Weibel[1,], A. Hoshino[1], T. Hirayama[1], M. Sakurai[2], and I. Arakawa[1]*

[1]Department of Physics, Gakushuin University, Toshima, Tokyo 171, Japan
[2]National Institute for Fusion Science, Chikusa, Nagoya 464-01, Japan
*Present address: Faculty of Chemical Sciences,
  University of Cordoba, Argentina

Abstract.   Photo-stimulated desorption of Ne metastables(Ne*) from a thin Ne film adsorbed on solid Ar, Kr, and Xe was studied using synchrotron radiation as an excitation source. The excitation which is corresponding to the first order surface exciton initiates the desorption of Ne* from the Ne submonolayer. The mean kinetic energies of Ne* desorbed from Ne/Ar, Ne/Kr, and Ne/Xe are 0.15, 0.13, and 0.10 eV, respectively, while that of Ne* desorbed from pure solid Ne is 0.18 eV.

The excitons in rare gas solid play an important role in DIET (Desorption Induced by Electronic Transitions) of neutral molecules[1, 2]. For the desorption of metastable Ar[3, 4] and Ne[5, 6, 7], clear correlations with the several types of excitons and the kinetic energy distribution of desorbed metastables have been clarified experimentally. Excitation of the first order surface exciton leads to the desorption of a metastable Ar* with the kinetic energy of about 0.04 eV[4] and Ne* with 0.18 eV[6, 7]. In the above desorption phenomena, an excited atom at the surface is thought to be ejected by the repulsive force which is due to an expanded electron orbit of the exciton and a negative electron affinity of the matrix. This is called the cavity-ejection (CE) mechanism as this repulsive interaction is known as the origin of the formation of a cavity around a self-trapped exciton in solid Ne and Ar. It has been thought that the CE mechanism does not work for solid Kr and Xe which have a positive electron affinity[8]. We present here the first experimental evidence of DIET of Ne* from the systems of Ne adsorbed on solid Ar, Kr, and Xe. The kinetic energy distribution of Ne*, which is initiated by the first order surface exciton, and the dependence on a thickness of Ne film are presented.

The detail of the experimental system and procedure has been described elsewhere [4, 9]. A small amount of Ne was adsorbed on the surface of a thick Ar, Kr, or Xe film which was condensed on a copper substrate at a temperature of 6K or less. An amount of Ne adsorbed was calculated from the exposure assuming the condensation coefficient to be unity. The base pressure of the experimental chamber was 10^{-8} Pa or less. In order to reduce the accumulation of impurities, the sample was surrounded by the cold shield which had only a narrow slit for the entrance of light and for the detection of desorbed species. An excitation source was monochromatic VUV light from the beam line BL5B of the UVSOR facility at the Institute for Molecular Science. The velocity distribution of the desorbed Ne metastables was measured by a time-of-flight (TOF) technique[9]. The pulse beam (a width of 16 µs and a cycle of 250 Hz) was obtained by a rotating disc with slits.

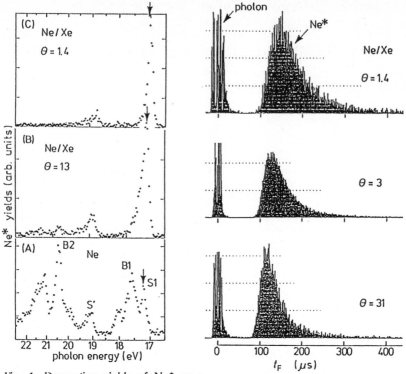

Fig. 1. Desorption yields of Ne* as a function of a photon energy. θ is a thickness of an Ne film on solid Xe.

Fig. 2. TOF spectra of Ne* desorbed from Ne/Xe by S1 excitation.

Desorbed metastables were detected by a secondary electron multiplier (Ceratron, Murata Corp.), which was placed at a distance of 155 mm from the sample.

The relations between the desorption yield of Ne* and the excitation photon energy are shown in Fig. 1 for pure solid Ne (A) and adsorbed Ne on solid Xe (Ne/Xe)(B and C). The thickness θ of an Ne film is given by a number of atomic layers. The arrow in Fig. 1(A) indicates the maximum yield of Ne* which is desorbed through the cavity-ejection mechanism after the excitation of the first order surface exciton (S1). The corresponding excitation in Ne/Xe systems also leads significant desorption of Ne* as indicated by the arrows in Fig. 1(B) and (C). Relative desorption yields of Ne* by the excitation of the other excitons, the first (B1) and the second (B2) bulk excitons and the higher order surface exciton (S'), change in reasonable accordance with the thickness of the Ne film. The peak to the left of B2 originates from the third and higher order excitons and the ionization. The photon energy E_m that gives the maximum Ne* yield by the S1 excitation shows a slight red shift in the adsorbed Ne: 17.00 ± 0.03 eV for Ne/Xe ($\theta = 1.4$) while 17.17 ± 0.03 eV for pure Ne. Figure 2 shows TOF spectra of Ne* desorbed from Ne/Xe by the S1 excitation. The kinetic energy E_k of Ne* at the peak of distribution shifts from 0.18 ± 0.01 eV at $\theta = 31$ to 0.10 ± 0.01 eV at $\theta = 1.4$. The TOF spectrum at $\theta = 31$ was

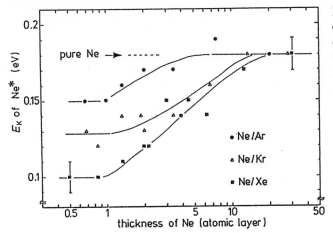

Fig. 3. Kinetic energies of Ne* desorbed by the S1 excitation from Ne adsorbed on Ar, Kr, and Xe.

identical with that obtained at pure Ne. The change in E_m and E_k observed in Ne/Xe is also the case with Ne/Ar and Ne/Kr though the magnitudes of the shifts are different in the three systems. The shifts ΔE_m, the difference of E_m between pure Ne and adsorbed Ne, were 0.09, 0.12, and 0.17 eV for Ne/Ar, Ne/Kr, and Ne/Xe, respectively. The kinetic energies of Ne* desorbed from the Ne films on Ar, Kr, and Xe are shown in Fig. 3 as a function of the Ne thickness θ. In the submonolayer region ($\theta < 1$), E_k seem to be almost constant and are 0.15, 0.13, and 0.10 eV for Ne/Ar, Ne/Kr, and Ne/Xe, respectively. E_k increases gradually as increasing θ and reaches the value 0.18 eV, which is E_k of Ne* from pure Ne, at around $\theta = 10$.

The desorption of metastables through the CE mechanism have experimentally been revealed at the surface of solid Ar[3, 4, 10] and Ne[1, 6, 7]. This mechanism has been expected to be less efficient in Kr and Xe[8]. Though the electron affinity of an isolated rare-gas atom is negative, both Kr and Xe have a positive electron affinity in bulk because of their large polarizability. The present experiment shows that the desorption of Ne* from Ne/Kr and Ne/Xe occurs. This result suggests that the electron affinity is negative at the surface of solid Kr and Xe if this desorption phenomena is discussed in terms of the CE mechanism. It is plausible that the smaller number of the surrounding atoms at the surface is not enough to change the electron affinity from negative to positive by their polarizability.

The photon energy E_m, which is deposited in the system, is divided into the following terms at the desorption[5]: $E_m = E_g + E_k + E_{coh} + E_l$, where E_g is the excitation energy of the corresponding state of an isolated Ne atom in gas phase, E_{coh} is the cohesive energy of an Ne atom on the surface, and E_l is the energy absorbed in the lattice after the desorption. It has been shown experimentally[5, 7] and by a molecular dynamic calculation[11] for Ne and Ar that 40 - 70 % of the lattice distortion energy, which can roughly be estimated by $E_m - E_g$, is transferred to the kinetic energy of the desorbing metastable. The shifts ΔE_m observed in the submonolayer Ne, namely, 0.09 eV(Ne/Ar), 0.12 eV(Ne/Kr), and 0.17 eV(Ne/Xe), lead to the shifts ΔE_k of the kinetic energy of Ne*, 0.03eV(Ne/Ar), 0.05 eV(Ne/Kr), and 0.08 eV(Ne/Xe), respectively. The ratios $\Delta E_k / \Delta E_m$ are in good agreement with the above value $E_k/(E_m - E_g)$. It means that the CE mechanism still works in the adsorbed Ne systems. Another factor which can affect the kinetic energy of Ne* is the cohesive energy due to the van der Waals interaction. Though it is difficult to estimate the true binding energies of Ne adsorbed on Ar, Kr, and Xe because the adsorption condition is not clear, the relative change can roughly be estimated by the geometric mean of the cohesive energies of each species[12]. They are calculated at 0.036(Ne/Ar), 0.042(Ne/Kr), and 0.050 eV(Ne/Xe), while the cohesive energy of pure solid Ne is 0.019 eV. It is found that the differences of the binding energy are comparable with ΔE_k and may have a significant effect in the desorption process.

In summary, the desorption of Ne metastables is initiated by the excitation corresponding to the first order surface exciton in the Ne films on solid Ar, Kr, and Xe, where the cavity-ejection mechanism seems likely to work.

Acknowledgements

One of the authors (D.E.W.) gratefully acknowledges the financial support from the Center for International Exchange of Gakushuin University.

References

1. F. Colletti, J.M. Debener, and G. Zimmerer. J. Phys.(Paris) Lett. 45, L467 (1984).
2. P. Feulner, T. Muller, A. Puschmann, and D. Menzel, Phys. Rev. Lett. 59, 791 (1987).
3. I. Arakawa, M. Takahashi, and K. Takeuchi, J.Vac.Sci.Technol. A7, 2090(1989).
4. I. Arakawa and M. Sakurai, *Desorption Induced by Electronic Transitions, DIET-IV* (Springer, Berlin, 1990) p. 246.
5. T. Kloiber and G. Zimmerer, Radiat. Eff. Def. Sol. 109, 219 (1989).
6. T. Kloiber and G. Zimmerer, Physica Scripta 41, 962 (1990).
7. I. Arakawa, D.E. Weibel, A. Hoshino, T. Hirayama, and M. Sakurai, to be submitted.
8. W.T. Buller and R.E. Johnson, Phys. Rev. B 43, 6118 (1991).
9. M. Sakurai, T. Hirayama, and I Arakawa, Vacuum 41, 217 (1990).
10. D.J. O'Shaughnessy, J.W. Boring, S. Cui, and R.E. Johnson, Phys. Rev. Lett. 61, 1635 (1988).
11. S. Cui, R.E. Johnson, and P.T. Cummings, Phys. Rev. B 39, 9580 (1989).
12. W.A. Steel, *The Interaction of Gases with Solid Surfaces* (Pergamon, Oxford, 1974) p. 34.

Dynamics of Multi-Electronic Photon Excitation in Multilayers of Argon Studied by Kinetic Energy Analysis of Photodesorbed Ions

G. Dujardin[1], L. Hellner[1], L. Philippe[1], M.J. Besnard-Ramage[1], R. Azria[2], M. Rose[1], and P. Cirkel[1]

[1]Laboratoire pour l'Utilisation du Rayonnement Electromagnétrique (LURE),
Centre National de la Recherche Scientifique, Commissariat á l'Energie
Atomique, Ministère de l'Education Nationale, Bât. 209 D,
Université Paris-Sud, F-91405 Orsay Cedex, France; and
Laboratoire de Photophysique Moléculaire, Centre National de la Recherche
Scientifique, Bât. 213, Université Paris-Sud, F-91405 Orsay Cedex, France
[2]Laboratoire des Collisions Atomiques et Moléculaires,
Bât. 351, Université Paris-Sud, F-91405 Orsay Cedex, France

Abstract : The kinetic energy distribution of photodesorbed ions from multilayers of argon is analyzed at two photon excitation energies, i.e. 40 eV and 100 eV, corresponding to the formation of satellite states and double photoionization, respectively. Surprisingly, the kinetic energy release is much smaller than what is predicted from the Coulomb repulsion model.

1. Introduction

Multi-electronic excitation processes are considered to play an important role in ion desorption from adsorbed atoms and molecules [1] although an unambiguous identification of these excitations is rather scarce in the literature. Recent studies on multilayers of argon clearly assigned the primary one photon two electron processes producing ion desorption, i.e. the formation of double excitation satellite states [2,3], the double ionization [2] and the formation of exciton pairs [4,5]. These results raise numerous questions about the electronic excitation processes themselves but moreover on the dynamics of the desorption, the migration of the excitation at the surface and in the bulk, the mobility of argon atoms and ions, etc... In order to get an insight into these dynamics aspects we have performed experiments on the kinetic energy analysis of the photodesorbed ions. There have been relatively few such experiments [6-8]. Kinetic energy measurements are of special interest in multilayers of rare gases because of the well identified nature of the primary electronic processes. This opens the possibility of a very detailed understanding of ion desorption mechanisms.

2. Experimental

We use a very high flux ($\simeq 10^{13}$ photons/s) undulator beam line at the Super-ACO synchrotron radiation source in Orsay as a photon source of variable energy in the 20-120 eV range. The experiments are performed in an ultrahigh-vacuum system (pressure below 10^{-10} mbar) equipped with a liquid-helium flow cryostat. The argon multilayers are prepared by condensing pure argon on a polycrystalline platinum substrate cooled to 15 K. The film thickness is estimated from photoabsorption measurements. The photodesorbed ions may be mass selected through a quadrupole filter and counted as a function of the photon energy. Ions may be also kinetic energy analyzed through a separate hemispherical electrostatic analyser. In this latter case ions are no more mass selected and therefore valuable measurements require that a single ion mass is detected. It is known [3] that dimer and trimer argon ions are also desorbed from argon multilayers. However their intensity remains small as compared to that of monomer argon ions as long as the sample thickness does not exceed 50 monolayers. We then checked, before and after each kinetic energy measurement, in the ion mass spectrum, that Ar^+ ions account for more than 90% of the total number of desorbed ions.

Reported kinetic energies of Ar^+ ions are relative to the vacuum level of the adsorbate. In order to get these values from the measured kinetic energies relative to the Fermi level [6], we measured the work funtion of the sample by photoemission experiments. This leads to an uncertainty of about 0.2-0.3 eV in the kinetic energy absolute values.

3. Results and discussion

Kinetic energy distributions of desorbed Ar^+ ions from 20 monolayers of condensed argon are shown in figure 1.

At a photon excitation energy of 40 eV, Ar^+ ions are desorbed via the formation of double excitation satellite states of Ar^+[2,3] whereas at 100 eV, the dominant process is the double photoionization [2,3].

The previously proposed mechanism [2] for ion desorption involves either the ionization of a neighbour Ar atom ($h\nu$ = 40 eV) or the charge exchange with a neighbour Ar ($h\nu$ = 100 eV). Within this model, the Ar^+ ion desorption is expected to be due to the Coulomb repulsion between positive charges in both cases. For such reactions occuring at the surface of the sample one would then predict a high kinetic energy of the desorbed ions. For example, doubly charged Ar^{++} ions of which the formation energy is around 39 eV [2] would decay into $Ar^+ + Ar^+$ (formation energy = 2x13.8 eV = 27.6 eV [2]) with a total kinetic release of 11.4 eV. Assuming that each Ar^+ ion takes half of this energy i.e. 5.7 eV, the desorbed ion would leave the sample with a reduced kinetic energy ($\simeq$ 3.7 eV) due to difference between the ionization energy of argon in the solid (= 13.8 eV) and in the gas phase ($\simeq$ 15.8 eV). This markedly contrasts with the observed kinetic energy distributions in figures 1a and 1b which have a maximum around 0.5 eV and do not extend beyond 2

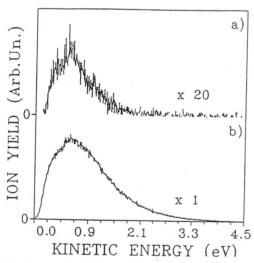

Figure 1 : Kinetic energy distribution of Ar^+ ions desorbed from condensed argon (thickness $\simeq$ 20 monolayers) at two photon excitation energies 40eV (a) and 100 eV (b).

eV and 4 eV respectively. This precludes to consider that satellite state or double ionization excitations in the bulk may migrate to the surface and then produce ion desorption directly via the

Coulomb repulsion mechanism. Several processes may explain the relatively low kinetic energy of the desorbed ions. If the Coulomb explosion occurs immediatly in the bulk, emitted Ar^+ ions may lose energy before leaving the sample by collision with Ar atoms. If the Coulomb repulsion mechanism occurs at the surface or close to the surface, the reduced kinetic enery of desorbed Ar^+ ions may be due to vibronic coupling with the vibrational modes of the surrounding medium. Therefore an important part of the initial Coulomb repulsion energy would be transferred to the evaporation of neutrals or to the excitation of phonons. We emphasize that other desorption scenarios such as the electron capture by a doubly charged Ar_2^{2+} ionomer

leading to repulsive Ar_2^+ ionomers [9] are not likely to occur since the Ar_2^+ potential energy curves [10] are not repulsive enough to allow the Ar^+ atoms to overcome the $\simeq 2$ eV energy barrier before desorption. In order to clarify the desorption mechanism, kinetic energy measurements as a function of the condensed argon thickness are required. This will enable us in particular to specify the rôle of surface and bulk processes. Such experiments are in progress in Orsay.

4. Conclusion

Ar^+ ions desorbed by photon impact at 40 eV and 100 eV on multilayers of argon have a kinetic energy which is much reduced as compared to what is expected from a direct Coulomb repulsion model. A large amount of the initial electronic excitation energy is considered to be transferred to the evaporation of neutrals or to the excitation of phonons due to charge separation and/or Coulomb repulsion occurring in the bulk or in the few top layers of the sample.

References

[1] D.E. Ramaker, J. Chem. Phys. **78**, 2998 (1983).

[2] G. Dujardin, L. Hellner, M.J. Besnard-Ramage and R. Azria
    Phys. Rev. Lett. **64**, 1289 (1990).

[3] G. Dujardin, L. Hellner, L. Philippe, M.J. Besnard-Ramage and F. Combet Farnoux
    to be published.

[4] Y. Baba, G. Dujardin, P. Feulner and D. Menzel, Phys. Rev. Lett. **66**, 3269 (1991).

[5] T. Schwabenthan, R. Scheuerer, E. Hudel and P. Feulner, Solid State
    Commun **80**, 773 (1991).

[6] J.A. Kelber, R.R. Daniels, M. Turowski, G. Margaritondo, N.H. Tolk and
    J.S Krauss, Phys. Rev. **B30**, 4748 (1984).

[7] H.H. Madden, D.R. Jennison, M.M. Traum, G. Margaritondo and N.G. Stoffel,
    Phys. Rev. **B26**, 896 (1982).

[8] J.A. Yarmhoff and S.A. Joyce, Phys. Rev. **B40**, 3143 (1989).

[9] D.Menzel, Appl. Phys.A **51**,163 (1990).

[10] H.V. Böhmer and S.D. Peyerimhoff, J. Phys. D 3, 195 (1986).

Photon Stimulated Desorption from Cryogenic Surfaces: Predicting the Vacuum Behavior of the Superconducting Supercollider

P.H. Lippel, Q.T. Jiang, S.P. Erichsrud, A.R. Koymen, and A.H. Weiss

Physics Department, Box 19059, University of Texas at Arlington, Arlington, TX 76019, USA

We describe planned experiments to measure photon stimulated desorption coefficients and related properties from copper-plated stainless steel and similar materials, at 4° K and 300° K. An accurate knowledge of the behavior of a vacuum system constructed of such materials, under intense synchrotron radiation, is neccessary to properly predict the vacuum behavior of the Superconducting Super Collider.

We have recently begun a new research program aimed at improving the understanding of synchrotron radiation induced desorption from technical surfaces, i.e., from surfaces which can routinely be achieved by normal commercial manufacturing methods. The immediate goal of this research is to obtain sufficient information to accurately predict the time development of the vacuum within the beam tubes of the Superconducting Super Collider (SSC). As discussed in more detail by A. Maschke elsewhere in this volume, the SSC presents an unprecedented combination of intense synchrotron radiation and liquid helium (LHe) temperature beam tube. To accurately model the vacuum behavior of the Collider, one must understand the photon- and electron- induced desorption coefficients, photon reflection coefficients, and photoelectron production crossections for candidate beam tube materials. While these parameters should not have strong temperature dependencies, the necessary surface conditions cannot be achieved except by performing the experiments directly at the relevant temperature, since the hydrogen adsorption isosteres on metal surfaces vary rapidly near 4° K[1]. The most likely material for the beam tubes is stainless steel electroplated with 100 microns of copper; the stainless steel is necessary to provide adequate mechanical strength should one of the superconducting magnets quench, while the copper provides the low electrical resistivity necessary for the image current to flow without heating the walls.

Previous research on photodesorption tends to fall into two categories. Fundamental studies of desorption mechanisms (e.g., most of the work reported at previous DIET conferences[2,3]) generally utilize well-characterized, carefully prepared single crystal surfaces. At the opposite extreme, a large body of work exists which attack the entire synchrotron-induced desorption problem at once, by installing technical surfaces directly in synchrotron lines and attempting to measure total desorption yields, photoelectron currents, etc. In this category we include numerous studies by Mathewson et al at CERN[4], previous SSC- sponsored work by Bintinger et al[5]., and a large body of work relating to the vacuum history of light sources such as the NSLS at Brookhaven[6].

In our effort to characterize the behavior of surfaces which can be obtained in over 150 kilometers of cooled, unbaked beam tube we must aim for a middle ground between these two extremes. We plan to measure room temperature and LHe temperature desorption coefficients, photoelectron production rates, and photon reflection coefficients in an ultrahigh vacuum environment from candidate beam tube materials, and to use single crystal samples to ensure that our results compare properly with prior work in the "fundamental" category. We are presently finalizing the design of the apparatus to perform these measurements, as indicated in Figure 1.

An XPS-type photon source will be used to provide X-rays with various discrete energies in the range from 200 eV to 1200 eV, through changes in anode material. Two different anodes materials can be installed in the gun at one time. The absence of sharp features in the energy dependence of previously reported neutral desorption rates[7]

Springer Series in Surface Sciences, Vol. 31
Desorption Induced by Electron Transitions DIET V
Editors: A.R. Burns · E.B. Stechel · D.R. Jennison   © Springer-Verlag Berlin Heidelberg 1993

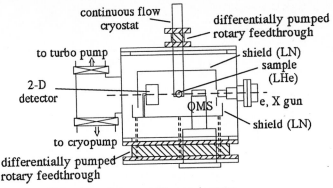

Fig. 1a Side view of the experimental set up.

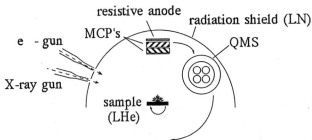

Fig. 1b Schematic drawing of the instruments inside our UHV chamber.

justifies the simulation of the broadband synchrotron radiation spectrum with a few discrete lines. A commercial electron gun will provide a 10-1000 eV beam with essentially constant current and diameter over much of this energy range; this gun includes a grid for rapid pulsing. A microchannel plate array with resistive anode readout will provide 2-dimensional position sensitive detection of photoelectrons and scattered photons; a quadrupole mass spectrometer will be employed to detect desorbed neutrals and, with its ionizer off, directly produced ions. Since the dominant desorption product from previously tested samples is H_2, the system must be optimized for the detection of H_2 in the presence of the H_2 background which dominates clean UHV systems. To minimize this background we will use a UHV compatible cryopump with a rated H_2 pumping speed of over 2000 l/s and an expected base pressure of 1x10-11 torr; chemical getters may additionally be employed. The incident electron or x-ray beam will be pulsed on and off, while the mass spectrometer is operated in the single particle counting mode. Subtraction of the beam off signal will remove the averaged H_2 background from the desorption signal. A glass cup surrounding the ionizer will substantially eliminate the variation of ionization probability on the energy of the desorbed particleand will allow for improved measurements of the ratio of atomic to moleculear hydrogen. Separate experiments using time of flight techniques will be necessary to measure the energy distribution of the desorbates. Both detectors will be mounted on a differentially pumped rotary flange at the bottom of the chamber, while the sample will be attached to a continuous flow cryostat mounted atop the chamber on a second differentially pumped rotary flange. The x-ray and electron sources will be mounted in fixed positions on the chamber sides. For LHe temperature experiments, a cryoshroud is needed to shield the sample from radiation from the room temperature walls; the ionization region of the mass spectrometer and the channelplate assembly will extend into this shroud.

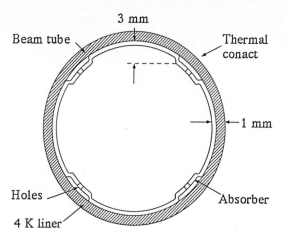

Fig. 2   4 K liner for H$_2$ pumping.

   The radiation load is characterized by a synchtron spectrum with a critical energy of 290 eV. The average power dissipated per linear meter of beam tube is 0.14 Watts. Only a narrow strip along the tube is illuminated by primary photons. The radiation comes in 30 ps pulses with a repition rate of 60Mhz; the peak flux in the area of primary illumination is of the order of 4 x 10^4 J/m^2. While many dedicated electron synchrotron facilities have comparable loads, such facilities have room temperature beam tubes and employ distributed ion pumping to remove desorbing gases. The beamline gradually "cleans up" and can reach UHV without baking. The baseline design of the SSC vacuum system relies on cryotrapping at the tube walls to attain UHV. Thus desorbed molecules are not actually removed from the vacuum system but are merely redistributed along the beam pipe. If a sufficient quantity of H$_2$ accumulates in weakly bound states, the pressure quickly rises to unacceptable values. This cannot be allowed even locally, as increased beam-gas interactions in any one region cause wall heating in that region, desorbing further gases and causing the pressure to run away. The accelerator must instead be shut down and warmed to room temperature, allowing H$_2$ (and other species present--usually CO, CO$_2$, and CH$_4$) to be slowly removed from the system by the true pumps. Our task is to understand how often this will be necessary, and whether the time between warmups will increase as the system cleans up, or will remain constant due to an essentially infinite supply of hydrogen. The origin of the desorbed hydrogen must be better understood to answer these questions; is diffusion from the metal bulk completely frozen out at 4° K, is H nonetheless mobile through tunnelling from site to site (at least in the oxide layer), or does most of the original hydrogen come from the adsorbed water layer?   Furthermore, we must verify and understand a complicating effect seen in the experiments of reference [7]. Here it appears that, rather than the walls cleaning up at LHe temperatures they get dirtier with time; i.e., the desorption coefficient increases with total photon dose. One possible explanation for this is the desorption of hydrogen from chemisorbed states with re-adsorption at low temperature into physisorbed states due to the presence of an activation barrier on the metal oxide surfaces. As hydrogen shifts from strongly to weakly bound states, the static base pressure and the desorption coefficient both rise. Can barrierless surfaces be achieved on suitable materials? Can practical coadsorbates be found which would circumvent these effects? Can the incident synchrotron radiation be absorbed in a small area, minimizing all these problems? If not, some form of distributed pumping must be considered. This approach involves the addition of a liner inside the beam tube, with holes or slots to allow gas desorbed from the liner to escape to the outer volume (Fig.2). The beam tube walls now become a true pump rather than a trap, in that gases desorbed from the liner can escape the region exposed to synchrotron radiation. This pumping might also be augmented with charcoal or some other sieve or getter placed at appropriate regions in the linertubewall space. While such a liner seems to simplify the vacuum problems of the collider, it complicates the mechanical and electrical problems, and is costly. If a liner must be employed, many of the same questions must be answered as to desorption behavior, except that the temperature of the liner now becomes an adjustable parameter; temperatures of 4o, 20°, or 77° K are feasible.

This work is sponsored by University Research Associates under Subcontract #92-Z-16301.

REFERENCES

[1] C. Benventi, R. S. Calder, and G. Passardi, J. Vac. Sci. Tech. 13, 1172. 1976.
[2] Desorption Induced by Electronic Transitions I, N. H. Tolk, M. M. Traum, J. C. Tully, and T. E. Oraday, eds, Springer-Verlag, Berlin, 1983.
[3] Desorption Induced by Electronic Transition II, W. Brenig, and D. Menzel, eds, Springer-Verlag, Berlin, 1985.
[4] A. G. Mathewson, E. Alge, O. Grorer, R. Sauchet, and P. Strubin, J. Vac, Sci. Tech. A5, 2512 (1987).
[5] D. Bintinger, P. Limon, and R. A. Rosenberg, J. Vac. Sci. Tech. A7, 59 (1989).
[6] C. L. Foaster, H. Halama, and C. Lanni, J. Vac. Sci. Tech. A8, 2855, 1990.
[7] P. Feulner, R. Treichler, and D. Menzel, Phys. Rev. B 24, 7427, 1981.

Erratum in DIET IV Proceedings

The editors wish to point out a printing error in the previous (DIET IV) proceedings (Volume 19, Springer Series in Surface Science) which occurred in the article by E. Matthias and T. Green, entitled "Laser Induced Desorption", pp 112-117. To correct this error, the reader should move the last four lines of page 114 and the first 12 lines of page 115 as a block to the top of page 114.

Index of Contributors